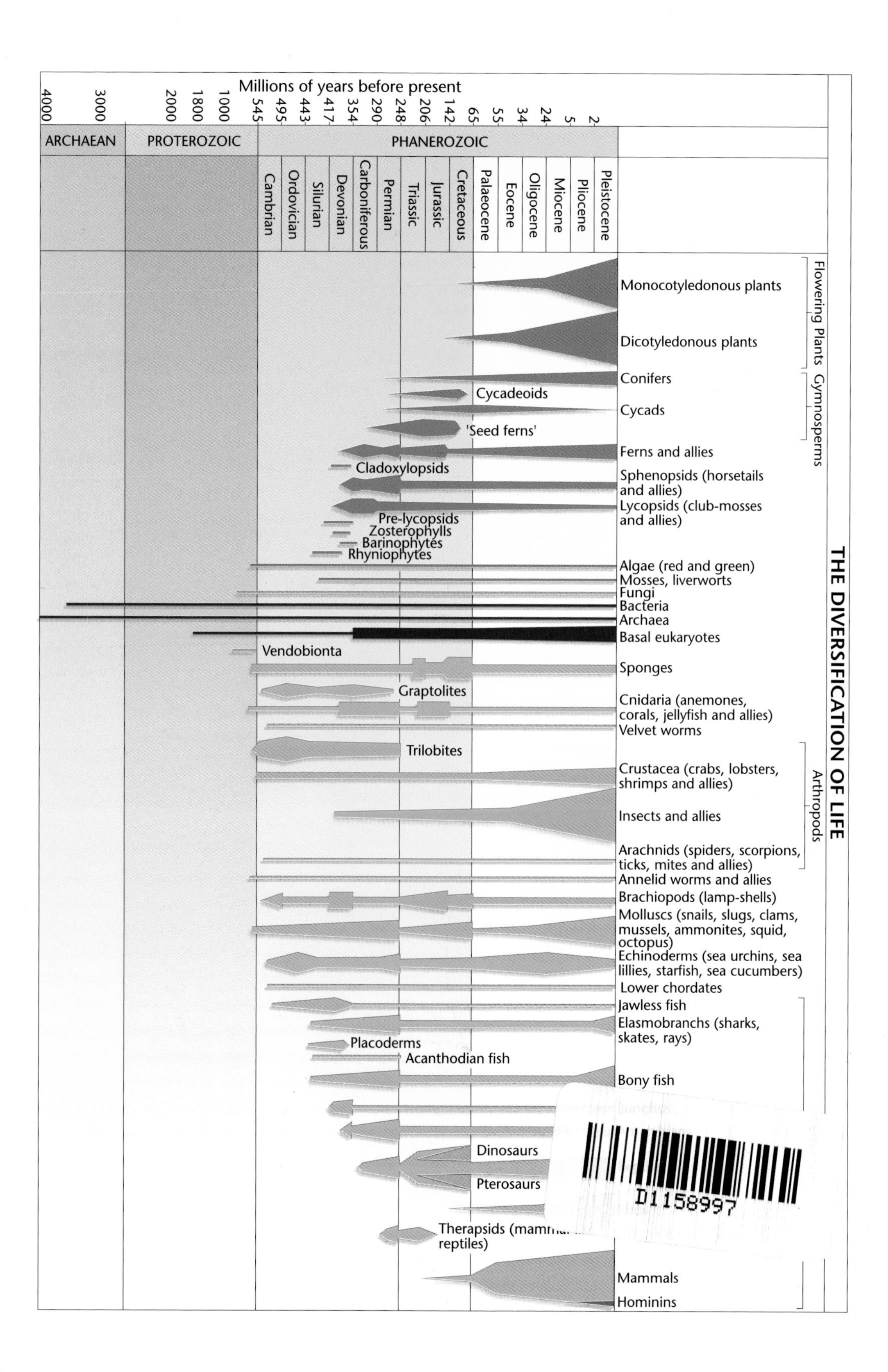

THE DIVERSIFICATION OF LIFE
Millions of years before present
4000
3000
2000
1800
1000
545
495
443
417
354
290
248
206
142
65
55
34
24
5
2
ARCHAEAN
PROTEROZOIC
PHANEROZOIC
Cambrian
Ordovician
Silurian
Devonian
Carboniferous
Permian
Triassic
Jurassic
Cretaceous
Palaeocene
Eocene
Oligocene
Miocene
Pliocene
Pleistocene
Flowering Plants
Gymnosperms
Monocotyledonous plants
Dicotyledonous plants
Conifers
Cycadeoids
Cycads
'Seed ferns'
Ferns and allies
Cladoxylopsids
Sphenopsids (horsetails and allies)
Lycopsids (club-mosses and allies)
Pre-lycopsids
Zosterophylls
Barinophytes
Rhyniophytes
Algae (red and green)
Mosses, liverworts
Fungi
Bacteria
Archaea
Basal eukaryotes
Vendobionta
Sponges
Graptolites
Cnidaria (anemones, corals, jellyfish and allies)
Velvet worms
Trilobites
Arthropods
Crustacea (crabs, lobsters, shrimps and allies)
Insects and allies
Arachnids (spiders, scorpions, ticks, mites and allies)
Annelid worms and allies
Brachiopods (lamp-shells)
Molluscs (snails, slugs, clams, mussels, ammonites, squid, octopus)
Echinoderms (sea urchins, sea lillies, starfish, sea cucumbers)
Lower chordates
Jawless fish
Elasmobranchs (sharks, skates, rays)
Placoderms
Acanthodian fish
Bony fish
Dinosaurs
Pterosaurs
Therapsids (mamm reptiles)
Mammals
Hominins

THE STORY OF EARTH & LIFE

A **southern African** perspective on a 4.6-billion-year journey

Terence McCarthy
Bruce Rubidge

Compiled by staff of the School of Geosciences,
University of the Witwatersrand, Johannesburg

Struik Publishers
(a division of New Holland Publishing
(South Africa) (Pty) Ltd)
Cornelis Struik House
80 McKenzie Street
Cape Town 8001

New Holland Publishing is a member of
the Johnnic Publishing Group

Visit us at **www.struik.co.za**

Log on to our photographic website
for an African experience

First published in 2005

3 5 7 9 10 8 6 4

Publishing manager: Pippa Parker
Managing editor: Lynda Harvey
Editor: Roxanne Reid
Design director: Janice Evans
Designer: Robin Cox
Illustrators: Colin Bleach, Rose Prevec
Proofreader: Pat Barton
Indexer: Cora Ovens

Reproduction by Hirt & Carter Cape (Pty) Ltd
Printed and bound by Kyodo Printing Co (S'pore) Pte Ltd, Singapore

ISBN: 1 77007 148 2

Contributing authors

Prof Carl Anhaeusser is a graduate of the University of the Witwatersrand. He is a former Director of the Economic Geology Research Institute at Wits and is a Professor Emeritus at Wits. His primary research interests centre on the mineralisation and evolution of Archaean granite-greenstone terranes and especially Barberton. He has published extensively and is an internationally acknowledged authority in these fields.

Dr Lucinda Backwell is a graduate of the University of the Witwatersrand, where she is a Research Fellow. Her primary research interests centre on early hominid behaviour and cognition.

Dr Marion Bamford is a graduate of the University of the Witwatersrand and a Senior Research Fellow at the Bernard Price Institute for Palaeontological Research at Wits. She is an authority on fossil woods of southern Africa, and is active in research in the Karoo basin and younger deposits, notably those of Laetoli, Olduvai and Sterkfontein.

Dr Lee Berger is a graduate of Georgia Southern University (USA) and the University of the Witwatersrand. He holds academic positions at Wits, Duke University and the University of Arkansas. His primary research is in the field of human origins and he has discovered several new hominid sites in the Cradle of Humankind.

Dr Bob (CK) Brain (guest author) is a graduate of the University of Cape Town and a former Director of the Transvaal Museum. He is a passionate naturalist, has published in numerous fields of natural science, and has won international recognition for his palaeoanthropological research at Swartkrans and particularly for the development of the discipline of cave taphonomy. His current research interests centre on the origin of metazoans.

Dr Dion Brandt is a graduate of the University of the Witwatersrand, where he is currently a lecturer. He has research interests in impact craters and environmental geology.

Prof Grant Cawthorn is a graduate of Durham and Edinburgh Universities. He is the Platinum Industry Professor of Igneous Petrology at Wits and a world authority on the geology of layered complexes.

Prof Guy Charlesworth, a graduate of the University of Natal, is an Associate Professor at Wits. His primary research interests are in structural geology and the influence of structure on mineral deposits.

Dr John Hancox, a graduate of the University of the Witwatersrand, is a Research Fellow at the Bernard Price Institute for Palaeontological Research at Wits. His primary research interests are in the geological evolution of the Karoo Basin and the evolution of its fauna.

Prof Roger Gibson is a graduate of the Universities of Natal and Cambridge. He is an Associate Professor at Wits. His main research interests lie in deformation and metamorphism of rocks and in meteorite impact structures.

Dr Andre Keyser is a graduate of the Universities of Pretoria and the Witwatersrand. After retiring from the Geological Survey of South Africa, where he researched Karoo mammal-like reptiles, he joined the University of the Witwatersrand's palaeoanthropology research initiative. He has been responsible for the discovery and development of the Drimolen fossil hominid site and the discovery of numerous hominid remains.

Dr Sharad Master is a graduate of the University of the Witwatersrand, where he holds a lectureship. He is an authority on the geological evolution of the African continent.

Prof Terence McCarthy is a graduate of the Universities of Cape Town and the Witwatersrand, and is Professor of Mineral Geochemistry at Wits, where he headed the Department of Geology for 17 years. He has wide research interests in the earth sciences and is a leading authority on the geology of wetlands, especially the Okavango Delta in Botswana.

Dr Mike Raath is a graduate of Rhodes University. He is a former Director of the National Museums and Monuments of Rhodesia, the Port Elizabeth Museum Complex and the Bernard Price Institute for Palaeontological Research at Wits. He is currently the curator of fossil collections at Wits. His research interests centre mainly on early dinosaurs.

Prof Uwe Reimold is a graduate of the University of Münster in Germany and is currently Professor of Mineralogy at Wits. He is a leading international authority on the geology of terrestrial impact craters.

Prof Laurence Robb is a graduate of the University of the Witwatersrand. He is Director of the Economic Geology Research Institute at Wits and an internationally acknowledged economic geologist.

Prof Bruce Rubidge, a graduate of Stellenbosch and Port Elizabeth Universities, is director of the Bernard Price Institute for Palaeontological Research at Wits. He is an authority on the formation of the Karoo Basin and the evolution of its fauna, especially the mammal-like reptiles.

Sue Webb is a graduate of the State University of New York (Binghamton) and Memorial University (Canada). She holds an academic position at the University of the Witwatersrand. Her research involves investigating the deep structure of Kaapvaal Craton using magnetic, gravity and seismic methods.

Sponsor's Foreword

Although the mineral wealth of southern Africa has been exploited to some degree for well over 1 000 years, the transformation of the subcontinent from an agrarian to an increasingly industrialised economy was triggered by the discovery of diamonds along the Orange River in the 1860s. This was followed during the succeeding decades by the discoveries of gold in Limpopo, Mpumalanga and Zimbabwe, and then, in 1886, by what was to become the world's richest gold repository, the Witwatersrand. Next came the delineation of some of the world's largest deposits of platinum, chrome, manganese, iron ore, uranium, coal, titanium, vanadium and fluorspar, creating the platform for a burgeoning mining sector that would underpin southern Africa's industrialisation throughout the last century.

This astonishing concentration of mineral resources, unequalled by any other region in the world, is the result of a unique combination of diverse geological environments, an immense history of geological stability and generally excellent exposure. As if this were not sufficient, the region also hosts a strikingly comprehensive palaeontological record – from among the oldest multicellular organisms yet identified, to the remains of some of our earliest ancestors at the Cradle of Humankind, a World Heritage Site.

Given this legacy of enormous mineral wealth, great geological diversity and an extensive evolutionary record, it is surprising that, until now, there has not been a publication that could offer both the interested lay reader and the entry-level tertiary student a concise, yet comprehensive, accurate, topical and readable account of our geological and palaeontological past.

This book aims to address this shortcoming: in clear language, accompanied by lavish illustrations, innovative vignettes and chronological 'route maps', it introduces the reader to the processes and events that led to the fascinating accumulation of strata that lie beneath our feet and to which we owe the rich diversity of our natural environment.

Kumba Resources is a leading South African-domiciled company that has been involved in the extraction and beneficiation of various minerals for almost 80 years. At the same time, we are committed both to the protection of the fragile environment in which we live, as well as to the education of our fellow citizens about our business and the context in which it operates.

It is therefore with great pleasure that we have had the privilege of becoming associated with the highly esteemed group of contributors and editors responsible for the publication of this much-needed volume. We hope that this book will be widely read and that it will become an essential reference text in the libraries of businesses, colleges, schools and homes across the country.

Con Fauconnier

CON FAUCONNIER
CHIEF EXECUTIVE,
KUMBA RESOURCES LIMITED
(AND PRESIDENT OF THE CHAMBER OF MINES OF SOUTH AFRICA, 2004/2005)

Preface

Southern Africa is a mineral-producing region of global importance and has been so for more than a century. It has a record of life preserved in its fossils that is more extensive than any other region in the world, a record that preserves not only the evolution of plants and animals, but humans too.

There is a growing awareness of the natural environment and in order to understand the landscape and ecosystems, a good grasp of local geology and geological history is essential because the basis of ecosystems is ultimately geological. Yet, ironically, information on the geological formations that host the region's mineral and fossil wealth, the story of life that the fossils tell, and the geological history of southern Africa are largely stashed away in technical publications that are accessible only to the specialist. With this book we hope to rectify this situation and share with non-specialists the exciting story of how southern Africa, and to some extent the world, came to be the way it is – how its mineral deposits formed, how its life evolved, and how the landscape of southern Africa was shaped.

Writing a book such as this requires a team effort, and we would like to express our appreciation to all of our contributing authors for their enthusiasm in the project, their on-time delivery and their good grace in the face of editorial reworking of their material. We would also like to thank our many colleagues who willingly provided illustrative material, as well as Robin Cox, Stanley Duncan, Andrew Terhorst, Philip Frost, Neil McKenna, Hein Pienaar, Glen McGavigan, Lew Ashwal and Marina Rubidge for assistance in sourcing images.

The subject matter of the book is extremely wide ranging, and as far as possible we wished to ensure that the content is accurate, or at least as accurate as such a work can be, given the divergent opinions that often exist in the earth sciences. We also wanted to ensure accessibility of the writing and clarity in the general approach to this vast subject. Many colleagues and associates assisted towards these ends: we thank Fernando Abdala, Carl Anhaeusser, John Begg, Bob Brain, Grant Cawthorn, Fred Daniel, Mike and Maarten De Wit, Doug Erwin, Nok Frick, Rob Gess, Roger Gibson, James Hersov, Judith Kinnaird, Pieter Kotze, Rodrigo Lacruz, Judy Maguire, Erna McCarthy, Jennifer Oppenheimer, Rose Prevec, Mike Raath, Uwe Reimold, Chris Sidor, Peter Tyson, Rob Veal, Richard Viljoen, Lyn Wadley, Lilith Wynne and Adam Yates. We nevertheless accept full responsibility for errors or omissions in the text and illustrations.

On the production end, we thank Pippa Parker (publishing manager), Robin Cox (designer) and especially Roxanne Reid for her editorial skills; Lynda Whitfield and Diane du Toit for the pioneering illustrative material, and Rose Prévec and Colin Bleach for the finished product. Henia Czekanowska undertook specialist photography for the book and Jacqui Thobois provided invaluable secretarial support.

Finally, we thank Kumba Resources, especially Richard Wadley and Trevor Arran, for supporting our book financially, thereby subsidising the price and increasing its accessibility.

TERENCE MCCARTHY

BRUCE RUBIDGE

CONTENTS

Adey, Shaen / IOA

Tugela Falls, Royal Natal National Park, KwaZulu-Natal.

THE STORY OF
EARTH & LIFE

1
INTRODUCTION

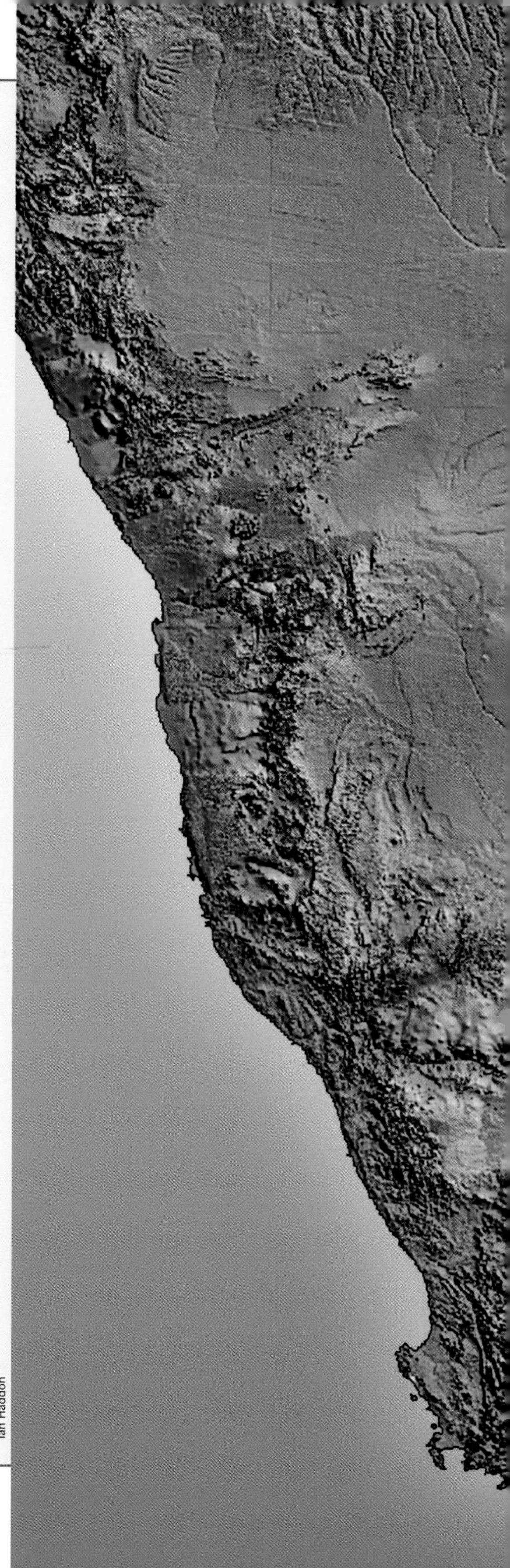

Ian Haddon

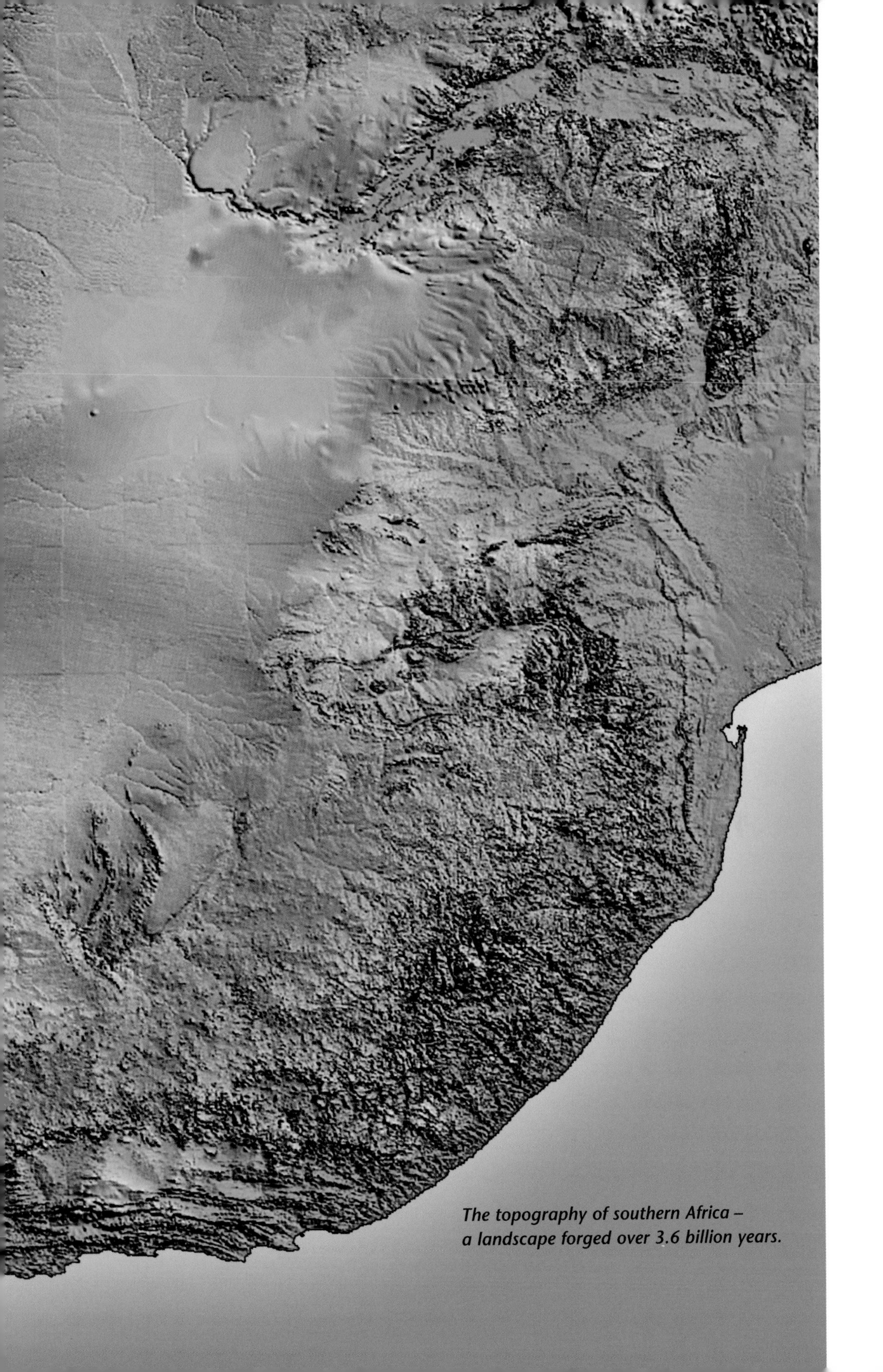

*The topography of southern Africa –
a landscape forged over 3.6 billion years.*

The ability to reason, to understand cause and effect, is a uniquely human trait. Evidence for its beginnings extends back more than one and a half million years in the form of carefully selected digging tools found in cave deposits containing hominin (human ancestors) remains, such as at Sterkfontein and adjacent sites in Gauteng. Slightly younger deposits contain primitive stone tools.

In later deposits dating from just over a million years ago, we see evidence of the ability to design and manufacture tools to order, in the form of uniform, beautifully fashioned, symmetrical hand axes of the Acheulian period (**figure 1.1**). Whether these ancient tool makers possessed a true culture is not known, and the earliest evidence of what might be called culture – the use of decorative symbolism and of ochre paint (**figure 1.2**) and shell beads – comes from deposits in Blombos cave in the Western Cape, which are about 70 000 years old.

By about 40 000 years ago, these cultural practices had become commonplace and sophisticated both in Africa and Europe. They included elaborate paintings on cave walls, the use of decorative jewellery and even musical instruments. Many of the cave paintings created by these people are shamanistic (spiritualistic) and indicate that human thought extended well beyond the simple needs of daily living; the human mind had begun to probe spiritual questions. It is likely that these people had developed some form of mythology, perhaps seeking explanations for common events such as the rising and setting of the Sun and Moon, the passing of the seasons, and even of death. Probably later, deeper questions related to the origin of the world began to be probed. We have no idea when these appeared, but all cultures possess some form of creation mythology, usually intimately intertwined with religious beliefs.

In Judaeo-Christian philosophy, the Creation is specifically described in the opening chapter of the Book of Genesis. So specific is this description and the subsequent narrative that in 1654 the Bishop of Armagh, James Ussher, was able to use the biblical lineages to calculate the date and time of the Creation: 09h00 on 23 October, 4004 BC.

Understanding of the world throughout the Dark Ages was based on a combination of the Bible and the writings of Greek natural philosophers and astronomers, especially Ptolemy and Aristotle. The Renaissance brought with it not only a revival of art and literature, but a more enquiring approach to the natural world, initially in the area of astronomy, and later in physics and chemistry. Serious geological enquiry that addressed the question of Earth history, however, only arose in the late 1700s.

Kathy Kuman

Figure 1.1 ***Examples of hand axes from the Acheulian period illustrate an ability to produce a standardised tool form of varying size and from different types of rock. These tools may be more than 500 000 years old.***

PROBING EARTH'S HISTORY

While matters related to the origin of the Earth fell into the realm of religious mythology, the more practical aspects of rocks and minerals received considerable attention. Stone was used since the earliest times, initially for tool manufacture and later for building. Minerals were mined and processed to extract metals such as iron, copper, lead, tin, silver and gold. Practical knowledge about mining and metallurgy was passed from generation to generation, and much of this accumulated knowledge was recorded in a famous book, *De Re Metallica* by Georg Agricola, published in 1556. Fossils, which are common in rocks over large parts of Europe, attracted considerable attention and were the subject of speculation since the earliest times. Greek and Roman scholars such as Pythagoras, Herodotus, Aristotle, Theophrastus, Strabo and Pliny developed ideas on their origins and their implications: they attributed the occurrence of fossil marine shells and fish at high altitude to alternate depression and uplift of the land.

Following the Dark Ages, writings again began to appear on geological topics. Fossils attracted the attention of theologians in particular, as they were considered to provide proof of the biblical flood. The rocks hosting fossils also began to attract attention, and their regularity was noticed – so much so that in 1684 Martin Lister proposed the concept of a geological map.

During the 1700s, the study of rocks and fossils began to grow in popularity among gentleman scientists. The French made important contributions. In his 1749 book on natural history, Georges Louis Buffon proposed that the Earth was of great antiquity and that its surface had experienced slow and gradual changes, but he was severely censured at the Sorbonne and the Faculty of Theology in Paris, and forced to withdraw his views. The study of rocks continued unabated, and in 1751 Jean Etienne Guettard published possibly the earliest geological map, showing the distributions of minerals and rocks in France. The term geology was coined in 1778 by JA de Luc.

In Germany, AG Werner, Professor of Mining at Freiburg, recognised the ordering of strata and the fact that rock strata could be characterised by the fossils they contained. His ideas were published in 1796. Similar observations were made in the United Kingdom by William Smith, whose geological maps began to make their appearance in the 1790s. Smith went further to produce the first geological column for the United Kingdom, which characterised and described the sequences of rocks and their fossils. His great work, a geological map of England, Wales and parts of Scotland, was published in 1815.

Meanwhile, in 1785, James Hutton published his extremely influential work *A Theory of the Earth*, which became a milestone in geological thought. Hutton argued that sediment accumulation that we can see taking place today, such as on mud flats and at river mouths, is very slow. Great thicknesses of sedimentary layers had been mapped by Hutton, Smith and others, and Hutton argued that these deposits must have taken immense periods of time to accumulate. The implication was that the Earth is very much older than suggested by the Old Testament. These arguments were expanded by Charles Lyell in his three-volume treatise *Principles of Geology* (1833).

Jean Baptiste Lamarck, a French biologist regarded as the father of invertebrate palaeontology, rekindled the ideas of Buffon regarding the great antiquity of the Earth in a series of books published between 1801 and 1822. From his knowledge of the changing fossil forms in successive rock layers, he developed the notion of the evolution of life. His idea was that life forms evolved as a consequence of changing environmental factors and that characteristics or adaptations acquired by an organism during life could be inherited by succeeding generations.

Charles Darwin and Alfred Wallace, who were both keen naturalists with an interest in geology, also made the connection between changing fossil forms, the probable immense time involved, and

Figure 1.2 *Engraved red ochre from Blombos cave on the southern Cape coast. The specimen is 77 000 years old, and provides the earliest evidence of cognitive abilities central to modern human behaviour.* ***A*** *shows the clay tablet as it appears to the camera, while* ***B*** *emphasises the red ochre scratches.*

South Africa's mineral reserves and production in a global context

Reserves are minerals or metals known to exist but not yet mined; **production** is the actual amount extracted annually. Percentages refer to South Africa's share of the world's reserves or production. The units are: t = tonnes; mt = million tonnes; bt = billion tonnes; mc = million carats. PGM denotes Platinum Group Metals (platinum, palladium, ruthenium, rhodium, osmium and iridium). The value of mineral production in 2001 was approximately R100 billion (b = billion; m = million).

Commodity	Reserves		World Rank	Production		World Rank	Rand Value
	%	Units		%	Units		
Manganese	80	4 bt	1	20	3.6 mt	1	3 b
Chrome	76	5.5 bt	1	45	6.6 bt	1	7 b
PGM	56	63 000 t	1	46	207 t	1	25 b
Gold	52	40 000 t	1	17	428 t	1	30 b
Vanadium	44	12 mt	1	57	18 000 t	1	780 m
Vermiculite	40	80 mt	2	45	210 000 t	1	132 m
Refractories				36	183 000 t	1	118 m
Zirconium	22	14 mt	2	28	0.25 mt	2	7 m
Titanium	20	146 mt	2	23	1 mt	2	600 m
Fluorspar	10	36 mt	3	5	213 000 t	3	1 000 m
Diamonds					11 mc	5	
Uranium	9	0.2 bt	4	2	860 t	9	215 m
Nickel	8	12 mt	6	3	37 000 t	9	2 b
Antimony	6	8 mt	4	3	3 700 t	4	29 m
Phosphate	7	2.5 bt	3	2	2.8 mt	9	900 m
Copper	2	13 mt	14	1	0.14 mt	13	1.6 b
Zinc	3	15 mt	5	1	63 000 t	18	310 m
Lead	2	3 mt	5	2	75 000 t	9	109 m
Iron	1	1.5 bt	9	4	34 mt	8	3 b
Coal	11	55 bt	5	6	224 mt	6	20 b

the diversity of life we see on the planet today, and simultaneously proposed their Theory of Evolution in 1858. They differed from Lamarck in that they noted there is always a range of characteristics (e.g. shorter or taller stature) within any individual species, and that environmental pressures can favour particular variants over others, allowing some variants to breed more successfully and therefore drive evolutionary change – a process Darwin termed natural selection.

Thus, in a brief period spanning the late 18th and early 19th centuries, the foundations of our present understanding of the Earth and its life were laid.

HOW CAN WE KNOW THE PAST?

We are a unique species. We possess the ability to manipulate natural materials to suit our purposes. We are capable of abstract thought and can communicate these thoughts to others. We can work in teams, executing complex plans. We can pass accumulated knowledge from one generation to the next. We alone among species can pose fundamental questions, including questions about our own origins and the origins of the world around us.

But posing such questions is one thing; providing the answers is quite another. How can we see back into the past, to a time before written records, before our species walked on this planet, or perhaps even to the time when our planet was born? There are some who turn to religious texts for answers, but there are others, scientists, who seek answers in the world around them, using the considerable mental abilities with which our species is endowed.

We are all familiar with archaeological methods, at least in general terms. Archaeologists dig at a site and recover artefacts (**figure 1.3**). These artefacts have an origin – they were made by early people. The deeper the archaeologists dig, the older the artefacts must be. Archaeologists can deduce what an artefact was used for by characteristics such as shape and wear patterns. They can also tell much about the lifestyle of its makers from the context of the artefact – how and where it was positioned in the site and what other artefacts and remains were associated with it. They can even measure its age by dating the object itself, or suitable material associated with it, using the radioactive decay of a kind of carbon atom present in organic matter, carbon 14 (^{14}C).

Archaeological sites are time capsules, having frozen during some event or succession of events in the past. By excavating and studying them archaeologists can reconstruct these past events. They have been very successful in piecing together the history of mankind, especially our more recent history. But as they go further back in time, the record becomes increasingly fragmentary and more difficult to interpret: fragmentary because of subsequent partial destruction by natural agencies such as erosion and decay; and more difficult to interpret because the artefacts become increasingly removed from our own experience.

We now understand that rocks are perfectly analogous to the ancient artefacts excavated by archaeologists. Rocks, too, are time capsules. All rocks are not of the same age, but formed at different times and in different ways. Some rocks, for example, were formed by solidification from the molten state – from lava erupted from volcanoes. Others were formed by accumulation of sediment, such as gravel in a riverbed, sand on a beach, or silt and mud on a river delta. Yet others may have been transformed by heat and pressure during deep burial to produce new kinds of rocks.

From the nature of a rock we can tell how it formed; from its context – the rocks with which it is

Lyn Wadley

Figure 1.3 ***An archaeological excavation at Rose Cottage Cave in the Free State, where the deepest layers provide information about the earliest occupants of the cave. The 6 m of cave floor deposits encapsulate a record of 100 000 years of habitation.***

associated – we can deduce the environment in which it formed. In this way we can reconstruct geographic environments and geological events of the past. And in the same way as archaeologists date their artefacts using ^{14}C, geologists can date rocks using not carbon but other radioactive elements.

Rocks are much more common than archaeological sites; they are everywhere beneath our feet. By mapping their distribution and establishing how they formed and in what type of environment, geologists can reconstruct the changing environmental history of a particular region. Every region has had a unique history and each must be individually determined, just as each country has its own, unique social and political history. As we go further back in time, the record becomes increasingly fragmentary, evidence having been obliterated by erosion or later geological events that often overprint the earlier record. Very ancient rocks are therefore extremely rare and difficult to interpret.

Dion Brandt

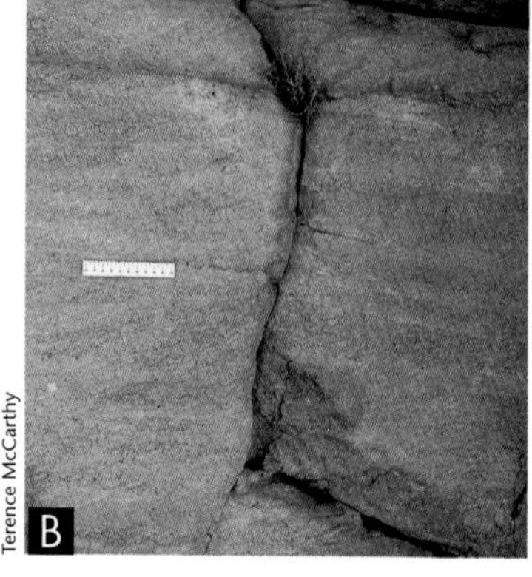

Terence McCarthy

Figure 1.4 ***Rocks deposited in southern Africa in ancient times are remarkably well preserved. A shows sand and gravel deposited by rivers a few thousand years ago (Rooisloot, Mokopane, Limpopo Province). B shows sand and gravel deposited 2 900 million years ago (Witwatersrand Supergroup, Gauteng).***

ABOUT THIS BOOK

South Africa is, geologically speaking, a very diverse and in many ways unique place, without equal on the globe. There are several factors that contribute to this uniqueness. South Africa is a treasure house of valuable minerals. Despite occupying only 1% of the Earth's land surface, the country is or once was (before some of the mines were exhausted) the world's largest producer of gold, chromium, diamonds, vanadium, manganese and platinum. It possesses very large reserves and is a world-class producer of iron, titanium, zinc, coal, fluorspar, refractory minerals and phosphorus, and also produces copper and lead. It has been said that hectare for hectare, the northern half of South Africa is the richest piece of real estate on earth. The table on page 12 shows just how spectacular the country's mineral wealth is.

South Africa has a very long geological history, its oldest rocks dating back some 3 600 million years. Rock-forming events extend from this ancient dawn virtually to the present, providing a long, albeit punctuated, geological history. The preservation of these ancient rocks is quite remarkable and many look little different today from the equivalents formed in very recent times (**figure 1.4**). The rocks record events during many crucial periods in Earth's history.

They also provide insight into globally important changes that took place in the past, such as the changing composition of Earth's atmosphere and the assembly and fragmentation of the supercontinents.

Finally, South Africa's rocks contain a very special and long record of life. The very earliest life forms are preserved as fossils in the rocks; the evolution of land plants and animals, and especially the origin of mammals and dinosaurs, are well preserved (**figure 1.5**). South Africa also has probably the best record of the origin of hominins. Truly an amazing record of Earth and its life.

In this book we will journey through time and examine the unfolding of that remarkable history. Although the logical place to start an historical account is at the beginning, we will begin our story at the end because it is not just an account of consecutive events, of seas that came and went, volcanoes that erupted and then became extinct. We also want to address the questions of how and why these events occurred. In other words, we want to put our geological history in context.

To do this, it is necessary to understand how the Earth works. In the last few decades, earth scientists have made remarkable strides in understanding the Earth, and it is believed that many processes we see operating today have operated since the earliest times. This knowledge has radically improved our understanding of past events, as it provides a context for these events. So our story begins with the modern Earth, and describes how the Earth works. Armed with this knowledge, we will be able to journey back in time to the birth of the Earth.

Readers unfamiliar with geology will be faced with many new concepts and terms, introducing essentially a new discipline that – like all disciplines – has its own special jargon. While we have made every effort to reduce jargon to a minimum, there are essential concepts and terms you need to know to appreciate South Africa's geological history.

For some, this may have the unfortunate effect of making what is basically a fairly simple story seem quite complicated. If you tend to forget the meanings of terms, it can be a source of immense frustration. Just as a map helps you recall place names when travelling in a foreign country, we have included route maps in our journey through time to help recall the meanings and significance of names and terms used in the text. In addition, we have included a glossary for quick reference to the meaning of these terms (*see* page 318).

We live in a four-dimensional space-time world, and our minds are attuned to certain natural scales in both space and time. Time we measure in seconds, minutes and hours, and we are familiar with scales of seasons and years. As we age, we begin to appreciate time scales of a generation or two. But that is where our experience and familiarity with time ends. In the same way, we measure distance in centimetres, metres and even kilometres. We can gaze on and appreciate mountains many hundreds of metres high. We think of local places in terms of their distance from where we live. But we have no appreciation for really great distances, distances on a global scale. We travel globally, but we express global distances not in terms of kilometres, but in terms of the travel time by jet airliner, because the time is more comprehensible.

Our natural, in-built space-time reference frame is too limited to cope meaningfully with Earth history; time and distances are simply too vast. There is no easy solution to this problem. In their training, earth scientists develop a familiarity with the numbers involved, a familiarity that makes them feel comfortable with very old or very large things. But like everyone else, earth scientists cannot really comprehend the vast time expanses or huge distances in and on

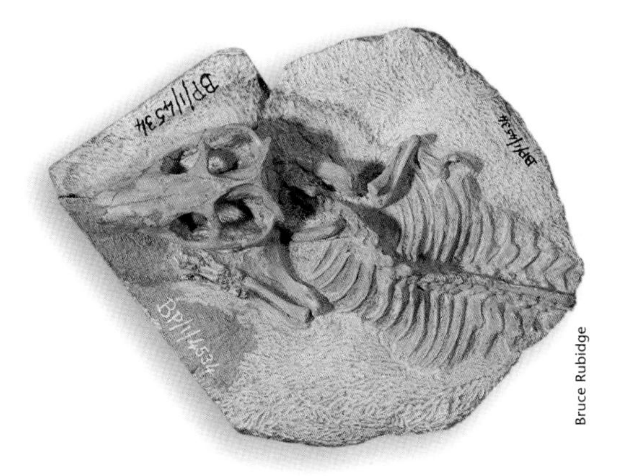

Figure 1.5 ***Well-preserved fossils from southern Africa, such as this skull of a reptile that lived in the Karoo region, provide insight into the origins of dinosaurs and mammals.***

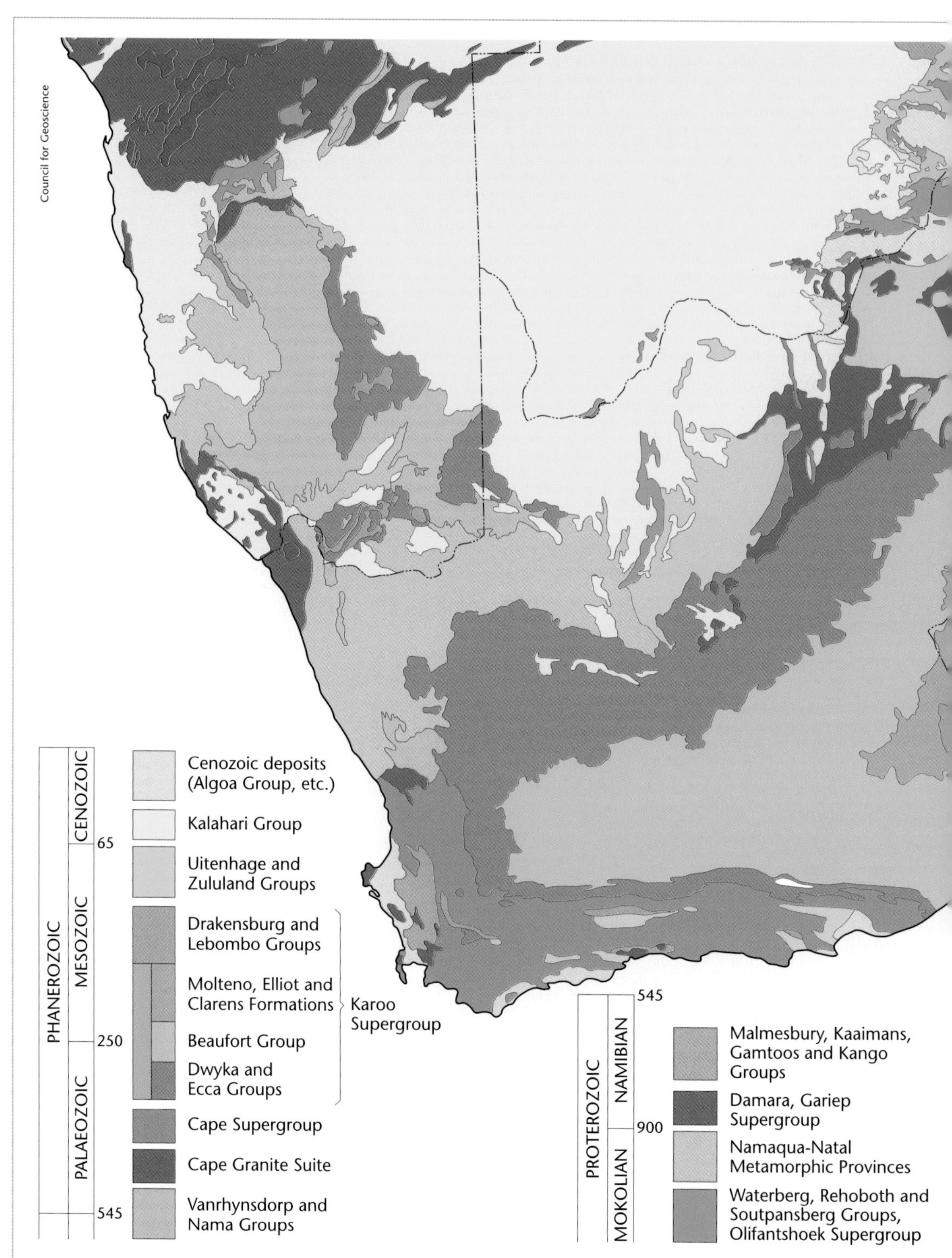

Figure 1.6 *In this geological map of southern Africa each of the colours represents a group of rocks formed during a specific period. These rocks record the geological history of the region, a history recounted in this book.*

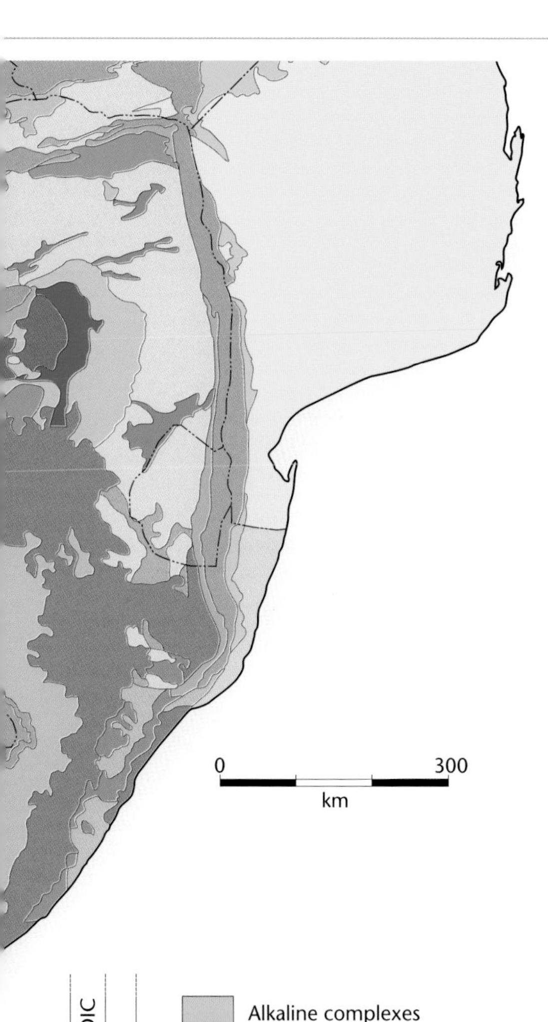

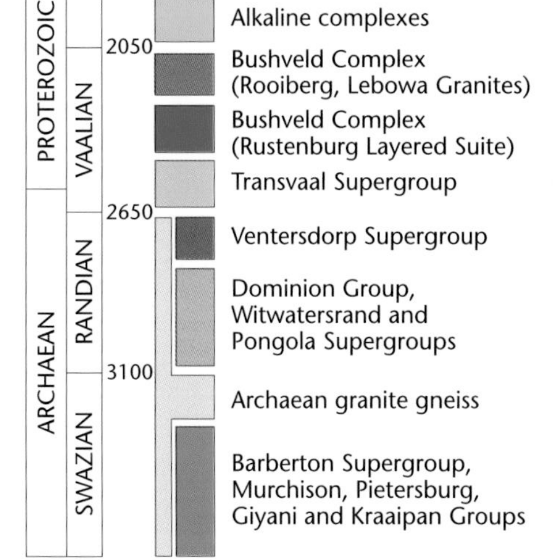

Earth. To really appreciate the story told in this book, you too need to develop a feeling for big numbers. Another useful skill is sensitivity to the Earth's three-dimensionality. Like architects, earth scientists operate in a world where the third dimension assumes a far greater importance than is required in everyday life. The ability to visualise a three-dimensional object from a two-dimensional sketch is therefore very useful.

It is inordinately difficult for most people to relate past events described by earth scientists to the world we see around us today. Reason and logic are strained when they describe mountains that were once seas, or seas where there were once mountains. It is hoped that this book will go some way to alleviating this kind of difficulty.

This book is about the rocks of southern Africa, shown in the geological map in **figure 1.6**, and the physiographic map on page 9, as well as the fossils they contain. Rocks are all around us, and they create the scenery we cherish – the Drakensberg, the Waterberg, the Cape mountains, the Karoo. When you gaze on these vistas, what are you seeing? Beautiful scenes to be sure, but there is another dimension to your view. Whether you realise it or not, you are looking back in time. This book is about that other dimension. How did the rocks form? And when? How did they get to be as we see them today? How did the mountains, valleys and rivers we see around us form, and how old are they? What animals and plants lived there in the past?.

Rocks in themselves are uninteresting to most people, and understandably so. But rocks are also time capsules because they encapsulate information about their surroundings when they form and preserve it. They have a story to tell. The story, however, is in code. Earth scientists are breaking the code and are steadily deciphering Earth's history. Study of the rocks of southern Africa has revealed the origins of the region's great mineral wealth. The most ancient rocks tell of a world very different from today, one with a crushing, toxic atmosphere and bacterial slime as its most advanced life. Rocks and their fossils also reveal the story of Earth's slow, at times life-threatening, journey to the present, and how southern Africa came to be the way it is.

THE STORY OF
EARTH & LIFE

2
HOW THE EARTH WORKS

Lava flowing from the Kitazungurwla volcano in Central Africa during the 1986 eruption.

Judith Kinnaird

ROUTE MAP TO CHAPTER 2

EARTH
(fig. 2.6)

CORE
(>5000°C)
(Ni-Fe alloy)

- **OUTER CORE**
(molten-source of magnetic field)
- **INNER CORE**
(solid)

MANTLE
(800° to 5000°C)
(magnesium, iron, silicon, oxygen)

- **MESOSPHERE**
- **ASTHENOSPHERE**
(Low Velocity Zone)
100-km-thick slippery layer on which plates move

PLATE TECTONICS
rigid plates of the lithosphere move relative to each other and interact at their margins

Plate interactions

SEPARATION
(fig. 2.10)
forms MID-OCEAN RIDGES where ocean crust is torn apart; shallow earthquakes occur; mantle melts, producing BASALT, which injects upwards and erupts on the sea floor, forming PILLOW LAVA

sea water circulates through the hot ocean crust, forming hot springs or HYDROTHERMAL VENTS

separation of continental plates forms RIFT VALLEYS (e.g. East African Rift Valley) (fig. 2.12)

Continued separation produces new oceans bisected by a mid-ocean ridge (e.g. Red Sea, Atlantic)

CONVERGENCE
(fig. 2.14)
forms SUBDUCTION ZONES where ocean crust is destroyed; forms trenches on the ocean floor (e.g. Mariana trench); shallow to deep earthquakes

ISLAND ARCS
(e.g. Japanese Islands)
oceanic crust descends (subducted) beneath oceanic crust; water released from descending slab causes melting of the overlying mantle, producing DIORITE and GRANITE, and explosive ANDESITE and RHYOLITE volcanoes that form the island arc; new continental crust is formed

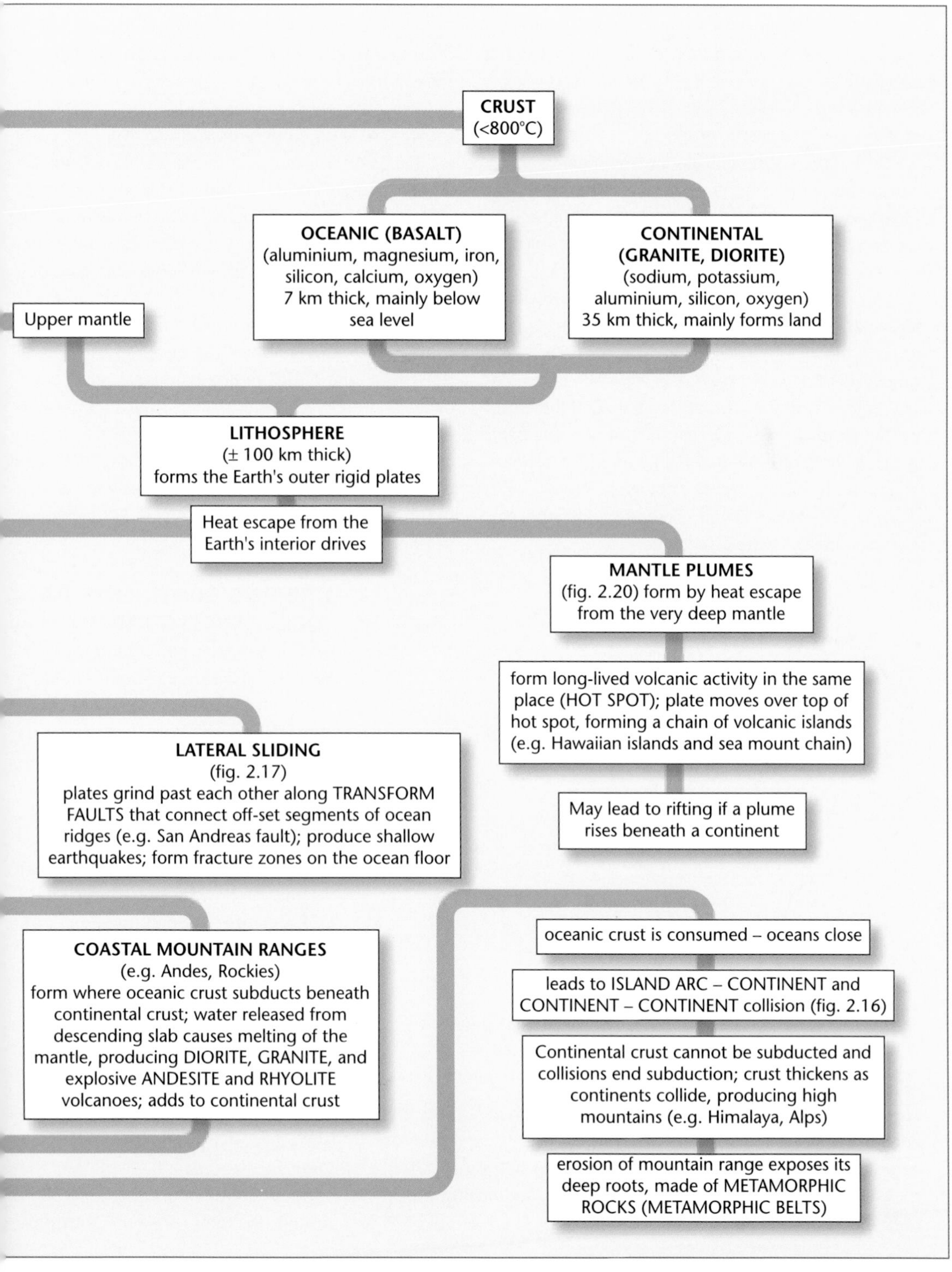

CRUST
(<800°C)

OCEANIC (BASALT)
(aluminium, magnesium, iron, silicon, calcium, oxygen)
7 km thick, mainly below sea level

CONTINENTAL (GRANITE, DIORITE)
(sodium, potassium, aluminium, silicon, oxygen)
35 km thick, mainly forms land

Upper mantle

LITHOSPHERE
(± 100 km thick)
forms the Earth's outer rigid plates

Heat escape from the Earth's interior drives

MANTLE PLUMES
(fig. 2.20) form by heat escape from the very deep mantle

form long-lived volcanic activity in the same place (HOT SPOT); plate moves over top of hot spot, forming a chain of volcanic islands (e.g. Hawaiian islands and sea mount chain)

May lead to rifting if a plume rises beneath a continent

LATERAL SLIDING
(fig. 2.17)
plates grind past each other along TRANSFORM FAULTS that connect off-set segments of ocean ridges (e.g. San Andreas fault); produce shallow earthquakes; form fracture zones on the ocean floor

COASTAL MOUNTAIN RANGES
(e.g. Andes, Rockies)
form where oceanic crust subducts beneath continental crust; water released from descending slab causes melting of the mantle, producing DIORITE, GRANITE, and explosive ANDESITE and RHYOLITE volcanoes; adds to continental crust

oceanic crust is consumed – oceans close

leads to ISLAND ARC – CONTINENT and CONTINENT – CONTINENT collision (fig. 2.16)

Continental crust cannot be subducted and collisions end subduction; crust thickens as continents collide, producing high mountains (e.g. Himalaya, Alps)

erosion of mountain range exposes its deep roots, made of METAMORPHIC ROCKS (METAMORPHIC BELTS)

RESTLESS EARTH

We tend to think of the Earth as constant, solid and unchanging, a notion that is enshrined in our language in expressions such as *terra firma*, or *solid as a rock*. But every now and then we are reminded that the Earth is not just a lump of dead, inert rock – reminders that come in the form of devastating earthquakes or volcanic eruptions.

Such things generally do not happen in South Africa, but other countries are less fortunate: the Japanese islands, home to about 150 million people, are prone to severe earthquakes and violent volcanic eruptions, while many Californians live in constant fear of earthquakes. This chapter examines why these forms of geological activity are unevenly spread around the globe. There is reason to believe the processes responsible have been operating throughout most of Earth history and have shaped the world we live in today. An understanding of how the Earth works is therefore essential background for appreciating South Africa's long geological history. The essential concepts you will encounter in this chapter are summarised in the route map.

Geology is a young discipline compared to, say, physics, mathematics or chemistry, and only began to emerge as a serious science in the late 18th century. Early geologists focused their attention on trying to understand the origins of different rock types, their relative ages and distribution. Nevertheless, even in those early years, there were individuals who thought on a global scale. As early as the 16th century, Elizabethan philosopher Sir Francis Bacon pointed out the complementary shapes of Africa and South America, while Dutch scientist Ortelius suggested in 1596 that these two continents were once joined, and subsequently became separated.

Shape alone is insufficient evidence on which to postulate movement of the continents. Most of the important early geological work was, however, confined to local studies, particularly the making of geological maps and the documentation and description of rocks and fossils. It was these data that were to provide a basis for examining global geology in a more scientific manner in later years.

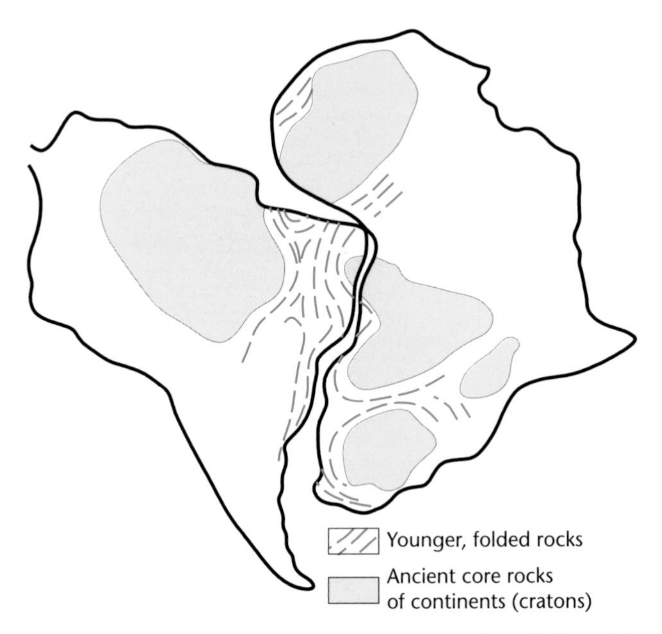

Figure 2.1 ***If South America and Africa are placed alongside each other there is a good fit between the coastlines (the edges of the continental shelves give an even better fit). The rock formations along the boundary match, suggesting that they were once joined.***

DRIFTING CONTINENTS OR AN EXPANDING EARTH?

The notion that the continents, particularly Africa and South America, were once joined continued to haunt the discipline. One of the first attempts to reassemble the continents using scientific evidence was made by Antonio Snider-Pelligrini in 1859, based on fossil evidence from Europe and North America. Other attempts followed, but it was Alfred Wegener, the German meteorologist, who in 1915 proposed the most significant hypothesis concerning the former unity of the continental landmasses, based on extensive field observations. He suggested that all the present-day continents were once assembled into a single landmass he called **Pangaea**, meaning all land. This landmass had split and the various continental fragments had drifted to their present positions.

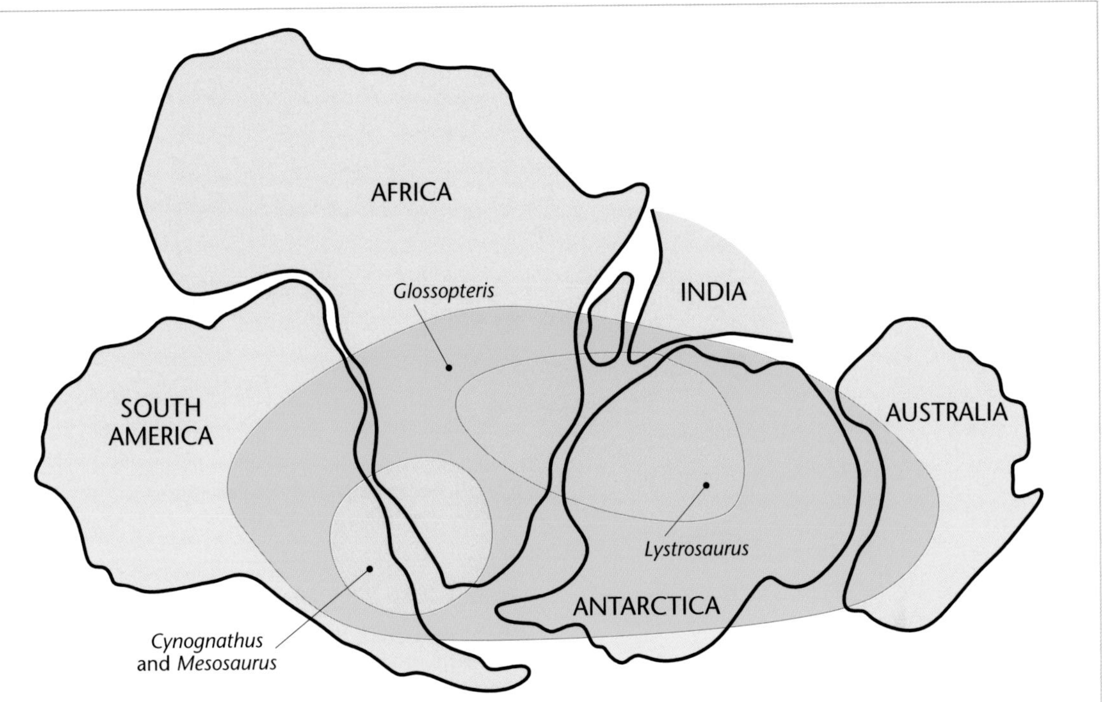

Figure 2.2 *The distribution of fossils of land-dwelling reptiles, such as* **Lystrosaurus,** *found in Africa, India and Antarctica,* **Cynognathus** *from Africa and South America, and the estuarine reptile* **Mesosaurus** *from Africa and South America – as well as the plant* **Glossopteris,** *which occurs on all of the southern continents and India – is inexplicable with the present arrangement of continents. But it makes sense if these continents were once joined together.*

Wegener's hypothesis of **Continental Drift** was supported by many geologists, among them prominent South African Alex du Toit in his work *Our Wandering Continents,* published in 1937.

The geological evidence that the continents were once joined is strong, particularly for the continents of the southern hemisphere. Old mountain belts and rocks of similar age in adjacent continents link up when the continents are reassembled like a picture on a jigsaw puzzle (**figure 2.1**). The reassembly of continents also explains the distribution of fossils of certain land-dwelling reptiles, as well as the fossil remains of a distinctive plant (**figure 2.2**). How could these terrestrial organisms have crossed the oceans if the continents had always occupied their present positions?

Other evidence is based on climatic conditions. Coal beds, which form in equatorial or temperate conditions, occur in Antarctica, indicating that Antarctica formerly occupied a position much closer to the equator. Glacial deposits, which formed about 300 million years ago, occur on all the landmasses of the southern continents, as well as India (**figure 2.3**). These glacial deposits can be observed in the Karoo region and in KwaZulu-Natal, and are discussed later (*see* page 195). The distribution of the deposits can best be explained if the southern continents had assembled as a large landmass or **supercontinent** centred on the South Pole some 300 million years ago. This supercontinent became known as **Gondwanaland**, subsequently shortened to **Gondwana** (Gondwana means land of the Gonds, a tribe in India, so to say Gondwanaland is tautology).

Although the hypothesis of Continental Drift had supporters, many earth scientists, especially from the northern hemisphere, were sceptical, because there was no acceptable mechanism to account for the splitting of large continents and their subsequent movement. After all, what conceivable force could be so powerful as to drive massive continents across the globe, especially

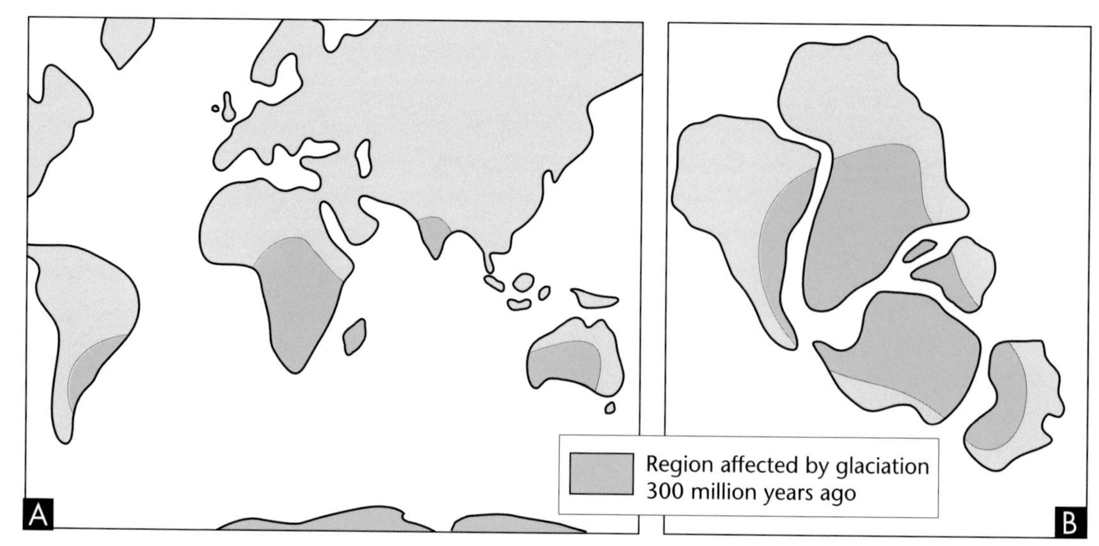

Figure 2.3A ***Glacial deposits of the same age occur on the southern hemisphere continents.***

Figure 2.3B ***If the southern hemisphere continents and India are united, the glacial deposits combine into a single entity, which could have formed when the combined continent lay over the South Pole.***

against the resistance provided by the rocks of the ocean floor? Sir David Attenborough, well known broadcaster and author, was a student during the 1940s when this debate was raging, and he provides an account of the attitudes of academics towards this new theory:

> *The Regius Professor of Geology in the University of Cambridge read our petition with care. We were a group of geology students. It was 1946. We demanded (in that way students are prone to do) that our lectures should examine the theory of continental drift. Our professor explained firmly that when someone could demonstrate to him the existence of the unimaginable forces in the earth that would be needed to make a continent move over the surface of the globe, he would lecture about it. But until then, the proposition was moonshine. So we went back to our lectures in this and other departments, and tried to make sense of the convoluted explanations we were being given of how, for example, dinosaurs managed to get across the oceans from North America to Australia, and why Australia's spectacular Banksia plants should be so extraordinarily similar to the beautiful Proteas of South Africa. And we were not allowed to dream about Gondwana.*

A rival hypothesis to explain the separation of the continents invoked expansion of the Earth. The **Expanding Earth Hypothesis**, championed by Australian Warren Carey, proposed that the present continents once covered the entire Earth. Expansion of the Earth resulted in the formation of the ocean basins, which separated the continental fragments. A simple analogy would be inflating a balloon covered with wet tissue paper. But this theory also suffered from the lack of a plausible mechanism to account for the expansion. The notions of Continental Drift and the Expanding Earth remained geological curiosities, and geologists continued their data gathering – mapping, collecting, describing and classifying rocks and fossils. The collection and classification of basic data has its place in science, and ultimately began to bear unexpected fruits.

A NEW VIEW OF THE EARTH

Since the start of space programmes in the late 1950s and early 1960s, views of Earth from space have become commonplace and have given us a completely new perspective of the Earth. Commonly referred to as the Blue Planet, some 70% of Earth's surface is covered by water, and satellite images (**figure 2.4**) typically show white swirls of clouds of the **atmosphere** and dark blue areas that are the ocean waters of the **hydrosphere**. The landmasses stand out in stark contrast to the oceans.

Extensive surveying and mapping of the topography of the ocean floor has been carried out since the 1950s, using sonar, seismic reflection and other techniques, ironically as a necessity for submarine warfare during the Cold War period. In addition, systematic geological investigation of the ocean floors commenced in the late 1960s. Knowledge of the ocean floor has had a profound impact on our understanding of the nature of the Earth's crust, the processes operating on it, and its evolution.

Maps of the ocean floor, created mainly by the military, combined with conventional maps of the continents, have enabled us metaphorically to peel off the atmosphere and the hydrosphere, providing a view of a naked Earth (**figure 2.5**). The two principal features on its surface are the **continents**, including the submerged **continental shelves**, and the **ocean basins**. Systematic surveys of the ocean basins over the past 40 years have revealed that the difference between continents and ocean floors is not just a matter of elevation – there are fundamental geological differences.

THE EARTH'S NAKED FACE

The Earth's radius is about 6 400 km, while the vertical distance between the deepest point on the ocean floor and the top of Mt Everest, Earth's maximum topographic relief, is about 20 km. To put this in an everyday context, suppose the Earth were shrunk to a ball 10 cm in diameter. The maximum relief would be just more than a tenth of a millimetre, whereas the average relief would be a tiny fraction of a millimetre. So the Earth is fairly smooth, but for us living on its surface, its relief is extremely important. For one thing, water occupies the lower ground, forming the oceans, while the higher ground forms the continents, providing habitat for terrestrial animals such as us.

Although really quite minor, the topography of the Earth nevertheless raises some fundamental questions. Why is the Earth divided into high and low areas – continents and ocean basins? Continents consist mostly of extensive flat areas, whereas the major mountains of the world form long, narrow belts such as the Cordillera of the Americas, extending from Alaska to the tip of South America, and the mountain chain of southern Europe and Asia, extending from the Pyrenees to the Himalaya (**figure 2.5**). Forces of erosion relentlessly attack and sculpt these mountainous regions, but why have they not been reduced to flat plains over geological time? And why do mountains occur in belts at all?

The topography of the ocean floor is in many ways even more remarkable than that of the continents. Perhaps the most striking feature is an almost continuous mountain chain (**figure 2.5**) that can be traced around almost the entire globe from the Pacific Ocean, through the Indian Ocean and across the Atlantic, a mountain chain about 70 000 km long, like the seam on a tennis ball. This is known as the **mid-ocean ridge** system, and it is approximately 1 500 km wide. It rises gently from the surrounding **abyssal plain** to reach heights of as much as 3 km in places, with a pronounced central valley.

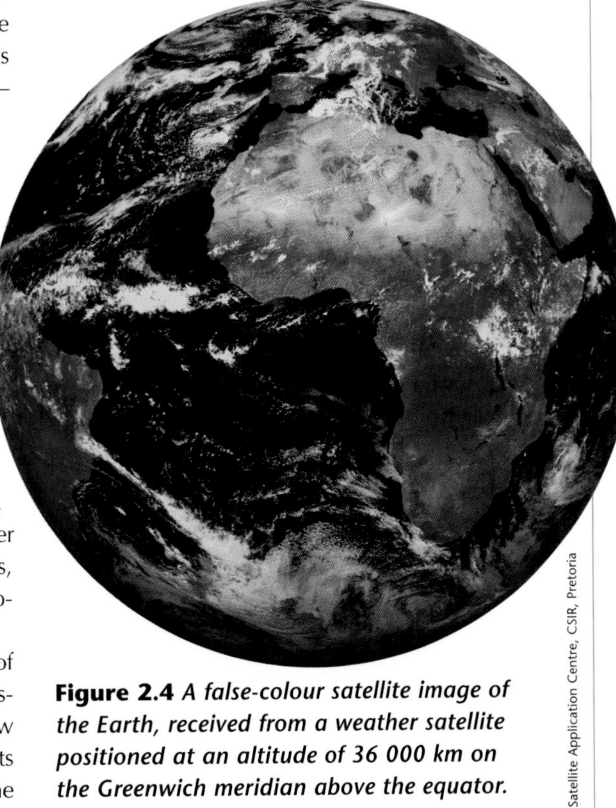

Satellite Application Centre, CSIR, Pretoria

Figure 2.4 *A false-colour satellite image of the Earth, received from a weather satellite positioned at an altitude of 36 000 km on the Greenwich meridian above the equator.*

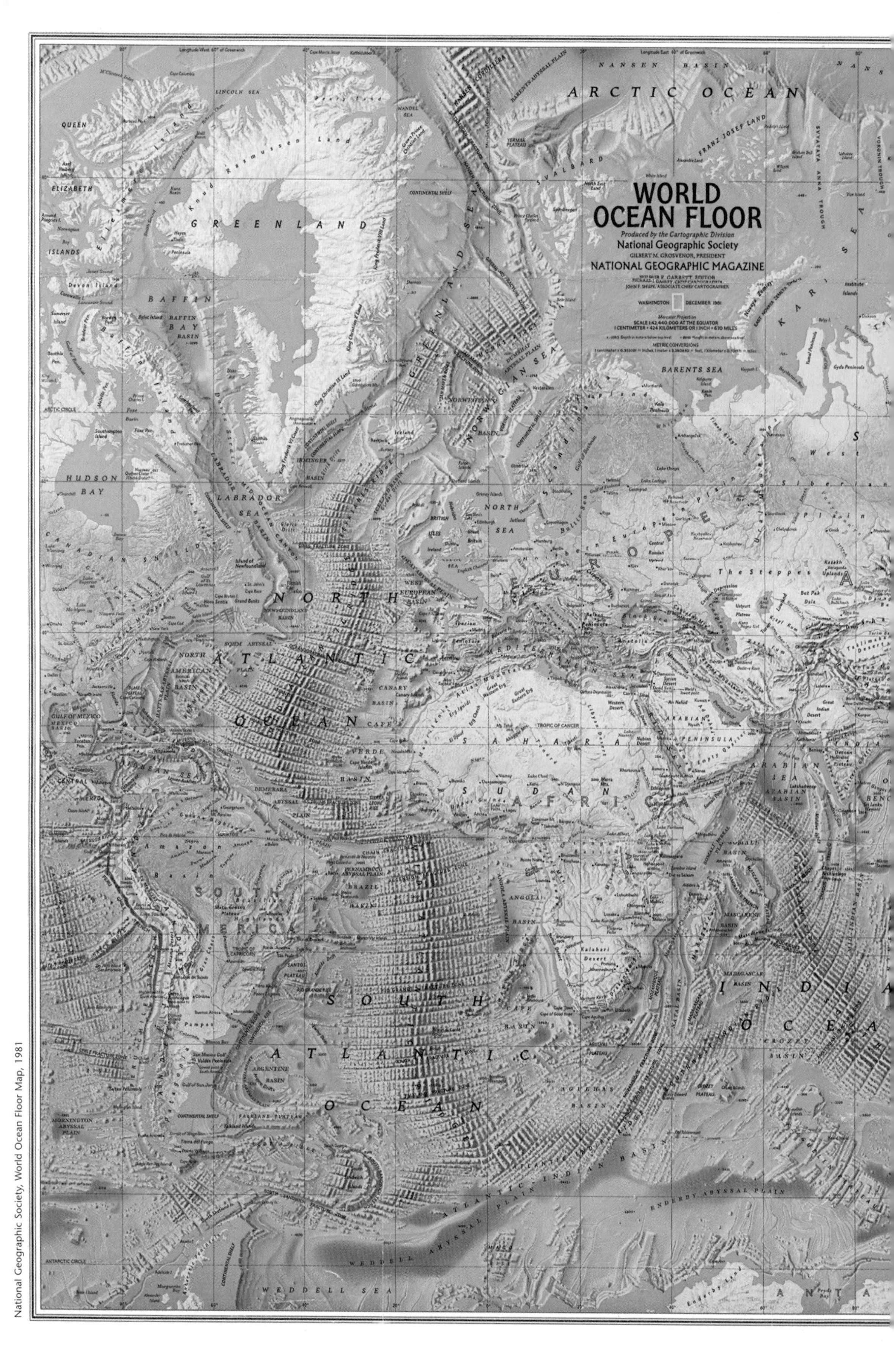

National Geographic Society, World Ocean Floor Map, 1981

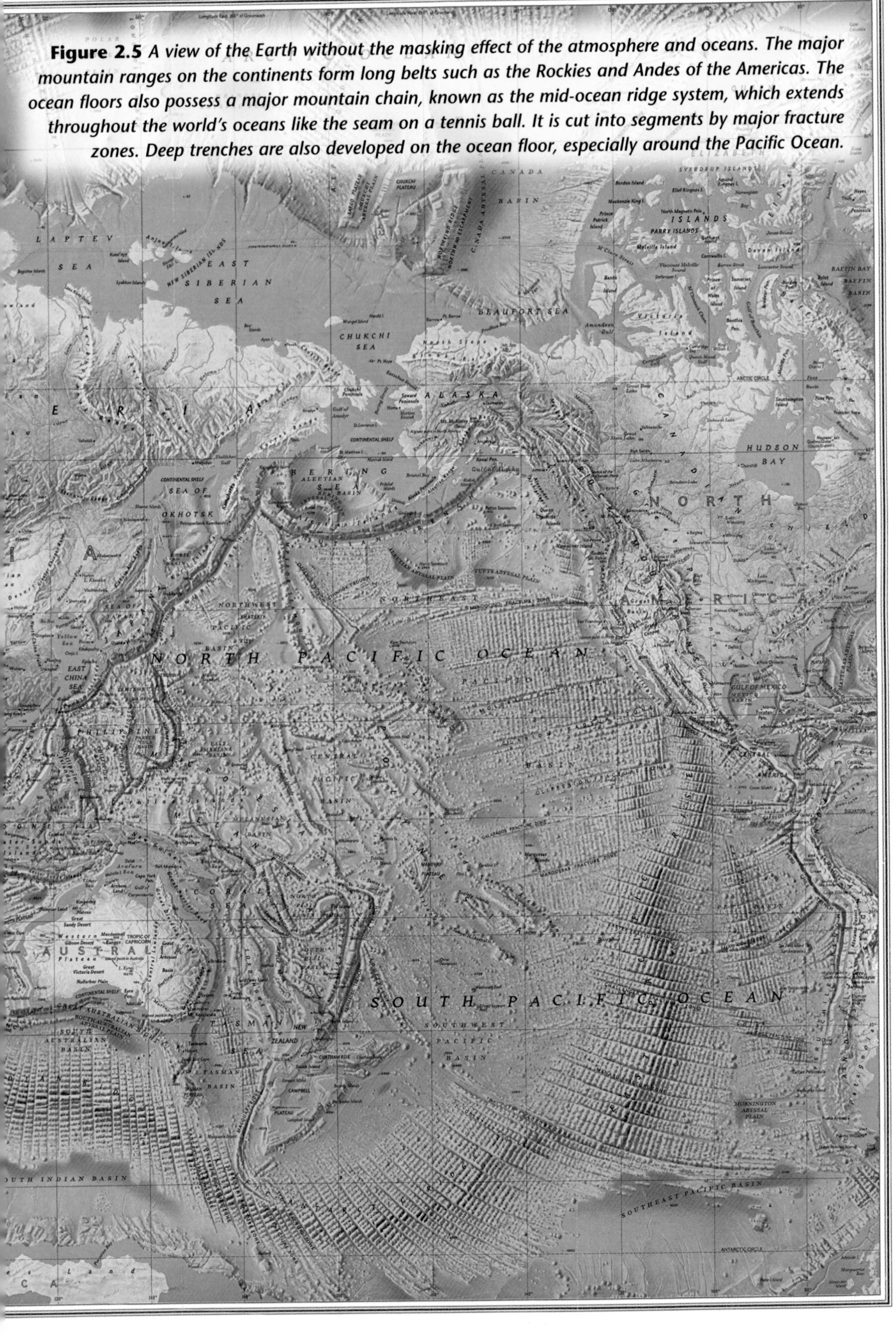

Figure 2.5 *A view of the Earth without the masking effect of the atmosphere and oceans. The major mountain ranges on the continents form long belts such as the Rockies and Andes of the Americas. The ocean floors also possess a major mountain chain, known as the mid-ocean ridge system, which extends throughout the world's oceans like the seam on a tennis ball. It is cut into segments by major fracture zones. Deep trenches are also developed on the ocean floor, especially around the Pacific Ocean.*

EARTHQUAKES

Earthquakes occur when rocks that have been subjected to sufficiently high stresses, fracture and in the process rebound and release energy in the form of seismic waves, much like the release of sound waves when a dry twig breaks. These seismic waves radiate outwards in all directions. Seismic waves behave in a similar way to light waves and can bend (refract) and reflect at boundaries between different rock layers.

The waves that travel through the interior of the Earth are called body waves, whereas those that travel along the Earth's surface are called surface waves. It is the body wave that is particularly useful in understanding the Earth's interior.

There are two main types of body waves, namely p-waves and s-waves. P-waves are compressional waves in that they compress and expand the material through which they pass. They can be transmitted through solids, liquids and gases. These waves are the fastest and arrive first at any measuring station. S-waves are shear waves and move particles from side to side at right angles to the direction in which the wave travels, like shaking a rope from side to side. S-waves are only transmitted through solids (liquids are unable to transmit shear) and travel slower than p-waves.

The time difference between the arrival of p- and s-waves is a measure of the distance to the source of an earthquake. The speed of seismic waves depends on the density and the rigidity or stiffness of the rock through which the waves travel. The speed at which the waves travel through the Earth gives an indication of the rock density and composition. The distribution of p- and s-waves received from distant earthquakes provides a view of Earth's interior (*see* diagram below).

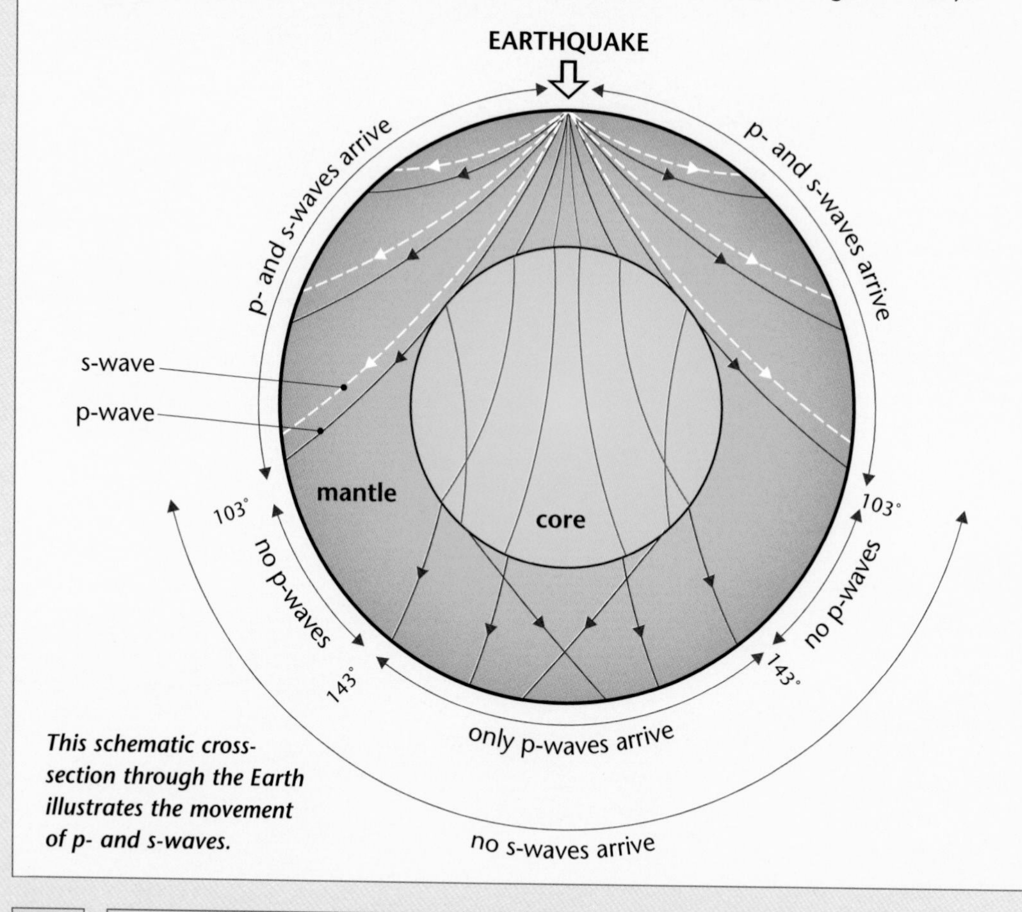

This schematic cross-section through the Earth illustrates the movement of p- and s-waves.

The range is cut into segments by **fracture zones** thousands of kilometres long. Virtually the entire length of this mountain range lies below sea level. Also hidden beneath the oceans are deep **trenches**, particularly around the Pacific Ocean, where the greatest ocean depths are attained, such as the 11-km-deep Mariana trench east of the Philippines. Also visible on the ocean floor are chains of **sea mounts**, or submerged islands. The most impressive is the chain extending northwards from Hawaii, via Midway Island, to the trench that borders the Aleutian Islands.

Erosion sculpts the mountain ranges of the continents, but there is no erosion on the sea floor. So how does this topographic variability on the ocean floor arise?

Before these questions about the topography of the Earth can be answered, we need to take a closer look at the Earth's deep interior.

INSIDE THE EARTH

Rocks on the Earth's surface are easily accessible for examination and we know a lot about them. We can also get a certain amount of information on rocks near the surface from deep boreholes (generally less than 5 km deep). Some indications of rock compositions down to depths of about 100 km in certain regions can be obtained by examining rock fragments transported to the surface in volcanoes.

But how do we know what materials occur at greater depths, towards the centre of the Earth 6 400 km below us?

The density of rocks at the Earth's surface averages about 2.8 g/cm^3, but the density of the Earth as a whole is 5.5 g/cm^3. This implies that there has to be much denser material in the Earth's interior, but we have no way of sampling it. In fact, our knowledge of the deep interior of the planet is largely derived indirectly, from the study of seismic waves generated by earthquakes. The way these seismic waves travel through the Earth gives an indication of the composition and internal structure of the planet – much like a CAT (or CT) scan used in medical examinations (*see* 'Earthquakes', left).

The study of the interior of the Earth using seismic waves has revealed that the Earth is a **differentiated** planet, consisting of concentric shells with different compositions and varying physical properties. The primary layers are called the **crust**, **mantle** and **core** (**figure 2.6**). The boundaries between these layers affect the velocity of seismic waves due to sudden density changes, and are known as **seismic discontinuities**.

The Earth's crust varies from about 7 km to 35 km thick. (Some notable exceptions will be dealt with later, *see* page 47). The boundary that separates the base of the crust from the underlying mantle is the **Mohorovičić Discontinuity**, or simply the **Moho**, where a substantial increase in density occurs due to a change in chemical composition. The mantle is solid, made up of rocky material consisting mainly of magnesium, iron, silicon and oxygen. The velocity of seismic waves increases with increasing depth as they pass through the mantle, a consequence of the rising pressure.

Near the top of the mantle, over a depth range from 100–200 km, the waves slow down by about 7% in a region called the **Low Velocity Zone (LVZ)**. In this zone, the rocks display more plastic behaviour because they are close to their melting point. At a distance of 2 900 km below surface is a major boundary, which is known as the **Gutenberg Discontinuity**, and which marks the mantle-core boundary. Seismic evidence indicates that the outer core is a very dense liquid; by analogy with certain meteorites, it is believed to be molten nickel and iron. The inner core is solid, but it is also thought to be composed of a nickel-iron alloy. The density of the core is about 10.8 g/cm^3, whereas the density of the mantle ranges from 3.3 g/cm^3 at its top to 5.5 g/cm^3 at its base. These higher density materials in the interior are responsible for the Earth's high overall density.

The region above the LVZ, consisting of the upper part of the mantle and the crust, is referred to as the **lithosphere**, which forms a fairly rigid outer carapace to the Earth. The LVZ, also known as the **asthenosphere**, is plastic. The mantle below the asthenosphere, termed the **mesosphere**, is more rigid than the asthenosphere, but less rigid than the lithosphere.

This second subdivision of the Earth into layers may seem confusing, especially as the lithosphere includes the upper part of the mantle, as well as the crust. The two subdivisions arise because different criteria are used: the crust-mantle-core subdivision is based on differences in chemical composition

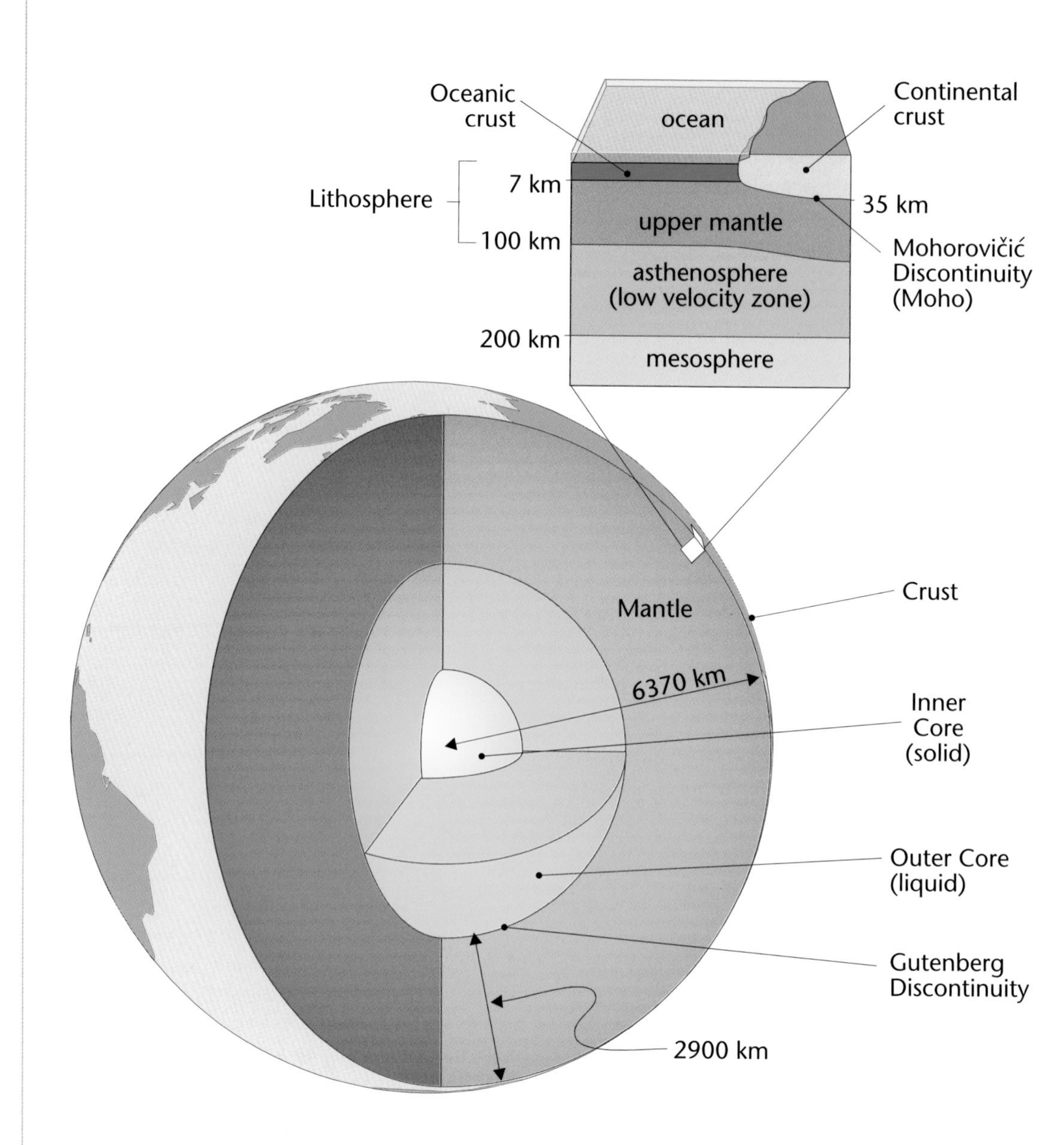

Figure 2.6 *A cut-away section through the Earth showing its component parts. The primary layers, which arise from major differences in chemical composition, are the crust, mantle and core. There are two kinds of crust, one forming the continents and continental shelves, and the other the ocean floors. These have different chemical compositions. The Earth can also be subdivided on a basis of the rheology of the rocks. The upper part of the mantle and the crust are rigid, and this combined layer is known as the lithosphere. Below it lies the 100-km-thick asthenosphere, which is very plastic. This layer is also known as the Low Velocity Zone because seismic waves slow down as they pass through it. Below the asthenosphere is the mesosphere, which is more rigid than the asthenosphere, but less so than the lithosphere.*

ROCK CLASSIFICATION

Rocks are aggregates of minerals (a mineral is a natural chemical compound). Rocks form in a variety of ways, but just two produce the vast majority of rocks. The first is by crystallisation from a molten state, producing **igneous rocks**, and the second is by accumulation of particles of sediment or by chemical precipitation, producing **sedimentary rocks.**

Pre-existing igneous or sedimentary rocks may become subject to pressure and temperature conditions very different from those under which they formed. As a consequence, the minerals in the rock invariably undergo change to form new minerals that are stable under the imposed pressure and temperature conditions. Such rocks are termed **metamorphic rocks.**

Each of these rock types is characterised by diagnostic mineral combinations and structures, and can be fairly easily identified by a trained observer.

Schist – metamorphic rock ***Granite – igneous rock*** ***Banded iron formation – sedimentary rock***

between the layers, while the lithosphere-asthenosphere-mesosphere subdivision is based on differences in rheological properties; the lithosphere is rigid, the asthenosphere very plastic, and the mesosphere slightly plastic.

Temperature increases with depth in the Earth, and in the upper lithosphere rises by about 25°C/km, but the rate of increase slows with depth. The temperature of the inner core is about 5 500°C. Iron-nickel alloy melts well below this temperature at the Earth's surface, but because of the high pressure in the centre of the Earth, the inner core is solid, as is the mantle.

Although the crust makes up only the thin outer part of the lithosphere, it plays an extremely important role in geological processes near the Earth's surface. Direct observation and seismic studies have revealed that there are two kinds of crust, which differ in important ways. One forms continents and continental shelves and the other the ocean floors.

The **continental crust** is about 35 to 40 km thick, but beneath high mountain ranges may be as much as 75 km thick. It is composed mainly of granite and related rocks rich in silicon, aluminium and oxygen (*see* 'Rock classification', above and 'Classification of igneous rocks' on page 34). These rocks are relatively light, with densities of about 2.6 to 2.7 g/cm^3, which may increase to about 2.9 g/cm^3 in the deeper parts of the continental crust. The continents also contain remnants of older rock masses that have been buckled and bent, and subjected to high temperatures and pressures (metamorphic rocks: *see* 'Rock classification', above) and include

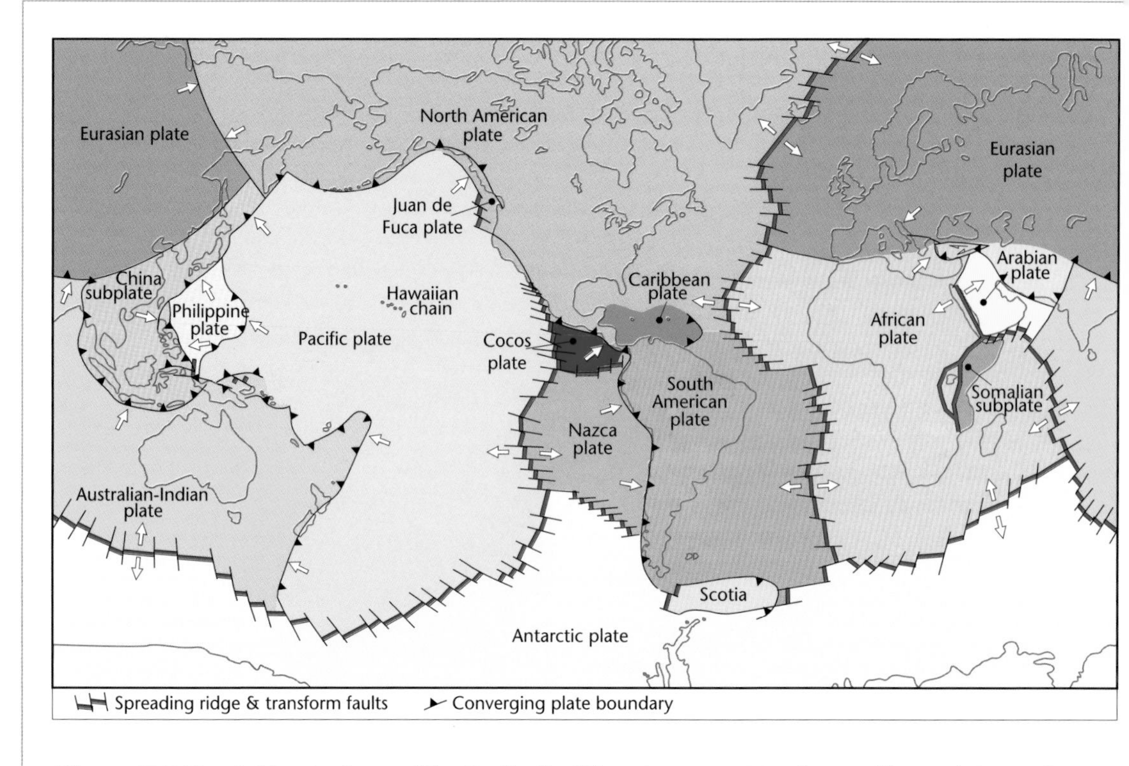

Figure 2.7 ***The rigid outer layer of the Earth, the lithosphere, consists of several large slabs or plates, which are moving relative to each other. The most intense geological activity on the Earth, in the form of earthquakes and volcanic eruptions, occurs along the boundaries between the plates.***

some of the oldest known rocks on the planet – up to 4 100 million years old. On many continents the granites and associated metamorphic rocks are covered by layers of sedimentary rock. The underlying granite and associated rocks are therefore often referred to as the **basement**, as they form the floor on which the sediments were deposited.

Oceanic crust, by contrast, is 6–8 km thick. It consists of basalt and related rocks (*see* 'Classification of igneous rocks' on page 34), which are dark igneous rocks, rich in magnesium, iron, silicon and oxygen, with a density of about 2.9 to 3.0 g/cm^3, and are denser than the continental crust. The rocks of the ocean floor are generally much younger than continental rocks, and the present-day ocean floor is not more than about 200 million years old.

THE THEORY OF PLATE TECTONICS

The geological evidence that certain continents were once joined together was already strong when Alex du Toit published his book in 1937 and it became stronger as more data on the geology of our continents accumulated. So strong was the evidence from the study of continental geology that there simply *had* to be a mechanism to account for the separation and dispersal of the continents, but knowledge of the nature of the ocean floors was lacking and there was a dearth of ideas.

Topographic maps of the ocean floors provided the critical clues. The mid-ocean ridges in particular attracted attention, especially the mid-Atlantic ridge, which neatly bisects the Atlantic Ocean (**figure 2.5**), and the idea of **sea-floor spreading** was born. It was proposed that the ocean floor was somehow created at mid-ocean ridges, causing spreading of the sea floor, which moved the continents apart like a conveyor belt. Military technology provided the tools to test the idea, in the form of sensitive instruments that could measure small variations in the magnetic properties of the rocks of the sea floor – instruments that had been developed to detect submarines. These revealed symmetrical magnetic patterns on either side of the ocean ridges. The concept of sea-floor spreading had passed the test (*see* 'Earth's magnetism' on page 36 and 'Palaeomagnetic studies and plate tectonics' on page 39).

The discovery of sea-floor spreading was good news for the proponents of the Expanding Earth Hypothesis, who proposed that the continents were being separated by the creation of new crust on the ocean floors as the Earth expanded. However, as information emerged on the rates of spreading at the ridges and on the ages of the ocean floors, many supporters of the Expanding Earth Hypothesis threw in their lot with the Continental Drifters.

It was improbable that the Earth could be expanding at the rapid rate indicated by the measurements. Their theory predicted rates of continental separation of about 0.5 cm/year, while the actual rates were several centimetres per year. Moreover, the ocean floors were found to be very young (the oldest is about 200 million years) compared to the age of the Earth (4 600 million years). This implied sudden, recent expansion of the Earth – a very unlikely scenario. Other discoveries that are discussed later in this chapter finally relegated the Expanding Earth Hypothesis to history.

The term sea-floor spreading was short lived. The broader implications were quickly realised, especially the fact that if ocean floor was being created at ocean ridges, somewhere else crust must be destroyed because the Earth's surface area was considered by most scientists to be more or less constant. Sea-floor spreading could thus be only half of the story. It was soon realised that ocean trenches were the places where crust was destroyed. The expanded concept of crust creation at ridges and destruction at trenches became known as the **Theory of Plate Tectonics**.

In brief, the theory postulates that the outermost shell of the Earth, the lithosphere, is divided into a number of separate, rigid slabs or **plates** that are moving relative to one another (**figure 2.7**). The zones where the plates make contact with each other are called **plate boundaries** and are the most geologically active regions on the planet. They are marked by earthquake and volcanic activity, and are usually associated with some topographic feature such as a mountain range or trench (**figure 2.8**).

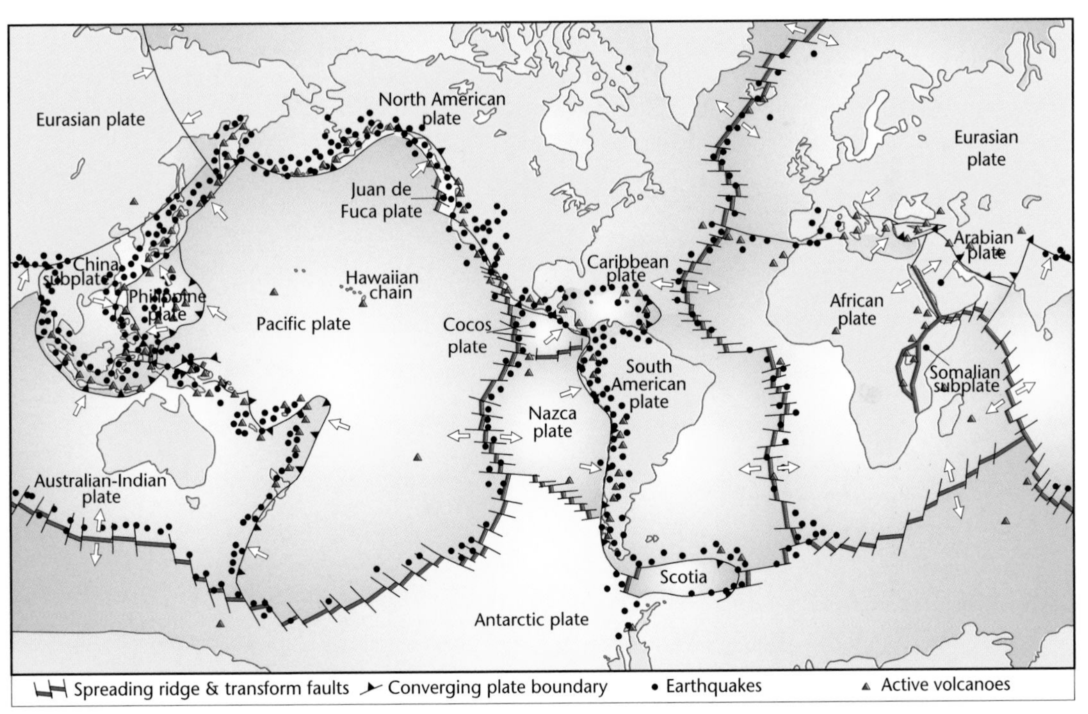

Figure 2.8 *The Earth's volcanic and earthquake activity forms belts, which outline the plates.*

CLASSIFICATION OF IGNEOUS ROCKS

Igneous rocks are those that form by solidification from the molten state. They consist of naturally occurring chemical compounds called minerals. Most of the minerals making up igneous rocks belong to the silicate mineral family, in which silicon and oxygen form a major component. The minerals present in an igneous rock are closely related to the chemical composition of the rock.

Igneous rocks form from molten material – called **magma** when below surface and **lava** when above. The texture of the resulting igneous rock depends on the rate at which the melt solidifies. Lavas, which are melts erupted from volcanoes, cool very quickly so they form rocks with extremely small crystals, and sometimes they even form glass (termed fine-grained rocks). Such rocks are referred to as **volcanic rocks**. Igneous rocks that form far below the surface where heat loss is slow because of the insulating effect of the surrounding rock crystallise very slowly and thus consist of large crystals – up to 1 cm in size (termed coarse-grained rocks). These are known as **plutonic rocks**. Rocks that form at intermediate depths (termed medium-grained rocks) are known as **hypabyssal rocks**.

There are many different types of igneous rock, each with its own name, but only a few of these are common, and only these will be discussed. Igneous rocks are classified according to the proportions and types of their constituent minerals, and their grain size. The important minerals are: iron and magnesium silicates, called **olivine** and **pyroxene**; calcium, iron and magnesium silicate, called **amphibole**; potassium, sodium and aluminium silicates, called **alkali feldspars**: calcium aluminium silicate, called **calcium feldspar**; and **quartz**, which is silicon dioxide.

The classification scheme based on minerals and grain size is shown in the diagram. The columns are determined by the minerals present and the rows by the grain size. Rocks in each column have exactly the same minerals, and differ only in their grain size. For example, **basalt, dolerite** and **gabbro** all contain about equal proportions of pyroxene and calcium feldspar. Rocks on the left-hand side, which consist mainly of quartz and alkali feldspars, are pale in colour (obsidian is an exception, being black), while rocks on the right, which have a large amount of pyroxene, are dark coloured.

Rock types intermediate between those shown in the diagram also occur. For example, **granodiorite** has mineral proportions intermediate between granite and diorite. **Dacite** is the volcanic (fine-grained) form of this rock. Basalt is the most common type of igneous rock, as it forms the upper part of the oceanic crust, followed by gabbro, which forms the deep oceanic crust. Next most common are granites and diorites, which form most of the continental crust.

Colour	Light grey to pink	Medium grey	Dark grey	Dark grey to black
Minerals	quartz, alkali feldspar ±amphibole	alkali feldspar, calcium feldspar amphibole	calcium feldspar pyroxene	olivine, pyroxene
Volcanic rocks (fine-grained)	RHYOLITE (obsidian when glass)	ANDESITE	BASALT	KOMATIITE
Hypabyssal rocks (medium-grained)	microgranite	microdiorite	DOLERITE	micro-peridotite
Plutonic rocks (coarse-grained)	GRANITE	DIORITE	GABBRO	PERIDOTITE

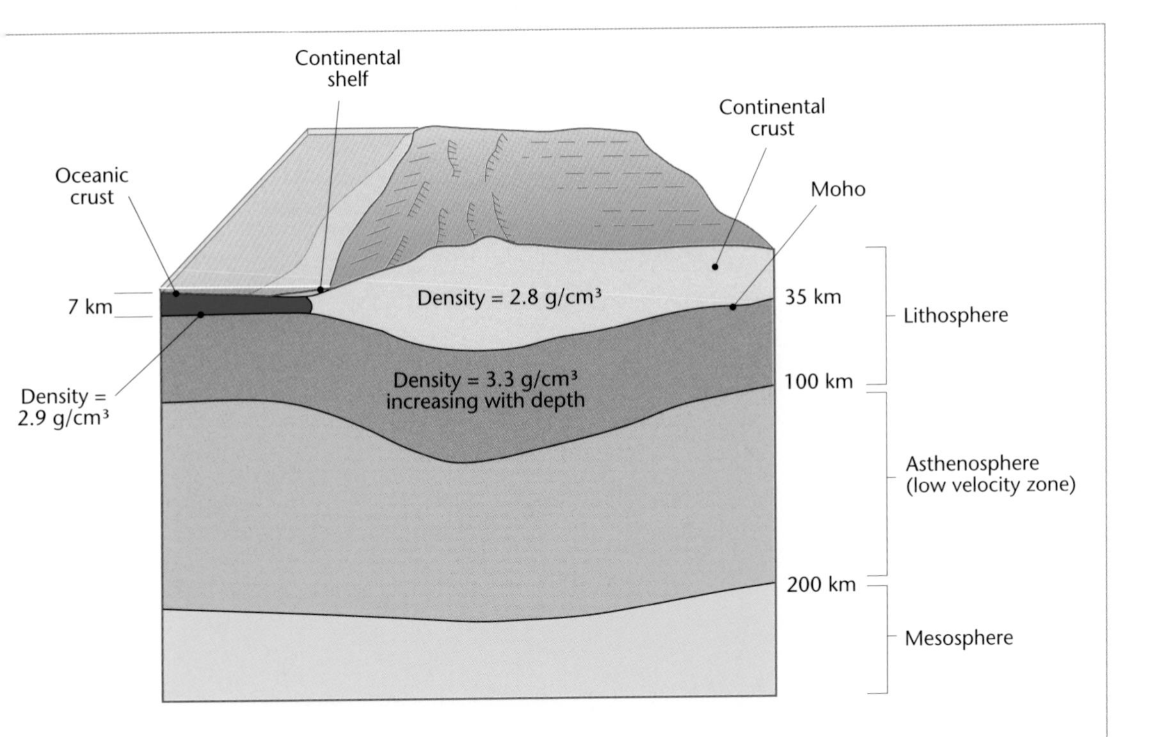

Figure 2.9 ***A cross-section through a typical plate. In this case, both continental and oceanic crust are developed. The continental crust has a slightly lower density than oceanic crust and is thicker, so rises above sea level. The continental shelf is also underlain by continental crust, but is thinner and submerged. It is usually buried beneath a layer of sediment eroded from the continent. Beneath the crust is rigid mantle material of relatively high density. Together with the crust this forms the lithosphere, which floats on the plastic asthenosphere below.***

Before looking at the processes of relative plate motion and plate boundary interactions, it is useful to examine a cross-section through a typical plate. A diagram showing a vertical slice through the outermost layers of the Earth is shown in **figure 2.9**. The lithosphere, which makes up the thin rigid plate, is about 100 km thick. It is underlain by the asthenosphere (or LVZ), which is plastic and acts like a lubricating layer on which the rigid lithospheric plates slide or, more correctly, creep. The lithosphere essentially floats on the asthenosphere. Plates can consist of either upper mantle plus continental crust plus oceanic crust (such as the African plate), or upper mantle plus oceanic crust (such as the Pacific plate) (**figure 2.7**).

Because continental crust is both thicker and less dense than oceanic crust, it stands (actually floats) higher on the Earth's surface than oceanic crust. Thinner continental crust around the edges of the continents is generally submerged, forming the continental shelves. The land areas of Earth therefore consist mainly of continental crust. Beneath mountain belts, the continental crust is thicker than average and this is why mountains are elevated (by analogy, large ice cubes float higher above the water surface than smaller ones).

Although the crust makes up only the thin outer skin of the lithosphere, the difference in composition and density between continental and oceanic crust fundamentally influences their behaviour and shapes the outer surface of the Earth.

JOSTLING PLATES

Plates are internally rigid, but are detached from and move relative to each other. The most important interactions occur at their boundaries. Here, plates may pull apart, collide – with one usually plunging beneath the other – or slide past each

EARTH'S MAGNETISM

Earth, like several other planets, has a magnetic field of internal origin. At the Earth's surface we can observe the presence of the magnetic field with a compass and we can measure its strength with an instrument called a **magnetometer**. The Earth's magnetic field resembles that generated by a simple bar magnet at the centre of the Earth, with magnetic North and South Poles located close to the geographic North and South Poles (the Earth's axis of rotation).

But the notion of a simple bar magnet inside the Earth is incorrect. If a magnet is heated above a critical temperature, called the **Curie Point** (550–600°C for most magnetic material), it loses its magnetism. Temperatures within the Earth increase with depth at a rate of about 25°C/km from the surface so the mantle and core are too hot to have a magnetic field generated from a permanent magnet. Just how the Earth's magnetic field is generated is still imperfectly understood, but the generally accepted theory is that the magnetic field is formed electromagnetically in the molten outer portion of the nickel-iron core as a result of convective motion.

It is well documented that the polarity of the Earth's magnetic field has switched repeatedly in the past in events known as **magnetic reversals**, when the magnetic North moves to the geographic South Pole, and *vice versa*. In the last 20 million years, 18 reversals have occurred, the most recent about 780 000 years ago. Normal polarity is taken when the magnetic North Pole is close to the geographic North Pole, as it is today. Reversed polarity is when the magnetic North Pole is located close to the geographic South Pole.

The South African National Research Foundation maintains a magnetic observatory at Hermanus in the Western Cape where the local magnetic field is constantly monitored. The strength of the field has declined by about 20% since observations began in 1940, which reflects a global weakening of the magnetic field. Satellites that have been launched since 1980 to study the Earth's magnetic field have revealed variations in the magnetic field to a degree of detail that was not previously possible. When the field is projected down to the Earth's core where it originates, it has been found that there is a patch below southern Africa where the field has actually reversed. The growth of this reversed patch may possibly be linked to the decrease in the magnetic field strength – a decrease that some people have suggested could eventually lead to a global magnetic field reversal.

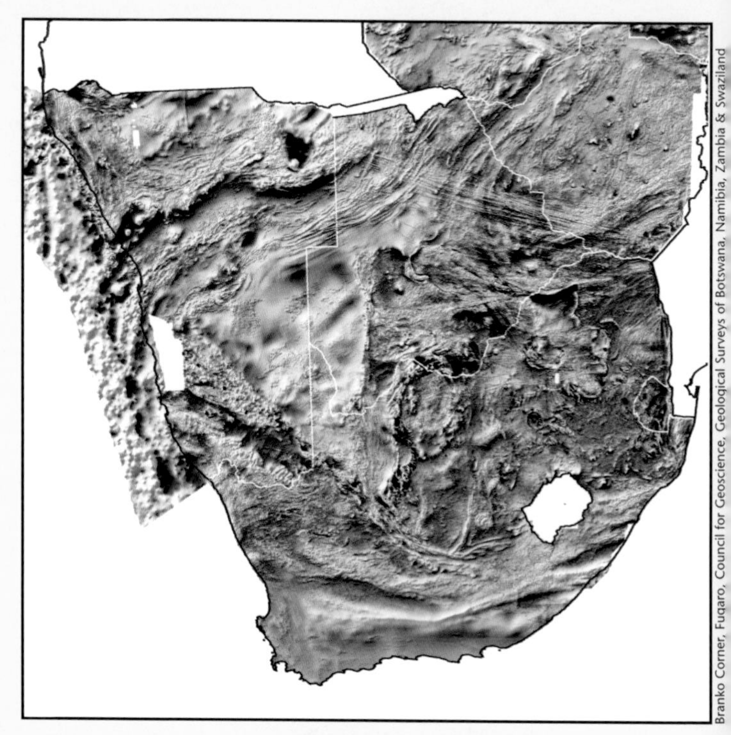

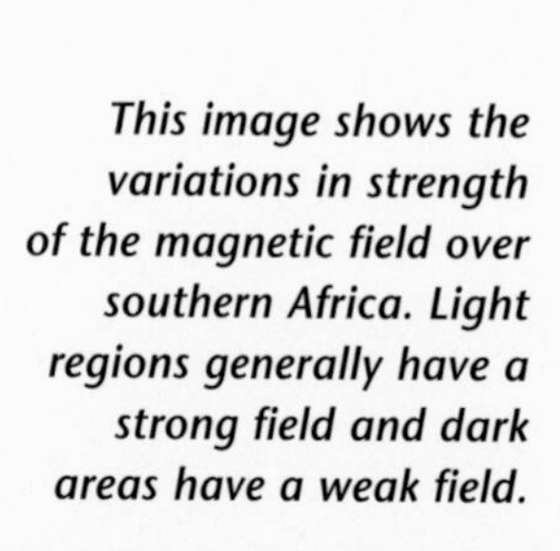

This image shows the variations in strength of the magnetic field over southern Africa. Light regions generally have a strong field and dark areas have a weak field.

Branko Corner, Fugaro, Council for Geoscience, Geological Surveys of Botswana, Namibia, Zambia & Swaziland

MAGNETIC MAPS

The magnetic properties of rocks vary as a consequence of their differing amounts of magnetic minerals. The strength of the Earth's magnetic field consequently varies slightly from place to place due to changes in the underlying rocks. Rocks rich in magnetite (a magnetic iron oxide) locally concentrate the field (induced magnetisation), producing a locally strong field. Rocks may themselves be weakly magnetised (remanent magnetisation), and this may enhance or oppose the Earth's field.

Magnetometers, often towed behind aircraft, are used to measure these small changes in magnetic field strength, and results are portrayed in the form of maps. Such maps, as well as the data themselves, are very useful as they provide an indication of the nature of rocks beneath the Earth's surface. They are especially valuable in exploration for mineral deposits and in geological mapping in general. Exploration for gold deposits in the Witwatersrand Basin, in particular, was greatly facilitated by the use of magnetic maps. The aeromagnetic map of southern Africa is shown in the accompanying figure (opposite), together with a partial interpretation (below).

PALAEOMAGNETISM

Study of the magnetic properties of rocks from both the continents and ocean floors indicate that most rocks behave like very weak magnets. Igneous and sedimentary rocks often contain very small particles of magnetic minerals (mainly magnetite). As igneous rocks cool down below the Curie Point the magnetite particles become magnetised in the orientation of the Earth's magnetic field at that time, and remain trapped in that orientation. In sedimentary rocks, fine particles and flakes of magnetite incorporated into the sediment at the time it was deposited are also aligned in the magnetic field. In this way a rock can acquire a weak magnetism, the orientation of which provides a record of the Earth's magnetic field at the time of the rock's formation, provided the rock was not subsequently heated above the Curie Point.

This old magnetism or **palaeomagnetism** is essentially a fossil magnetism, and its orientation is easily measured. Studies of these orientations can yield: **direction to the magnetic pole** at the time of rock formation; **magnetic inclination**, which provides information about the latitude where the rock formed (magnetic inclination changes with latitude: at the poles, a compass needle is inclined vertically and at the equator a compass needle is horizontal); and **polarity**, whether the magnetic field was normal or reversed at the time of formation. Palaeomagnetic information can be used to track past movements of continents.

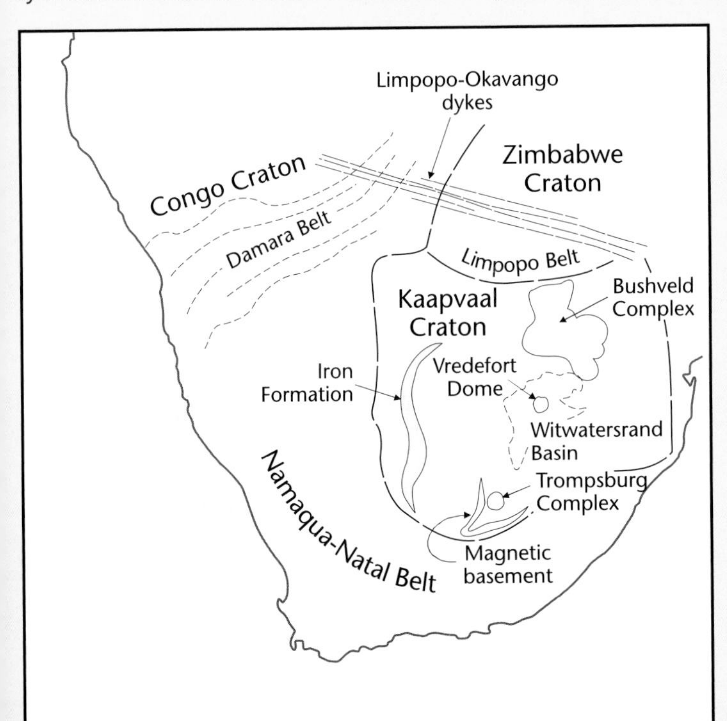

This map shows the geological formations responsible for some of the variations in magnetic field strength.

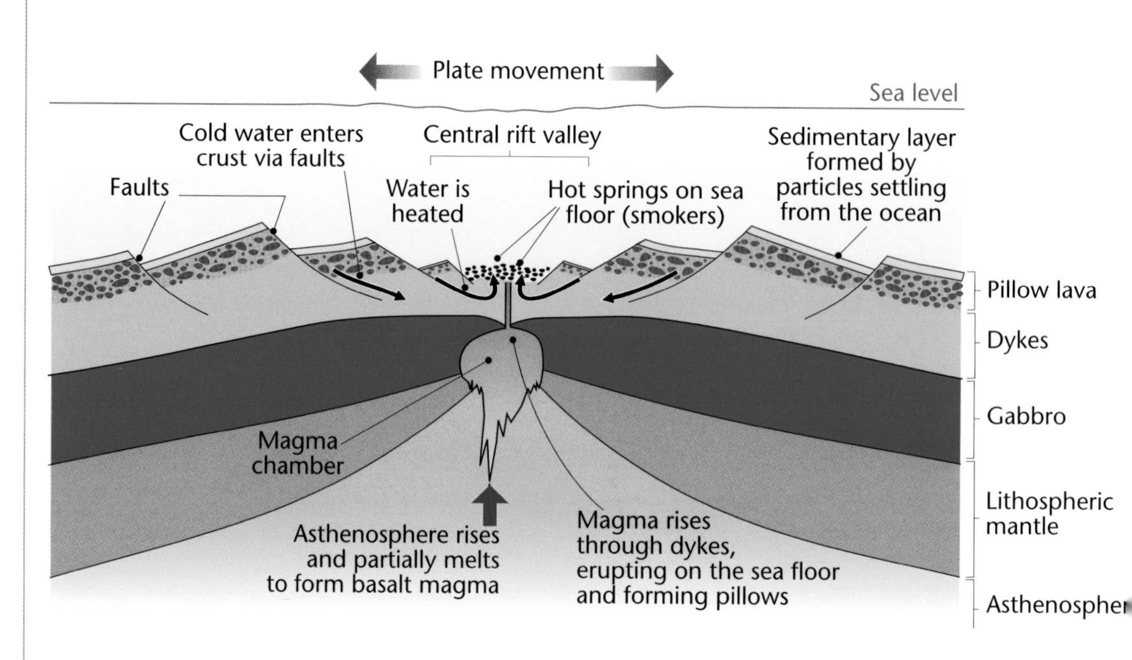

Figure 2.10 *This cross-section through a mid-ocean ridge illustrates the primary geological processes taking place. As the plates on either side of the ridge move apart (arrows), the crust and the underlying mantle (lithosphere) become thinner due to stretching and faulting. The faulting produces steps in the ocean floor, which culminate in a central rift valley. Hot, plastic mantle rises up beneath the thinned lithosphere, causing bulging, which produces the overall rise along the ridge. The rising hot mantle partly melts (due to pressure reduction), releasing basalt magma, which collects in a chamber below the central rift. Slow crystallisation of magma on the walls of the chamber produces gabbro. Periodic rupturing of the roof causes magma to inject upwards, forming dykes and piles of pillow-shaped blobs of lava on the sea floor. These processes continue as the plates separate, and result in the continuous formation of new oceanic crust at the ridge axis. Residues of mantle melting, as well as cooling of the hot mantle, thicken the mantle portion of the lithosphere as the plates separate. Sea water circulates through cracks in the rock, emerging as hot springs on the ocean floor.*

other. Each of these is associated with characteristic topography and volcanic and seismic activity, depending on the nature of the interaction. Although there are some notable exceptions, most plate boundaries occur in oceanic regions. The separation of plates occurs at ocean ridges, convergence usually occurs at ocean trenches, and relative sliding occurs at major fracture zones.

Plate separation

A boundary where two plates separate is marked by a linear topographic feature, usually an ocean ridge. As the plates move apart, the lithosphere becomes stretched. As it thins, the underlying hot asthenosphere rises, bulging the Earth's surface to form a broad arch, which is the mid-ocean ridge.

Stretching of the cold, brittle crust over the arch produces many faults (*see* 'Folding and fracturing of rocks' on page 41). These are parallel to the ridge axis and form a series of steep-sided steps and a central valley, a **rift valley**, like a collapsed keystone in a widening Gothic arch (**figure 2.10**). Shallow earthquakes (less than 30 km deep) dominate this type of boundary, and arise from movements on the faults as the crust and the lithosphere stretch.

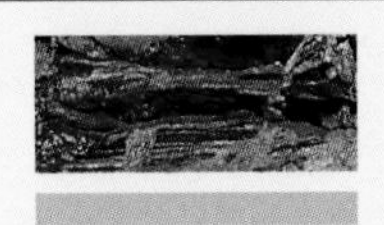

PALAEOMAGNETIC STUDIES AND PLATE TECTONICS

Palaeomagnetic studies have provided powerful evidence in support of Plate Tectonic Theory. Magnetic field reversals or switching of North and South Poles are preserved in the fossil magnetic record. From studies of layered lava flows, it has been shown that some lava flows indicate a normal polarity as it is today, whereas other lava flows indicate a reversed polarity, i.e. the magnetic North Pole is near the geographic South Pole. From combined magnetic polarity measurements and precise dating of lavas, a time scale of magnetic reversals or a magnetic stratigraphy has been established. This is known in detail for the past 50 million years and has been tentatively extended back some 300 million years. During the past five million years Earth's magnetic field has reversed on average once every 500 000 years. The cause of field reversals is not well understood.

Magnetometers measure two things: the main planetary or geomagnetic field and the local magnetic disturbance or **magnetic anomaly** due to the magnetised rocks on the ocean floor. Rocks strongly magnetised in a normal field show a positive anomaly and those in a reversed field show a negative anomaly. A symmetric magnetic banding of normal and reversed polarity is recorded in the sea-floor rocks on either side of a mid-ocean ridge.

Magma rises at the ridge and solidifies to form new oceanic crust. As the crust spreads away from the ridge, new magma rises and is added onto each plate. The magma extrudes at a temperature of 1 200°C. As it cools down to below the Curie Point (600°C), it takes on the polarity of the Earth's magnetic field at that time. If there is a subsequent change in the magnetic field (a magnetic reversal), then that will be recorded in the next lavas to be erupted. The net result is that the sea floor on either side of the mid-ocean ridges has acted like a tape recorder, displaying a record of normal and reversed polarity through time. Precise age determinations carried out on ocean-floor basalts across the ridges show that the age of the sea floor increases away from the ridges on either side. These magnetic patterns on the sea floor are important time markers and assist in the reconstruction and reassembly of landmasses.

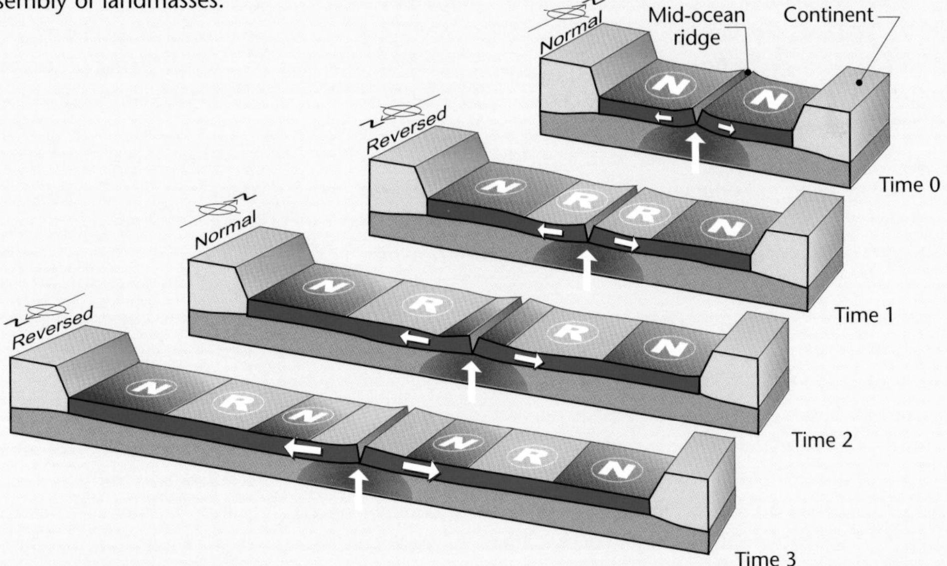

As ocean crust forms and cools along mid-ocean ridges, the magnetic minerals in the rocks align in the prevailing Earth field. Periodic reversals in the Earth field are recorded in the rocks and form symmetrical patterns astride the mid-ocean ridges.

Figure 2.11 *When molten lava comes into contact with water at the mouth of an undersea volcano, the outer surface of the lava freezes, forming a thin skin, while the lava inside remains molten. The skin stretches and tears as more lava is forced in, exposing more molten lava, which immediately freezes. Continued flow of lava into the expanding skin results in a bulbous, pillow-like shape. As the skin thickens due to cooling, it can no longer be further inflated. The pillow stops growing and new pillows begin to form around it. Undersea volcanic activity produces huge quantities of lava of this type, which is known as 'pillow lava'. Pillows make up the uppermost layer of ocean crust, and form the floor of the oceans. Pillow lavas are only visible near ocean ridges, though, as they eventually become buried beneath a layer of sediment. This photograph of pillow lava was taken off the coast of Hawaii.*

New oceanic crust is generated at these ridges. As the plates move apart, the oceanic crust and lithosphere are ruptured while the hot mantle material rises from below and partially melts, releasing molten rock or **magma** of basaltic composition. The magma accumulates in an elongated chamber under the ridge axis 2–7 km below the Earth's surface. Some of this magma crystallises slowly in the chamber, forming **gabbro** (*see* 'Classification of igneous rocks' on page 34) to add to the lower parts of the oceanic crust. The roof of the magma chamber is continually being ruptured as the plates move apart, and some magma rises into the cracks where it may solidify to form steep **dykes** (*see* 'Igneous intrusions' on page 46), or erupt on the sea floor, forming small volcanoes along the ridge axis.

The temperature of basaltic magma (termed **lava** when it erupts onto the surface) is about 1 200°C. When it comes into contact with cold sea water, its outer surface is immediately chilled, forming a black, glassy skin, while molten lava continues to flow into the skin, causing it to expand. The overall effect of this process is the formation of bulbous, sausage-like accumulations of lava on the sea floor, called **pillow lava** (**figure 2.11**). Pillow lavas, dykes and gabbros constitute new oceanic crust. Thus, as the plates move apart, new crust is created to fill the gap between them (*see* 'Palaeomagnetic studies and plate tectonics' on page 39).

The many faults and cracks along the ocean ridge allow cold sea water to penetrate deep into the crust, possibly as much as 6 km. Here it comes into contact with hot rock and is heated to temperatures up to 450°C. This hot water reacts chemically with the surrounding rocks, changing their composition. They retain some water, which is incorporated into minerals in the rocks. The hot water in turn leaches silica (silicon dioxide), sulphur and metals such as copper, zinc, lead, iron, manganese, gold and silver from the rocks, and takes them into solution. The resulting hot, metal-charged water, termed **hydrothermal fluid**, rises upwards through cracks and emerges on the sea floor to form hot springs, while more cold sea water is drawn into the surrounding rocks in order to replace it.

Although the temperature of the hot fluid may exceed 400°C, it does not boil because of the high pressure exerted by the overlying 2–3 km of water. As

FOLDING AND FRACTURING OF ROCKS

The earth's crust is continually subjected to various forces or stresses as a result of movement of tectonic plates. Sometimes rocks are stretched, at other times compressed. Rocks respond to these forces in different ways. The rate of application of the stress also affects how the rock responds. Rocks near the surface of the Earth are cold and generally brittle. When stress is applied, such rocks will fracture.

The way they fracture depends on the applied stress. If a slab of brittle rock is squeezed it will break in such a way as to reduce its length, thereby relieving the stress (**A**). The resulting plane of fracture is called a **fault**, and as the sense of movement across the fault serves to reduce the length of the slab, it is termed a **reverse fault**. In contrast, if a slab of rock is stretched, it will break in such a way as to increase its length (**B**). This is termed a **normal fault**. In some situations, slabs of rock slide past each other along a fault (**C**). These faults are termed **wrench faults**, and when developed along the margins of tectonic plates, they are referred to as **transform faults**.

Rocks that are deeply buried beneath the Earth's surface become hot as a result of the escape of heat from the earth's interior. Under these conditions they can become plastic. Rather than breaking under applied stress, they change shape. If the stress is rapidly applied, however, they may still break. The behaviour of hot rocks is much like that of warm toffee. Hot, plastic rock that is stretched simply becomes longer. When compressed, it usually buckles and becomes **folded** (**D**), which serves to reduce the length. Sedimentary rocks, which are deposited in almost horizontal layers, are particularly prone to folding. Folding in this way thickens the pile of rock, and this is why folded rocks occur in mountain belts, such as the Cape Mountains.

The collision of continents produces extreme cases of folding. As two continents are brought together by plate movement, the sediments that had collected on the floor of the ocean that formerly separated them are compressed and fold into elaborate structures, producing high mountain belts. The Alps and Himalaya formed in this way. Extreme folding of this type is often accompanied by a type of reverse fault, along which one slab of folded rocks slides over another (**E**). These faults are termed **thrust faults**, and the upper folded slab is termed a **nappe** (pronounced nap).

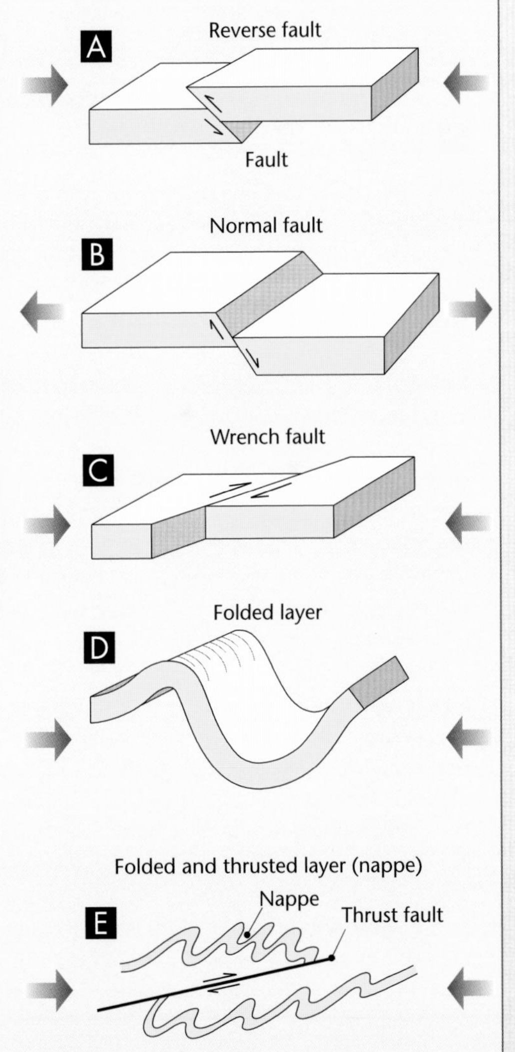

A: ***Brittle rocks under compression will fracture in a way that will reduce their length.***
B: ***Brittle rocks under extension will fracture and increase their length.***
C: ***Rock masses may slide past each other in response to horizontal stresses.***
D: ***Rocks that are hot and hence plastic will respond to compression by folding.***
E: ***Under extreme compression, folding may be associated with gently inclined reverse faults, called thrust faults.***

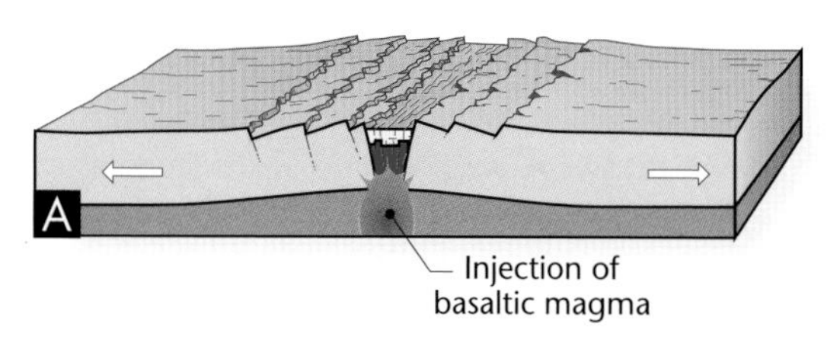

A very early stage of continental rifting: the crust and lithosphere have been thinned by stretching, and hot mantle has risen below, forming an elevated tract. Faulting in the rocks of the crust has produced a rift valley along the uplifted tract. The East African Rift Valley is an example of this stage of continental rifting.

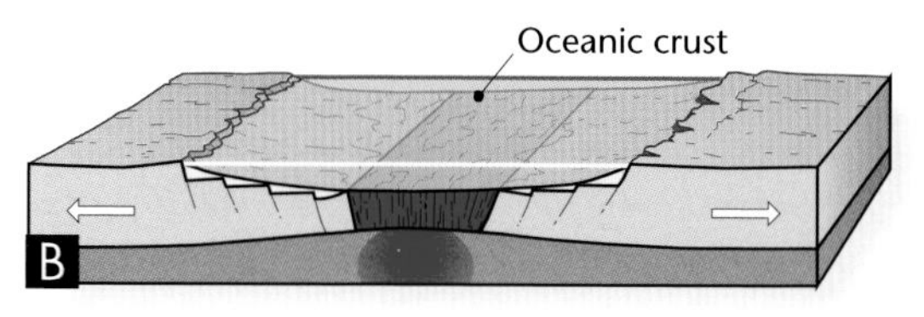

A more advanced stage: basalt magma rising from the hot mantle has been injected between the separating continental masses, forming oceanic crust. The crust has subsided because the ocean crust is thinner and denser than continental crust, and the sea has invaded the rift valley. The Red Sea is an example of this stage of rifting.

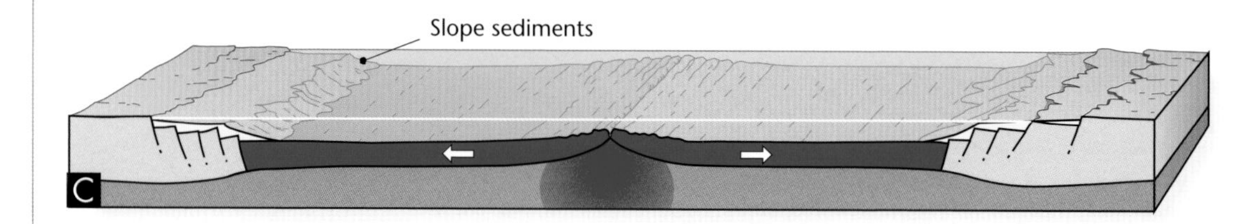

An even more advanced stage: here, the continents are widely separated by ocean crust and an ocean basin and a mid-ocean ridge have formed mid-way between them. The Atlantic Ocean is an example of this advanced stage.

Figure 2.12 *The rifting of continental crust can create new ocean basins.*

the hot fluid meets the cold sea water, metal sulphides and other compounds (particularly silica) precipitate, forming clouds of fine particles. As a result, these hot springs have been termed **smokers**. Some of the particles stick together to form a chimney around the hot spring, but most settle on the ocean floor in the surrounding area. These give rise to local concentrations of metals, and it is believed that certain types of base metal mineral deposits were formed in this way.

Hot springs on the ocean floors provide homes for unique forms of life, including giant tube worms and clams, which have adapted to this harsh environment. These animals form part of a complex ecosystem fed ultimately by bacteria that derive their energy from the chemical imbalances between the hot, mineral-rich spring water and the surrounding sea water. The bacteria include some of the most primitive life forms known, and it is believed by many scientists that life on Earth may in fact have originated at such hot springs (*see* Chapter 6).

Separation can also occur across plates containing continental crust, and can lead to the formation of new ocean basins (**figure 2.12**). Rifting of continents occurs when the lithosphere beneath the continental crust becomes stretched, allowing upwelling of hot mantle material. The continent then bulges upwards and stretches, producing faults that create steps down into a central rift valley. Because continents stand high on the Earth, these topographic features associated with rifting occur above sea level.

The East African Rift Valley is a classic example of a newly forming rift on a continent (**figure 2.12A**). Sediments eroded from the high flanks of the rift accumulate in the rift valley, and lava may erupt and spread over the valley floor. As rifting continues, the continental blocks stretch further and become thinner while the rift valley deepens, eventually allowing the ocean to flow into the rift. Basaltic magma injected into the central rift now begins to form new ocean floor. The Red Sea is an example of this stage of continental rifting (**figure**

2.12B). Separation of the plates may continue, forming an ocean basin between the two continents. The Atlantic Ocean is an example, formed by the separation of Africa and South America (**figure 2.12C**).

Plate convergence

The Earth is a sphere with essentially constant surface area, so creation of new crust by plate separation must be compensated for by destruction of crust elsewhere. This happens at convergent plate boundaries, where plates move towards each other, one plate sliding down under the other in a process known as **subduction**. The upper plate may contain either continental or oceanic crust. Convergent plate margins occur particularly around the edges of the Pacific Ocean (**figure 2.13**), which is steadily being consumed as the Atlantic and Indian Oceans expand. The topographic expression of convergence in ocean basins is that of a trench (**figure 2.13**), and these form the deepest parts of the ocean.

There are two main settings for ocean trenches: the first is adjacent to continental margins (**figure 2.14A**), such as along the west coast of the Americas; the second is alongside curved chains of oceanic islands – termed **island arcs** (**figure 2.14B**) – such as the Aleutian or Japanese Islands. These convergent margins are characterised by a zone of earthquake activity that is close to the surface near the trench but deepens on the landward side of the trench to a maximum depth of about 700 km. This seismic activity arises from the slow, jerky, downward slide of the subducting plate into the mantle (**figure 2.15**). Typically, the subducting plate is inclined at an angle of approximately 40–60° to the horizontal, but both shallower and steeper angles may occur.

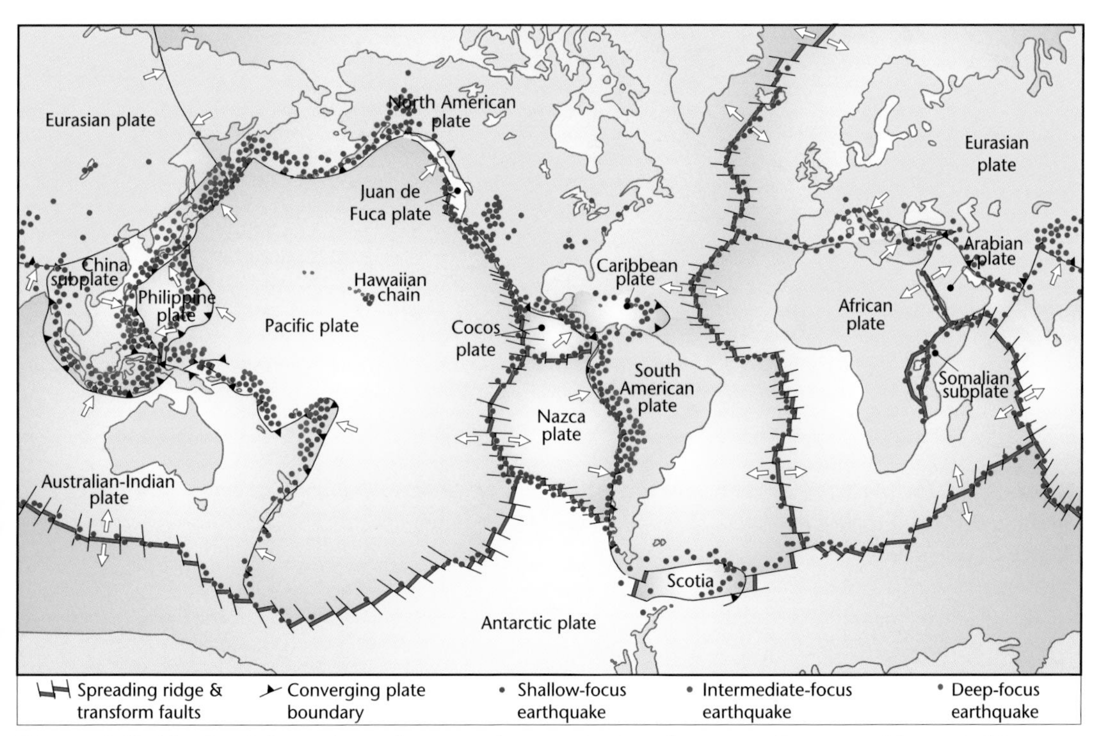

Figure 2.13 *The Earth is a sphere of more or less constant surface area. If new crust is created in one place by the divergence of plates, an equivalent area of crust must be destroyed elsewhere. This occurs where plates converge. This map shows the regions of the world where plates are converging. They are characterised by both earthquake and volcanic activity. Convergent zones in the oceans are also characterised by deep trenches in the ocean floor. Earthquakes are confined to one side of the trench and become progressively deeper away from the trench. The Pacific Ocean is largely rimmed by convergent margins, and the active volcanoes around its margin have been described as the ring of fire.*

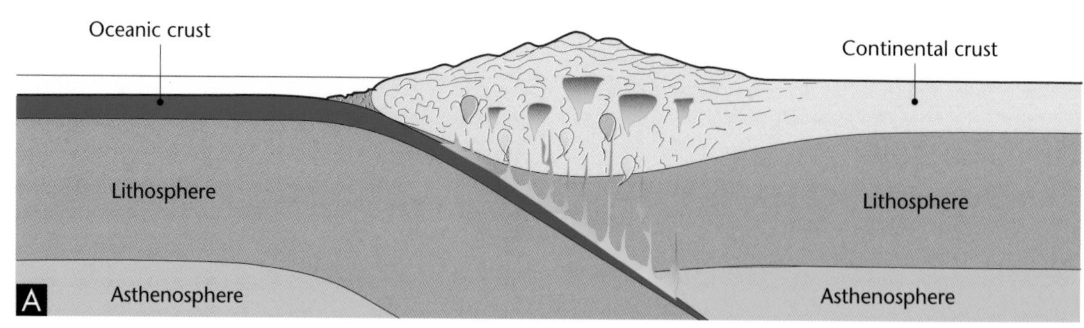

This diagram illustrates a plate containing oceanic crust subducting beneath a plate containing continental crust. This type of boundary occurs along the west coast of the Americas. The oceanic plate bends as it begins its descent, forming a deep trench on the ocean floor (e.g. the Chile-Peru trench). As the cold oceanic plate descends it is slowly heated and water is driven off. The water rises into the hot, overlying mantle, where it induces melting, producing magmas of diorite and granite composition. These magmas rise upwards, and most crystallise in the crust as large masses called batholiths. Some magma erupts at surface, forming explosive volcanoes.

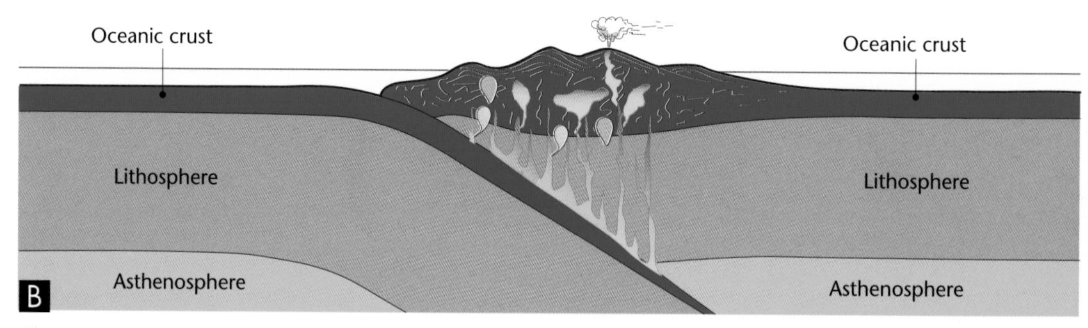

The convergence of two oceanic plates is shown here. An example is the Japanese Islands, where the rising magma crystallises as batholiths in the overlying oceanic crust, while volcanoes form on surface. The volcanoes produce a line of islands above the down-going slab, known as an island arc.

Figure 2.14 ***The diagrams above demonstrate the main processes taking place at convergent plate margins. The convergence of plates causes one plate to slide down beneath the other – a process that is known as subduction.***

By the time oceanic crust (and its underlying lithosphere) begins its descent at a trench, it has travelled some distance from the ocean ridge where it was generated millions of years before. The upper part of the slab is thus cold and dense, and takes time to heat up as it plunges into the mantle. It remains cool and brittle to great depths (**figure 2.15**), accounting for the deep earthquakes at subduction zones.

The oceanic crust attached to the subducting slab is charged with water. In addition, some water-bearing ocean floor sediment may also be dragged down by the subduction process. As the descending slab is gradually heated and compressed, the water is released (along with other substances such as sulphur and carbon dioxide) and rises, entering the hot mantle overlying the slab. The water acts as a flux, lowering the melting temperature of the mantle, which begins to melt (**figure 2.14**).

Details of the processes involved remain unresolved, but the magmas produced are different from those found at ocean ridges, and are rich in calcium, silicon, aluminium, sodium and oxygen. They are known as **diorite** and **granite** (or **andesite** and **rhyolite** when they erupt as lava on surface, *see* 'Classification of igneous rocks' on page 34). Intermediate magma types known as **granodiorites** (**dacite** is the term used for lava of this composition) also occur. These melts gradually work their way towards the Earth's surface, where they may erupt to form andesite, dacite or rhyolite volcanoes.

Magmas formed at subduction zones are physically quite different from the basalt magmas of ocean ridges. Whereas basalt is very fluid, andesites and rhyolites are extremely viscous, and do not flow readily from volcanoes. In addition, they are charged with fluids such as water, carbon dioxide and sulphur dioxide.

Rather than spewing out lava that rapidly flows away from the volcano, as basaltic volcanoes do, the volcanoes above subduction zones are violent and explosive, and they eject glowing rock fragments and especially avalanches of hot rock and dust known as **pyroclastic flows** (*pyro* = fire, *clastic* = fragmental). This style of eruption produces the typical steep-sided shape of their volcanoes (**strato-volcanoes**).

In addition, the volcanoes discharge enormous volumes of sulphur dioxide and carbon dioxide, as well as millions of tonnes of fine dust (**volcanic ash**), which rises high into the atmosphere. Occasionally, they may explode cataclysmically, destroying part or all of the volcano in seconds, as happened at Mt St Helens in 1980 and the island of Krakatoa in 1883. These volcanoes are responsible for hundreds of thousands of human deaths and have destroyed many towns and villages.

One of the worst recent cases occurred in 1902 on the island of Martinique in the West Indies when a pyroclastic flow avalanched through the capital, St Pierre, killing all but two (who were in prison) of the 28 000 inhabitants and destroying all buildings. The capital of Montserrat in 1995 suffered a similar fate, but fortunately had been evacuated in time. The Minoan civilisation on the island of Crete in the Mediterranean Sea is also believed to have been destroyed by a volcanic explosion on the island of Santorini some 120 km to the north.

Another danger associated with these volcanoes is rising heat from volcanic activity, which creates huge storms over these volcanoes. Torrential rain

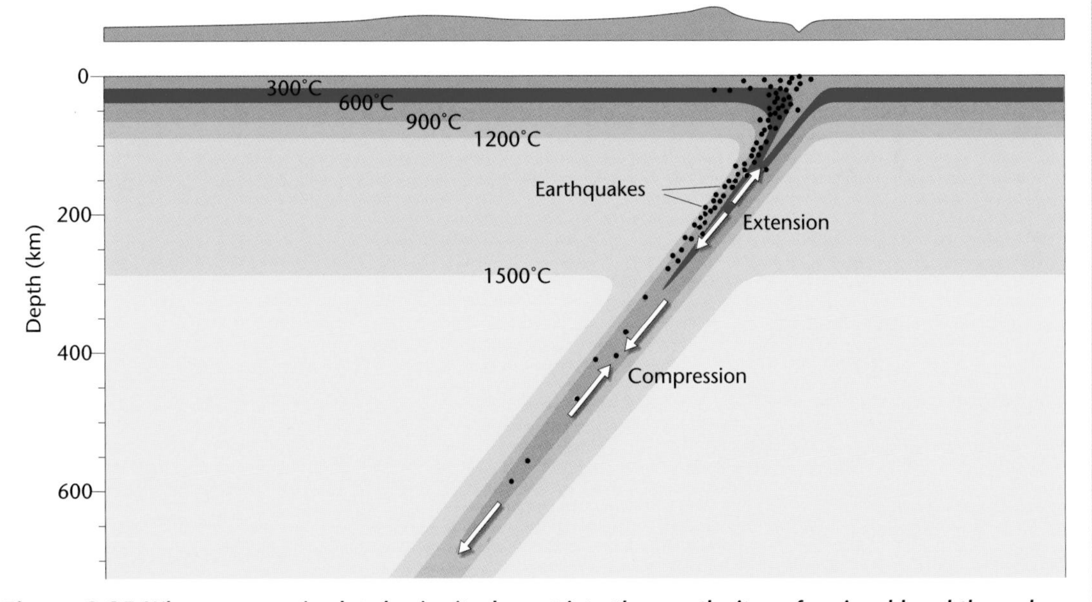

Figure 2.15 ***When an oceanic plate begins its descent into the mantle, its surface is cold and the rock is brittle. As the slab descends, it gradually heats up, as illustrated. The plate moves downward in a jerky fashion, producing shock waves (earthquakes). This results in earthquakes becoming deeper with increasing distance from the ocean trench.***

IGNEOUS INTRUSIONS

Molten rock or magma is generated far below surface in the Earth's mantle or the deep roots of mountain belts. Magma rises because it is slightly less dense than the surrounding rock and may erupt on surface, forming volcanoes. Near the site of generation, the rocks are very hot and soft, or plastic. Here, the magma rises as a blob, much like a hot air balloon. If the magma solidifies at this position, the resulting body is referred to as a **pluton**, or if very large, a **batholith**.

At higher levels in the crust, rocks are cooler and much stiffer or even brittle, so magmas move in a different way. In these regions, magmas move upward by injecting into cracks, often forcing the sides apart. Magma may solidify in the crack to form what is termed a **dyke**. Dykes vary in width from centimetres upwards. Magma injecting into layered sedimentary rocks may be forced between the layers to form a **sill**. Sills can reach several kilometres in thickness.

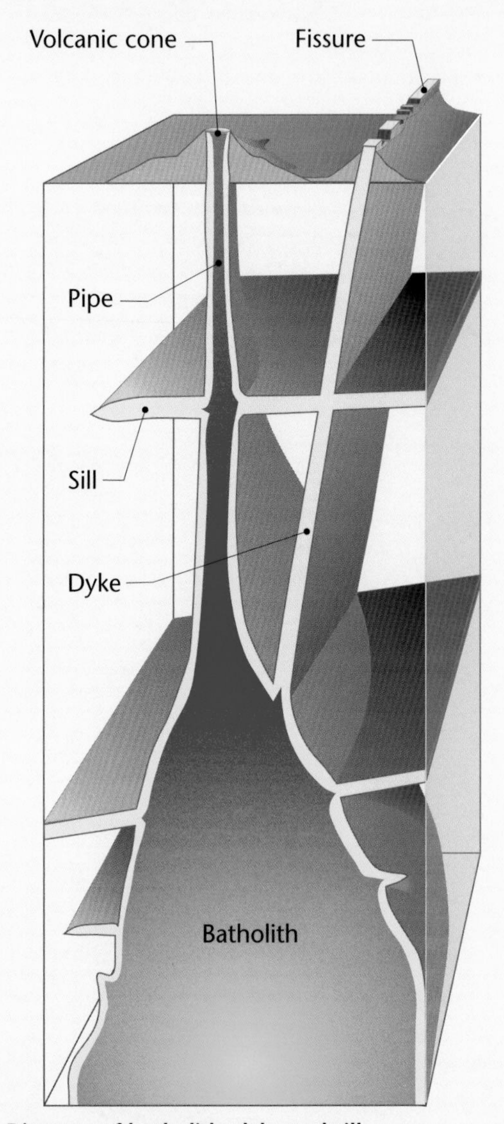

Diagram of batholith, dyke and sill.

Dyke (black) intruding granite (pink).

Sills cap many flat-topped hills of the Karoo.

erodes the loose volcanic ash and boulders on the steep slopes, resulting in very destructive mud-flows known as **lahars**.

When a plate containing oceanic crust is subducted beneath another also containing oceanic crust, the rising magmas produce a chain of volcanic islands along the trench's edge (**figure 2.14B**), known as an **island arc** (because the island chains are normally arcuate in shape). Examples are the Aleutian and Japanese Islands. When subduction occurs beneath a continent, volcanic activity occurs along the continental margin, such as along the west coast of North America (**figure 2.14A**). The distance between the volcanoes and the trench depends on the angle of subduction: the shallower the angle, the greater the distance. However, if the subduction angle is very shallow, no volcanic activity occurs because of the overlying mantle's limited volume and lower temperature.

Only a small proportion of the magmas generated above the subducting slab erupts at surface; most remain in the crust where they cool and solidify to produce huge bodies of igneous rock called **batholiths** (*see* 'Igneous intrusions', left). The magmas produced at subduction zones have the same composition as continental crust, so the process of subduction actually creates new continental crust. Ocean floor sediment scraped off the subducting slab further adds to this new crust.

Continents in collision

Sometimes, both of the converging plates contain continental crust. As subduction proceeds, the ocean separating the two continents is consumed and the continental masses draw closer until finally they meet (**figure 2.16**). Sediment on the ocean floor is scraped off as the ocean closes and becomes compressed between the converging continents. Continental crust has a relatively low density and is

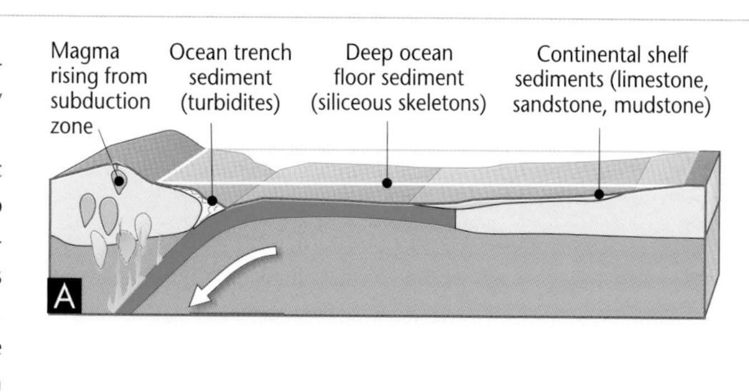

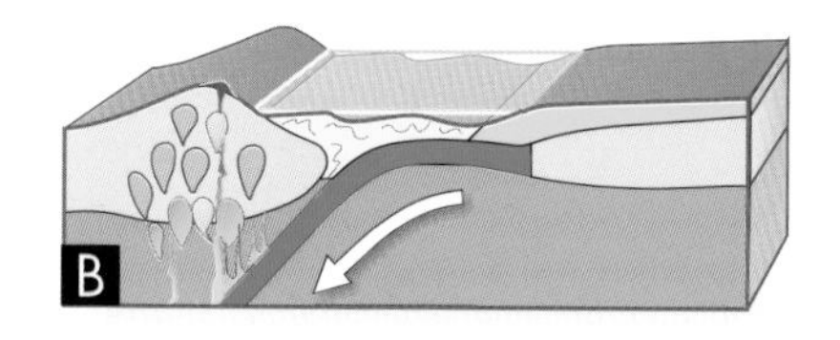

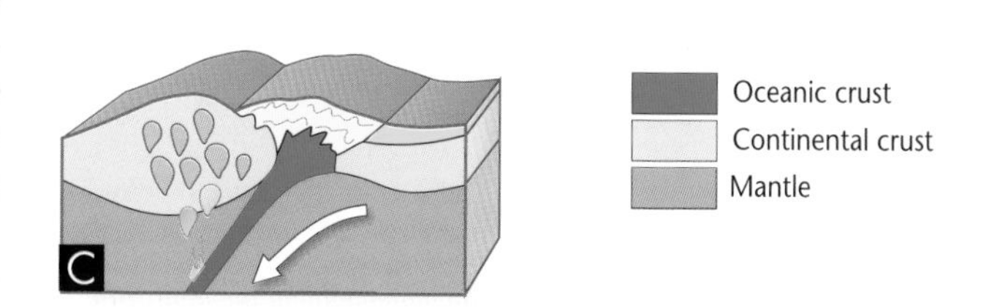

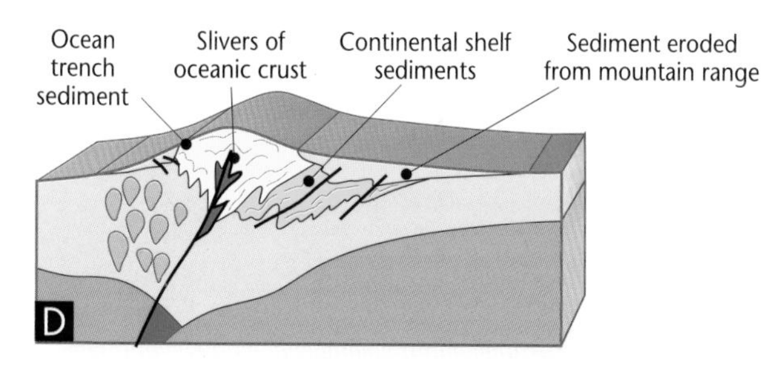

Figure 2.16 *The convergence of plates may bring two continents together, as illustrated by this sequence of diagrams. As the plates converge, the ocean separating them is gradually consumed (**A** and **B**), and the sedimentary deposits on the ocean floor are compressed into a thick pile (**C**), forming a mountain range. Continental crust has a relatively low density, and cannot be subducted to any significant extent. A portion of one continent will under-ride the other for a short distance (**D**), but then subduction ceases and the two slabs of continental crust remain welded together. The crust is greatly thickened by the collision process, resulting in a high mountain range. The Alps and the Himalaya formed in this way.*

buoyant in the context of dense mantle rocks. This prevents it from being subducted to any significant extent. Thus, when continental crust on the subducting slab meets continental crust on the upper slab, the former is subducted only for a short distance, producing a double thickness of continental crust. This thickening causes uplift of the double crust, producing a mountain belt. Thereafter, subduction ceases and the two continents remain welded together.

One of the most spectacular examples of this type of collision in the geological record is the collision of the Indian and Asian continents, which has produced the Himalayan Mountain chain. Collision commenced about 50 million years ago, but is not complete, and India and Asia are still converging at a rate of about 5 cm/year. The Tethys Sea, which formerly existed between India and Asia, was closed by the subduction process. The sediments that had accumulated in that ancient sea were squeezed between the colliding continents and crumpled, producing a thick pile that is still rising due to the thickened crust beneath. Sedimentary rocks (mudstones and limestones) deposited on the bed of the Tethys Sea now form the upper slopes and summit of Mt Everest, the highest point on Earth. Likewise, the collision of the African and European continents produced the Alps and associated mountains.

A similar process occurs during collision of an island arc and a continent. When a continent collides with an island arc, the island arc becomes welded or **accreted** onto the leading edge of the continent. Crustal thickening is generally less than during continent-continent collision. Subduction ends, but may resume behind the accreted island arc. By this process, the area of continental crust has increased over time.

Most sediment that accumulated in the ocean basin between two colliding continents becomes compressed as the continents converge (**figure 2.16**) and is intensely folded, as in the Himalaya, described above. This, plus the small amount of subduction of one continent beneath the other, leads to abnormal thickening of the continental crust – by up to 75 km in some cases. As mentioned earlier, the crust and underlying mantle (lithosphere) float on the asthenosphere. As the crust thickens, it rises, forming a mountain belt. The folded sedimentary rocks that form the mountain belt become heated by intrusions of magma rising from the subducting slab and by heat conducted from below.

The result is that these rocks, which formed at low temperature and pressure, now experience both high temperature and pressure. Consequently, they undergo **metamorphism**, developing new and usually distinctive mineral compositions appropriate to the temperature and pressure they experience (*see* 'Rock classification' on page 31).

The manner in which rocks respond to increasing temperature and pressure is well known from laboratory experiments in which these conditions have been simulated. Some of the sedimentary material may undergo partial or even complete melting, producing granite magma that forces its way upwards in the buckled and metamorphosed sediments. While these processes are taking place far below the surface, the mountain belt is undergoing erosion and is gradually being worn down.

As fast as the mountains are denuded, the root of the mountain belt rises, in the same way as slicing off the top of a floating block of ice will cause it to rise in the water. Hot rock is brought closer to surface, where it cools and any magma present solidifies. Eventually, the mountain belt is completely removed by erosion. By this stage rocks that once formed its deep root lie exposed on surface. The crust has now returned to its normal thickness of about 35 km (*see* 'Why the continental crust is 35 km thick' on page 51). The metamorphic rocks that once formed the root are distinctive and easily recognised, and form linear tracts known as **metamorphic** or **mobile belts**. These mark the boundaries between fused continental masses, the sites of once mighty mountain ranges.

Lateral sliding of plates

Major fracture zones, where two plates slide past one another, are known as **transform faults**. At such boundaries, crust is neither created nor consumed. The most common type is a ridge-ridge transform that links one segment of a ridge with another (**figure 2.17**). These transform faults develop at right angles to the ridge, and result from variations in the spreading rate along the ridge. Only that section of the fault between the two ridge segments is active; the remainder, which may be extremely long, is simply a scar left from previous tearing. For example, the great fracture

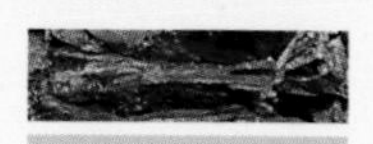

THE SAN ANDREAS FAULT

The San Andreas Fault forms part of a major transform fault zone almost 3 000 km in length along the west coast of the United States of America (*see* map, right). Here the Pacific plate is sliding northwestwards against the North American plate at a rate of about 6 cm/year. The fault zone links the East Pacific mid-ocean ridge in the Gulf of California with the Juan de Fuca mid-ocean ridge in the Pacific Ocean. Spreading on these two ridges is accommodated (in part) by slip along the fault zone.

The fault zone consists of a number of interconnecting fractures that form an overall braided network, in effect a whole zone of fractures of which the San Andreas Fault itself is the largest, more than 1 000 km in length. This zone is characterised by shallow (<15 km deep) earthquakes that can occur throughout the entire length of the fault zone. As the plates move past each other, they do so in a jerky, stick-and-slip manner.

If the strain energy is released gradually, the rocks will creep past one another and the earthquakes will be numerous but very minor in size. However, if the system locks up, an enormous amount of strain energy can be stored and when fracturing eventually does occur, a huge amount of energy is released in microseconds with devastating results to man-made structures, such as occurred in the San Francisco earthquake of 1906.

Movement can occur on any of hundreds of minor faults developed along the fault line at any time. Because of this complexity, predicting fault movements continues to elude scientists. The stakes are high because San Francisco and Los Angeles are situated on the fault zone.

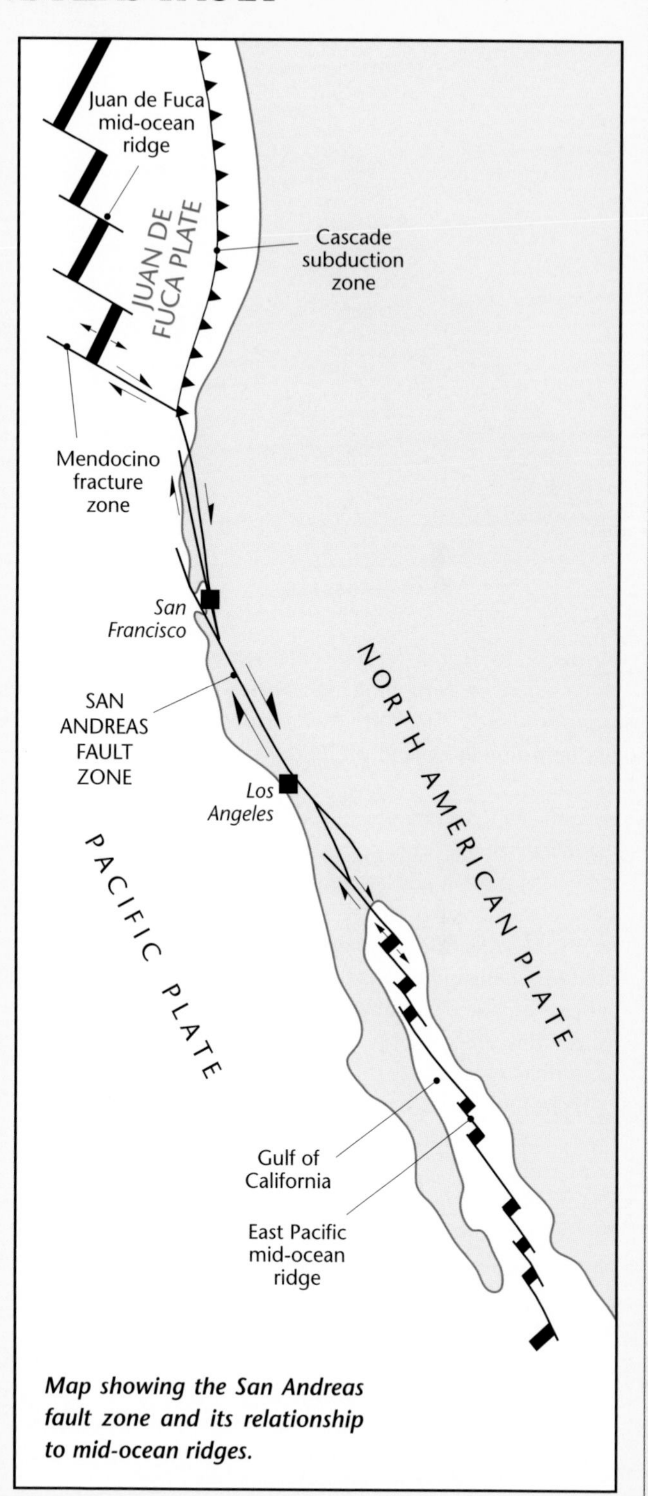

Map showing the San Andreas fault zone and its relationship to mid-ocean ridges.

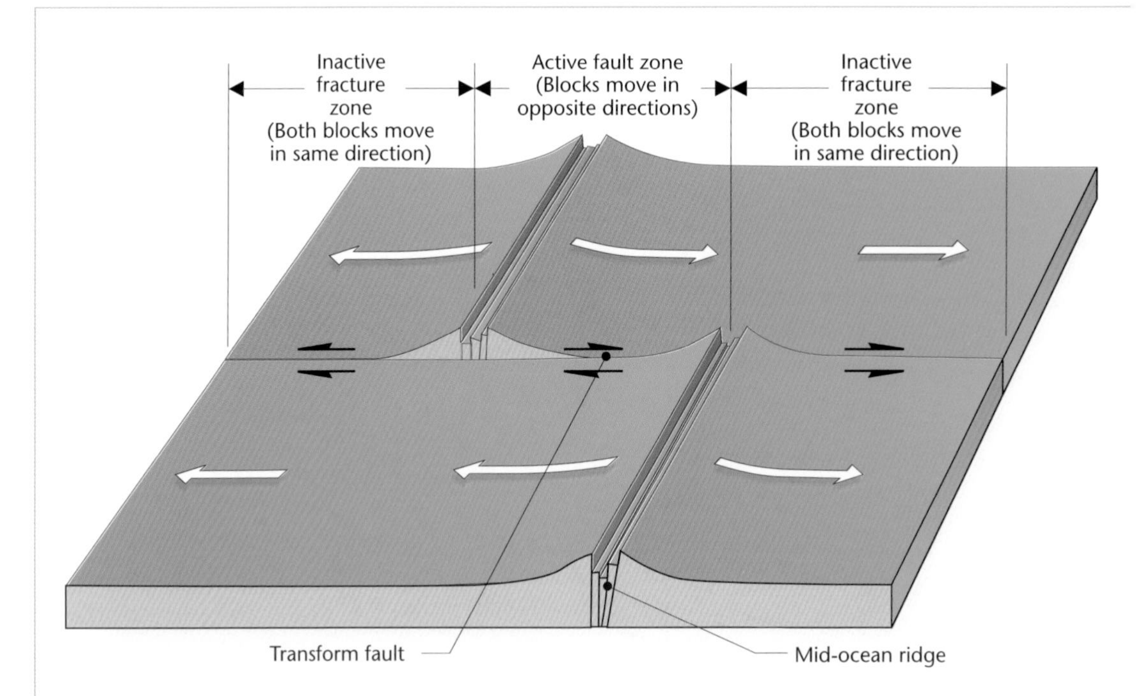

Figure 2.17 ***Transform faults link segments of a mid-ocean ridge. Arrows indicate the spreading of the ridge segments. Note that rock masses between the two ridges move in opposite directions along the transform fault. Transform faults usually occur on the ocean floors along the mid-ocean ridge system, but occasionally appear on land. An example of this is the San Andreas Fault.***

zones of the east Pacific Ocean extend over half the width of the ocean (*see* **figure 2.5**). They appear as distinct gashes on the ocean floor, with steep cliff-like valley sides. Transform faults develop in the direction of plate motion and can be used to establish the direction of plate movement. When combined with the dating of ocean floor rocks, this information allows geoscientists to track plate movements through time.

Transform faults normally occur on the ocean floor, but occasionally also involve continental crust and appear on land. The classic example of this phenomenon is provided by the San Andreas Fault in California (*see* 'The San Andreas Fault' on page 49). Less infamous examples are the Alpine Fault, which extends the length of South Island, New Zealand, and the Dead Sea transform fault, which extends from Turkey to the Red Sea, and gave rise to the Sea of Galilee, the Jordan Valley, the Dead Sea and the Gulf of Aqaba. The biblical prophet Zechariah warns us to expect continuing movements on the latter fault system in the future:

On that day, his feet shall stand on the Mount of Olives which lies before Jerusalem on the east; and the Mount of Olives shall be split in two from east to west by a very wide valley; so that one half of the Mount shall withdraw northward, and the other half southward . . . and you shall flee as you fled from the earthquake in the days of Uzzi'ah king of Judah. (Zechariah 14:4,5)

HOW FAST DO PLATES MOVE?

The directions of movements and the relative velocities of the major plates have been measured, and they range from less than 2 cm/year to more than 20 cm/year (**figure 2.18**). The plate that contains the African continent is stationary on the Earth's surface and has been so for the past 30 million years.

WHAT DRIVES THE PLATES?

The energy that drives plate movement is the Earth's internal heat that comes from two main sources: firstly, primordial heat generated during the formation of the Earth (*see* Chapter 3); and secondly, heat

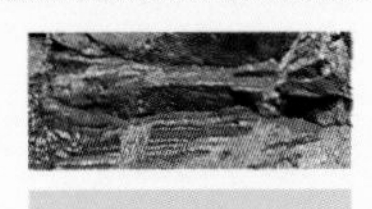

WHY THE CONTINENTAL CRUST IS 35 KM THICK

The continental and oceanic crust essentially float on the underlying denser, plastic mantle. Continental crust is both thicker and less dense than the oceanic crust and so stands higher on the Earth's surface, a phenomenon known as **isostasy**. The oceans occupy the less elevated areas of the Earth and hence cover the oceanic crust. So in ocean basins, we have a layer of oceanic crust overlain by about 4 km of sea water, and on continents, a thick layer of continental crust that rises above sea level.

These two systems are not independent, but are linked together by the process of erosion. Given sufficient time, and in the absence of geological processes that thicken the crust, continents will be thinned by erosion, the eroded material being dumped in the sea around the continental margins. As a continent is being reduced in thickness by erosion, it will float up until it is in isostatic equilibrium with the ocean crust plus the overlying sea water, and has been eroded down to sea level. In this way, equilibrium is established between oceanic crust and its overlying column of sea water on the one hand, and the thickness of the continental crust on the other.

This equilibrium is generally referred to as the continental freeboard problem, and has been explored by several well-known earth scientists, among them the late Harry Hess of Princeton University and Ross Taylor of the Australian National University. It turns out that the equilibrium can be described by a simple equation:

Thickness of the continental crust (in km) = 12.3 + 5.75 x depth of the oceans (in km)

Thus, since the average ocean depth is 4 km, the continental crust is 35.3 km thick.

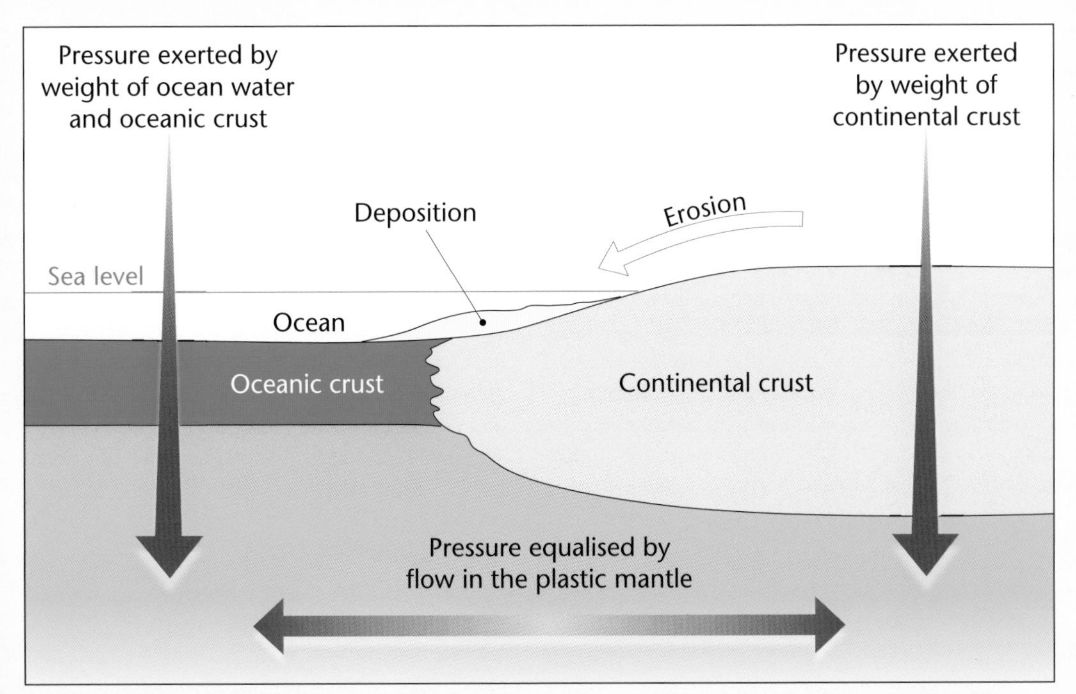

The thickness of continental crust is linked to the thickness of oceanic crust and the depth of the oceans. Pressure exerted by the two systems is equalised by flow in the plastic mantle. Assuming that the thickness of oceanic crust is fixed, the thickness of the continental crust is then dependent only on ocean depth.

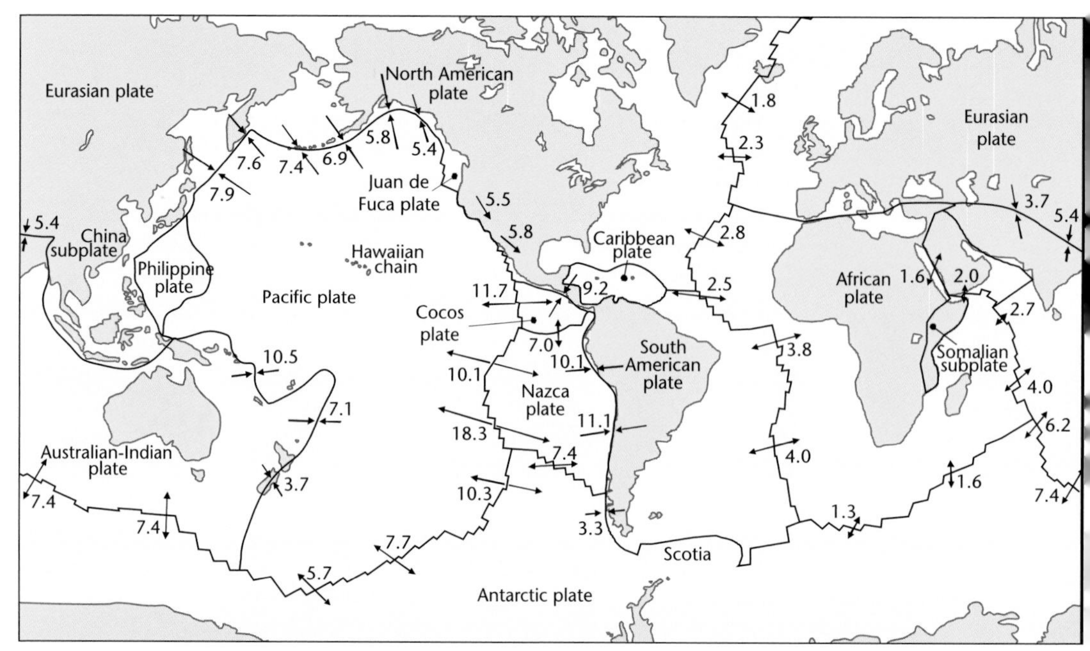

Figure 2.18 *All of the Earth's plates are moving relative to each other, the velocities and directions of motion being quite variable, as illustrated here. The velocities are expressed in cm/year.*

produced by radioactive decay of uranium, thorium and potassium within the Earth. The most efficient way heat reaches the surface is through convection in the mantle. Although solid, this is able to flow very slowly because of its high temperature, like tar on a road on a hot day. But how does convection in the mantle move the plates at the Earth's surface?

Two main theories or models have been proposed to explain plate movement: a **thermal model** and a **mechanical model**. In the thermal model, the cool, rigid plates are considered to be dragged about by the slowly convecting mantle. Implicit in this model is the assumption that hot material on rising convection cells would come to surface below ocean ridges and cool material would descend at subduction zones (**figure 2.19**).

The alternative mechanical model views the plates as active participants, with plate movement being driven by plate-edge forces, especially subduction. In this model, subduction occurs because the subducting plate is colder and more dense than the underlying asthenosphere, and the down-going plate pulls the remainder of the plate with it.

Ocean ridges are simply tears in the crust resulting from the subduction pull. But the tears also actively contribute to the movement: the uplifted lithosphere at the ocean ridges slides down the ridge slopes, providing a pushing force on the plate. The asthenosphere provides a slippery, lubricated surface on which the lithosphere can move freely.

It seems likely that plate movement is best accounted for by a combination of convection in the deep mantle and plate-edge forces. The pull created by subduction is probably the most important and can enhance convective motion in the mantle. Ocean ridges, self-inflicted tears in the outer skin of the Earth, are long-lived and assist plate movement and mantle turnover. How subduction is initiated is still unknown, and indeed, much of the detail of plate tectonic processes remains to be discovered.

VOLCANIC ISLANDS

Not all volcanic activity occurs at plate margins. Important exceptions are volcanic islands like Hawaii or St Helena, which occur well within plates (*see* **figure 2.5**). The existence of these islands has led to the realisation that convection processes are not the only way heat is lost from the Earth's interior. A second means of escape is via **mantle plumes**. These are narrow, cylindrical

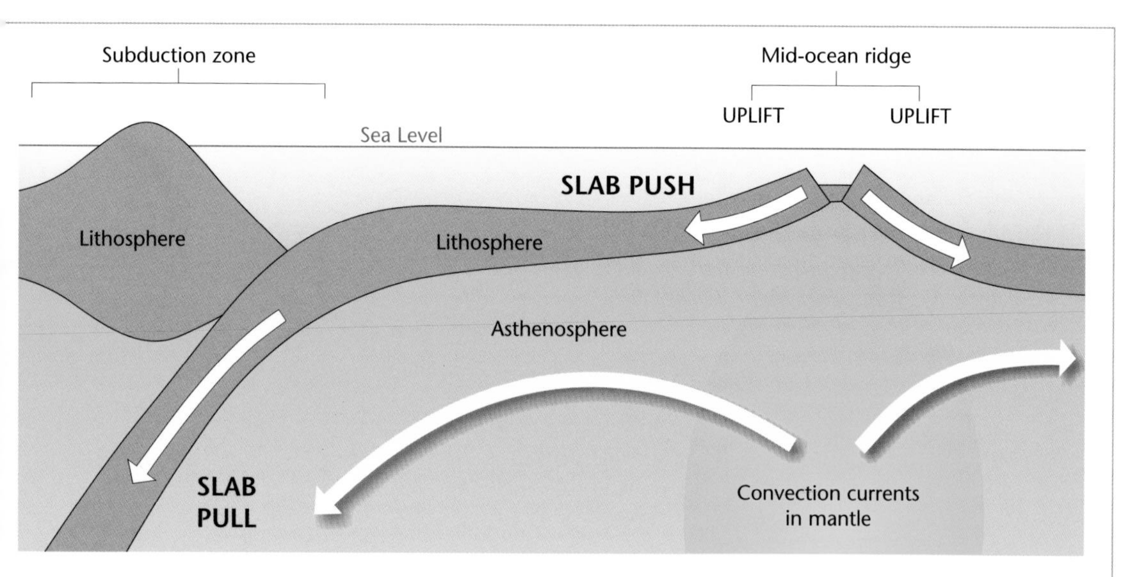

Figure 2.19 *Two theories or models have been proposed to explain plate motions. The thermal model proposes that plates are dragged along by convection currents in the underlying hot mantle. The mechanical model proposes that cold lithosphere descends at subduction zones because it is denser than the surrounding hot mantle material. In doing so, it drags the lithosphere trailing behind it towards the subduction zone. The mid-ocean ridges are simply resultant tears in the Earth's surface. The uplifted lithosphere flanking mid-ocean ridges slides away from the ridge axis under gravity, assisting plate movement. Movement is made possible by the slippery layer afforded by the asthenosphere.*

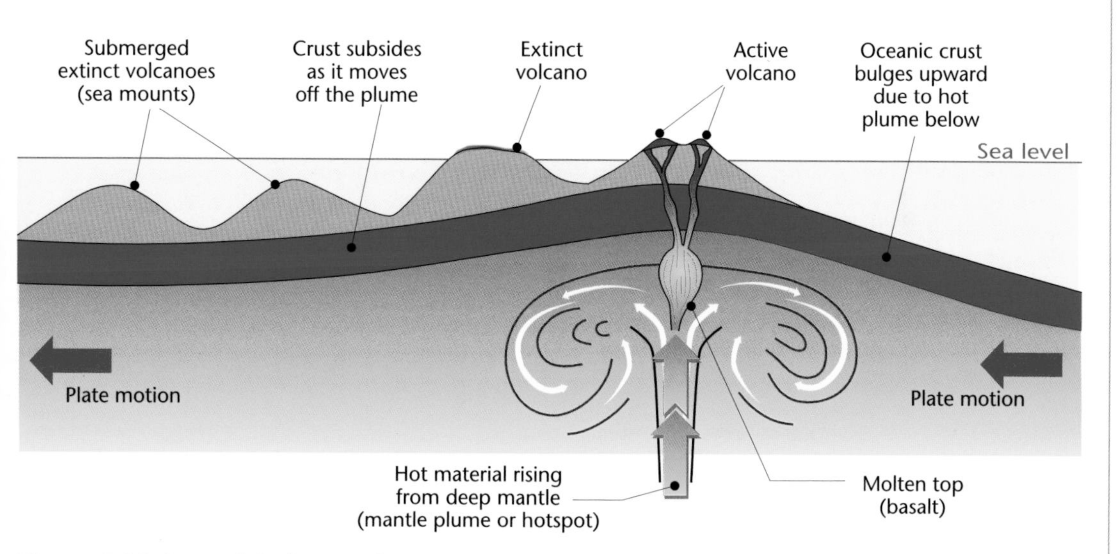

Figure 2.20 *Some of the heat in the Earth's deep interior escapes by rising as a column of hot rock, known as a mantle plume. These are long-lived features that remain stationary beneath the moving plates. The heat of the plume causes the upper mantle to melt and the crust to bulge upward. Large volumes of lava are produced above these plumes. As the plate moves across the position of the plume, new volcanoes form; the old volcanoes become dormant, cool and subside. Mantle plumes usually rise beneath oceanic plates and form volcanic islands such as Hawaii. The movement of an oceanic plate across the mantle plume results in a chain of sea mounts (drowned islands) across the ocean floor.*

columns of hot mantle rock that rise from the core-mantle boundary towards the surface, like steam rising from the bottom of a pot of stiff, hot porridge on a stove.

At the surface plumes express themselves as so-called **hot spots** with abundant volcanic activity, such as on the Hawaiian Islands. The plumes rise upwards because the material is hot and thus lower in density than the surrounding mantle. As with rising hot mantle beneath ocean ridges, rising plumes also generate basaltic magmas as they approach the surface. The hot head of the plume lifts the crust and bores through, releasing large amounts of fluid, basaltic lava. This lava flows rapidly away from the volcano, building a very broad, flat mountain known as a **shield volcano** because of its resemblance to an ancient warrior's shield.

Plumes most commonly occur beneath oceanic crust, where they form volcanic islands far from plate boundaries. They appear to remain stationary beneath a moving plate and are long lived. As the plate passes over the plume, new islands are formed, while the lithosphere beneath the old island cools and sinks, submerging the crust, drowning the old island and leaving a track of **sea mounts** on the ocean floor (**figure 2.20**).

Plumes may also rise under an ocean ridge – Iceland is an example – and beneath continental crust, as at Yellowstone in the western United States. Mantle plumes may be responsible for occasional outpourings of extremely large volumes of basalt on continents, such as the 180-million-year-old lavas of southern Africa's Maluti Mountains.

It is believed that plumes may sometimes initiate rifting of continents. When a large plume rises through the mantle beneath a continent, its large, hot, bulbous head lifts the lithosphere and weakens it by heating, so that it tears apart, rupturing the continent along three rift arms arranged at 120° to each other. This process is believed to have occurred at the junction of the Red Sea, the Gulf of Aden and the Ethiopian section of the East African Rift Valley (**figure 2.21**).

Figure 2.21 *Mantle plumes beneath continents may cause rifting along three axes arranged at 120° to each other. Such rifting is believed to have produced the Red Sea, the Gulf of Aden and the Ethiopian section of the East African Rift Valley, shown in this satellite image.*

NASA

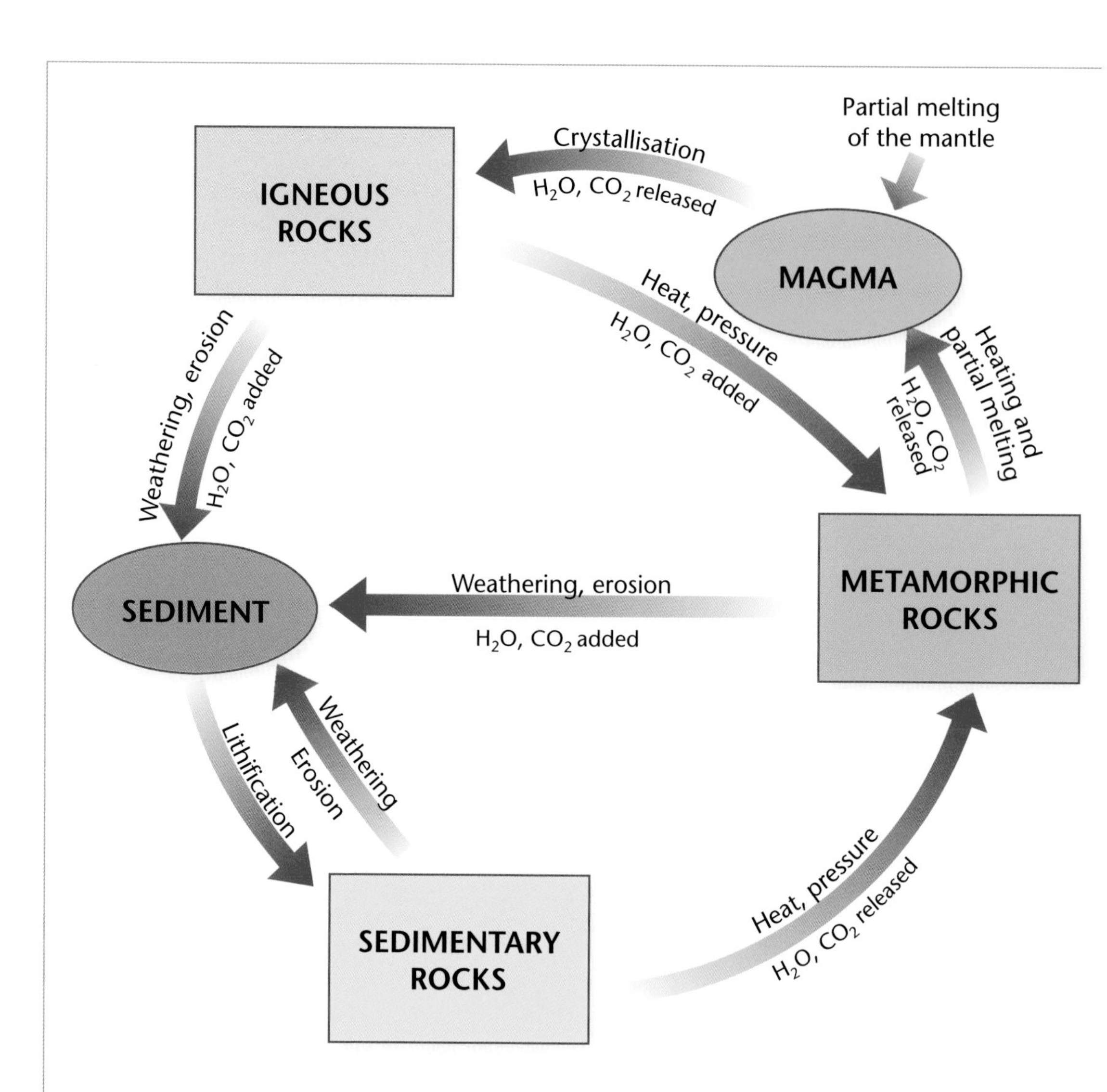

Figure 2.22 ***The Earth's crust, atmosphere and hydrosphere are involved in an endless cycle, known as the Rock Cycle, schematically shown here. The cycle is driven by plate tectonics, and involves the reconstitution of rock-forming compounds in response to changing environmental conditions, especially temperature and pressure. Water and the atmosphere (carbon dioxide and oxygen) form an integral part of this cycle.***

THE SIGNIFICANCE OF PLATE TECTONICS

Local effects related to plate interaction – such as earthquakes and volcanic eruptions – remind us of Earth's ceaseless activity, but we are completely unaware of its long-term consequences because human life spans are so short in relation to the speed of plate motion. Even the entire span of recorded history is short compared to plate motion. For example, the Atlantic Ocean has widened by only 150 m since the building of the Great Pyramids of Egypt. But in the context of geological time, plate motion is very fast. The Earth is 4 600 million years old and the oldest known rocks formed 4 100 million years ago (*see* Chapter 3), yet all of the ocean floors are younger than 200 million years. Therefore, 60% of the Earth's surface (the ocean basins) has been formed in the last 5% of Earth's history. Clearly, the Earth is extremely active.

As a unifying theory, plate tectonics provide a framework to interpret and understand geological processes and even the evolution of life. We now know that the continents are moving, and have been for thousands of millions of years. Continents are long-lived because they are not subducted, and they contain the oldest rocks on the planet. Moreover, the total area of continental crust has grown over time by the creation of new crust at subduction zones. We now also understand that continents not only move laterally, but also rise and fall with respect to sea level, and occasionally rupture and split apart. They rift, drift apart and collide with other continental fragments, only to rift apart again, drift and re-collide in the endless **Wilson Cycle**.

Most geological activity occurs at plate boundaries and is therefore very unevenly spread around the globe. Regions where subduction occurs are particularly prone to natural disasters in the form of both earthquakes and extremely violent volcanic eruptions. In contrast, there is little activity in the interior regions of plates, such as in southern Africa. Movement of the continents affects global climate, because Earth's climate is closely coupled to the oceans and ocean circulation is affected by the positions of the continents. The elevations of continents, and especially mountain ranges, impact on atmospheric circulation, and thus also influence climate.

The changing Earth environment constantly places animal and plant species under stress, thereby driving evolutionary change. In addition, the rifting of continents breaks up populations of animals and plants and the separated populations become subject to differing environmental pressures as the continental fragments migrate through different climatic zones. This results in divergent evolutionary paths on different landmasses. For this reason, each of Earth's major landmasses is populated by different species of animals and plants. Indeed, much of the diversity of life is a product of plate tectonics.

Plate tectonics also drives immense chemical cycles in the Earth. Water and carbon dioxide are incorporated into the crust at mid-ocean ridges and released at subduction zones, inducing mantle melting; mountain ranges formed along subduction zones are eroded and the products deposited as sedimentary rocks or dissolved in the oceans. The sedimentary rocks are reconstituted at subduction zones into metamorphic and igneous rocks and form mountain ranges, which are re-eroded in a cycle that is almost as old as the Earth – the **Rock Cycle** (**figure 2.22**).

What would the Earth be like in the absence of plate tectonics? Any mountainous terrain that may have existed would long ago have been eroded by wind, rain, rivers, glaciers and waves, and there would be no dry land on the Earth. Without the continual supply of chemical elements resulting from weathering, erosion and volcanic activity, the biosphere – the envelope of life that surrounds the Earth – would have been deprived of essential resources and would have died of malnutrition. In short, we would not be here to marvel at the wonders of our dynamic Earth.

When viewed in the context of geological time there is little that is constant about the face of the Earth, except the fact that it is constantly changing. A record of this constant change, extending back almost 4 000 million years, is extremely well preserved in the rocks of southern Africa. It includes not only information on the physical environment, but also on the Earth's changing atmosphere and life.

That history, revealed to us by the rocks, forms the subject of the remainder of this book.

SUGGESTED FURTHER READING

Hamblin, WK and Christiansen, EH. 1995 (8th Edition). *The Earth's Dynamic Systems*. Prentice Hall, New Jersey.

Johnson, MR, Anhaeusser, CR and Thomas, RJ. 2005. *The Geology of South Africa*. Council for Geoscience, Pretoria.

Marshak, S. 2001. *Earth – Portrait of a Planet*. W N Norton and Co, New York.

Skinner, BJ and Porter, SC. 1999 (4th Edition). *The Dynamic Earth*. John Wiley and Sons, New York.

Skinner, BJ, Porter, SC and Botkin, DB. 1999 (2nd Edition). *The Blue Planet – An Introduction to Earth Systems Science*. John Wiley and Sons, New York.

THE STORY OF
EARTH & LIFE

3

THE FIRST CONTINENT

Carl Anhaeusser

The rocks of the Barberton Mountain Land record the formation of the Earth's first continent.

ROUTE MAP TO CHAPTER 3

Age (years before present)	Event
13 700 million	• Big Bang: time began. Energy was transformed into hydrogen and helium as the Universe expanded. • Clumping of gas started galaxy formation. • Clumping of gas within early galaxies started star formation, in which the synthesis of heavy elements commenced. • Stellar explosions (supernovae) mixed heavy elements into the galaxies. Later generations of stars further modified composition of the galaxies.
4 600 million	• Gas and dust in the outer region of the Milky Way galaxy collapsed to form the Sun and planets: the Earth grew by gravitational accumulation of smaller bodies. • The Earth melted, heated by bombardment of smaller bodies, and segregated into layers of different density (core, mantle). A major collision knocked off a portion of the mantle, forming the Moon. • The Earth's surface cooled and an early crust formed. Bombardment declined, but destroyed the crust. The mantle also solidified: carbon dioxide, water vapour and nitrogen degassed from the Earth's interior, forming the atmosphere. • As the Earth cooled, water vapour started to condense to form the oceans.
4 100 million?	• Plate tectonics started; the oldest known rock (a granite in Labrador, Canada) formed.
3 900 million	• The oldest known sedimentary rocks formed (Amitsoq terrane, Greenland).
3 500 million	• The oldest known oceanic crust, consisting of komatiite, formed (the lower Onverwacht Group, Barberton).
3 400 million	• Subduction of oceanic crust created the oldest known island arc (upper Onverwacht Group, Barberton), with associated batholiths.
3 300 million	• Island arcs were eroded, shedding sediment into ocean trenches (the Fig Tree Group, Barberton).
3 200 million	• Island arcs started to amalgamate to form the first micro-continent. Erosion of continental rocks produced conglomerates, sandstones and mudstones (the Moodies Group, Barberton). • Granites formed by melting of thickened crust. • Micro-continents continued to accrete to form the Earth's oldest known continent, the Kaapvaal Craton.
3 100 million	• The continent finally stabilised.

WHAT IS SO SPECIAL ABOUT THE EARTH?

Earth is the third planet from the Sun, at a distance of about 150 million kilometres – an incomprehensible distance, at least in terms of human experience. Light from the Sun takes about eight and a half minutes to travel the distance to Earth.

Earth is, of course, unique in the Solar System in that it supports advanced forms of life and is mantled by a gaseous atmosphere that presently contains a significant amount of oxygen. One of the main reasons why Earth supports life is that it is positioned at the right distance from the Sun (the so-called habitable zone) to support the long-term existence of surface water in liquid form, mainly in the oceans. Any closer and the Sun's radiation would be so intense that water would boil and exist largely, or entirely, in the vapour state. The amount of heat energy reaching planets further from the Sun diminishes to the extent that water is entirely in the solid state (ice). The fortunate coincidence that places the Earth at just the right distance for much of its surface water to be liquid is a critical factor in the existence of life.

The prevailing view is that life not only originated in the oceans, but has existed entirely in water throughout much of Earth's history. The first amphibians emerged from the oceans only 350 million years ago, following plants that colonised land possibly 500 million years ago, and insects, about 430 million years ago. For the preceding several billion years life evolved entirely in the oceans. Earth is also just big enough to retain most of its gases in the atmosphere, whereas smaller bodies with less gravitational attraction (like Mars or the Moon) have long since lost their volatile envelopes.

In addition, Earth has a fairly strong magnetic field. This creates an encircling shield that deflects highly energetic particles ejected from the Sun, thus protecting the atmosphere, oceans and life. The oldest trace of life on Earth is found in cherts (a chemical sediment deposited on the ocean floor – *see* 'The formation of sedimentary rocks' on page 64) such as those at Barberton (**figure 3.1**), where microfossils believed to be cyanobacteria have been preserved. The story of the origin of life is dealt with in more detail in Chapter 6.

But how did Earth form, and when?

Carl Anhaeusser

Figure 3.1 *Chert layers from Barberton contain traces of early life.*

THE DAWN OF TIME

The late Stephen Jay Gould, well-known geologist-palaeontologist from Harvard University, made the point that the most profound contribution geology has made to human thought is the concept of Deep Time. This term, originally coined and popularised by American author John McPhee, refers to the immensity of geological time and the problem that man has in conceptualising the several-billion-year time span over which geological processes on the Earth have been operating.

Geological time – difficult though it may be to conceptualise – provides a sense of security in that there was a beginning, and that time's passage has been regular and marked by familiar cycles such as day-night, the lunar month and the seasonal year (*see* 'Measuring the age of a rock' on page 68 and 'The geological time scale' on page 71). Physicists who grapple with the origin of the universe have a much greater conceptual problem: their equations dictate that time is not a constant but varies as one approaches the speed of light. Stephen Hawking's book *A Brief History of Time* introduces this concept for the general public and explains that the Universe is finite in volume but expanding. Therefore, time and matter must have been formed at a fixed point in the past, which physicists call a singularity.

The concepts of relativity and quantum theory are complex. To keep things simple we will assume that time is constant and can be measured by the regular passage of events encapsulated in standard units which humankind has defined, such as the second and the year. Let us, however, accept the notion that time and the Universe (and therefore geological processes) commenced at a singularity

Figure 3.2 *The Solar System is embedded in the large Milky Way galaxy, which contains billions of stars and huge clouds of dust and gas. It probably resembles the M74 (or NGC 628) galaxy in the constellation Pisces, shown here. Galaxies began to form by mutual gravitational attraction of hydrogen and helium gas shortly after the Universe was born in the Big Bang about 13 700 million years ago. Local concentrations of gas within these early nebulae collapsed under gravity to give rise to the first stars, which converted hydrogen and helium into heavier elements. Many of these stars exploded as supernovae, distributing heavy elements into the galaxy, and subsequent generations of stars were made from a more complex mixture of elements.*

and that the Universe had a beginning and will, in all likelihood, have an end. The event that marks the beginning of time, as well as matter in the universe, is popularly referred to as the **Big Bang**.

THE BIG BANG

In 1929 American astronomer Edwin Hubble discovered that the universe is expanding. The evidence for this is provided by a shift towards the red (i.e. lower frequency) end of the light spectrum in virtually all galaxies for which these data have been obtained. A red shift in the spectrum of a galaxy is a consequence of the Doppler effect, which relates how the frequency of light (i.e. its colour) emitted from one galaxy is lowered relative to the observer galaxy as the two move apart. The same phenomenon causes the change in pitch of the sound made by a motor vehicle as it passes. If virtually all galaxies are moving apart relative to one another, then the Universe must be expanding away from a single point. The implication of this discovery is that all matter, as well as time, was created by the explosion of a point source of almost infinite density during an event that is believed to have occurred about 13 700 million years ago.

Protons and electrons were formed from other subatomic particles in the first few instants after the Big Bang. They combined to create neutrons that then amalgamated with other protons and electrons to form atoms of hydrogen and helium. The newly formed matter was ejected away from the singularity to create an expanding Universe. As this expanding cloud of gas began to form clumps under the influence of gravity, the Universe started to evolve towards its present form, with regions of concentrated matter, known as **nebulae**, separated by near total emptiness in between.

After the initial flash of the Big Bang, the Universe darkened and remained dark for millions of years. Light was only emitted once matter had locally concentrated to sufficiently high densities under gravity within the nebulae to heat up and trigger nuclear fusion reactions in the first generation of stars about 200 million years after the Big Bang. These reactions not only emitted light and other forms of radiation, but also synthesised heavier chemical elements by the nuclear fusion of lighter elements. These early stars evolved and died, many in spectacular explosions known as **supernovae**, in the process spreading their newly formed mixture of heavier elements into the nebulae.

New stars were born from this debris, modifying it further and eventually ejecting it into the nebulae. Over thousands of millions of years, these nebulae, the galaxies of today, evolved from simple clouds of only hydrogen and helium to gigantic discs containing billions of stars and huge clouds of gas and dust laced with heavy elements, such as are required to form a rocky planet like the Earth. Current theories on the formation of heavy elements in stars suggest that the atoms that make up our world, and us for that matter, have passed through at least two supernovae events, so we have a somewhat intimate association with these celestial extravaganzas.

The Solar System is embedded within one of these galaxies – which we see edge-on in the night sky as the Milky Way – and is a huge amalgamation of stars, dust and gas arranged in a flattened disc with spiral arms, very similar to the M74 galaxy shown in **figure 3.2**. It is about 100 000 light-years across and about 20 000 light-years thick at its centre. (A light-year is the distance light will travel in one year – 9 460 000 million kilometres). The galaxy is rotating, and completes a full revolution every 186 million years. It is believed to have grown to its present size by the amalgamation of smaller galaxies, and today forms part of a cluster of more than 50 galaxies. Two nearer neighbouring galaxies are the Magellanic Clouds, which are visible to the naked eye and appear as two detached portions of the Milky Way.

BIRTH OF THE SOLAR SYSTEM

One rather ordinary star in the outer portion of the Milky Way galaxy, which we call the **Sun**, began to nucleate from the gas and dust of the galactic nebula about 4 600 million years ago, most probably as a result of the gas being compressed by the shockwave of a nearby supernova explosion. Once compressed, the hydrogen and helium, together with other gases and dust, began to collapse under gravity, forming an ever-shrinking disc – the Solar Nebula. As collapse progressed, temperatures in the disc rose, eventually reaching several million degrees at its centre and creating a fireball that ignited nuclear fusion of hydrogen to form our Sun. The Sun is in essence a continuously exploding hydrogen bomb protected from disintegration by the force of gravity.

The young Sun remained embedded in the disc of gas and dust. In the hot, inner region of the disc, only materials with very high melting and boiling points – mainly silicate compounds and nickel-iron alloys – formed solid particles (other materials being in the vapour state). These provided the raw materials for the formation of the inner planets. In the outer, cooler regions of the disc, carbon compounds and frozen water and ammonia formed the bulk of the solid particles. The grains began to clump together to form larger bodies or **planetesimals**, kilometres in diameter. Continued collision and accretion of planetesimals gave rise to Moon-sized bodies that amalgamated under the force of gravity to form complete planets.

The planets closer to the Sun are denser (average 5 g/cm^3), because of the higher temperatures prevailing in the region where they formed. In contrast, more distant planets contain a high proportion of gases and have low densities (average 1.2 g/cm^3), reflecting the cooler conditions that prevailed in the outer regions of the disc during planetary formation. The results of a computer simulation of the process are illustrated in **figure 3.3.** With the passage of time, radiation from the young Sun (the **Solar Wind**) swept away gas and dust that had not collapsed to form planets, producing the dust- and gas-free Solar System we observe today.

Nine planets formed from the gas and dust cloud orbiting the Sun. The four closest to the Sun – Mercury, Venus, Earth and Mars (the so-called inner or terrestrial planets) – are relatively small and are

THE FORMATION OF SEDIMENTARY ROCKS

Rock exposed at the Earth's surface is subject to chemical attack by the atmosphere, a process called **weathering**, which leads to disintegration of the rock. Rocks may also be exposed to mechanical agencies that cause disintegration. Whereas mechanical processes produce only rock fragments, weathering produces a range of products, including: components of the original rock that are not susceptible to weathering; new minerals, particularly clay minerals; and soluble substances.

The products of weathering or mechanical disintegration are transported away from the site of generation by flowing water, wind or ice, in a process known as **erosion**. They will ultimately accumulate elsewhere as sediment. Sediments that are accumulations of fragmental material (clay minerals, sand grains, rock fragments) are known as **clastic sediments**, whereas those that originate by precipitation of material from solution are known as **chemical sediments**. Sediment accumulations may ultimately become converted into sedimentary rocks in a process called **lithification**, which basically involves cementation of the particles.

CLASTIC SEDIMENTS

Flowing water separates material according to size: large particles such as pebbles only move in rapidly flowing water, sand in slower flowing water and silt in even slower flowing water, whereas mud (mainly made of clay minerals) requires hardly any flow to keep it dispersed in water. Sediments deposited by flowing water are therefore differentiated by size into gravel, sand, silt and mud.

Lithification converts gravel into a sedimentary rock called **conglomerate**, sand into a rock called **sandstone**, silt into **siltstone** and mud into **mudstone** (*see* **A–D**, right). Sand usually consists of the mineral quartz (silicon dioxide) because it is hard, chemically resistant and fairly common (quartz is a major constituent of the igneous rock granite, which forms most of the continental crust). Quartz sand may become lithified to the point where it forms an extremely hard, resistant rock type called **quartzite**. Sand sometimes consists mainly of shell fragments, in which case the term **limestone** is used to describe the resulting rock type, although not all limestones are shell accumulations. Some form as a result of photosynthetic bacterial activity.

In contrast to flowing water, flowing ice (i.e. glaciers) does not sort material according to size. As ice melts at the end of a glacier or from an iceberg, the rock material contained in the ice is dumped in an unsorted manner. This results in sediment consisting of a random mixture of material ranging in size from boulders to mud. When lithified, this is referred to as **tillite**. Material consisting of such a random mixture of different-sized particles can also be produced by a mud-flow, such as an avalanche resulting from heavy rain. Distinguishing between deposits formed by mud-flows and tillite is often difficult, and can only be done by examining the context in which the material occurs. Because it is often difficult to tell whether an unsorted rock resulted from glacial activity or from mud-flows, the general term **diamictite** has been coined to describe such a rock, whatever its origin.

Wind is also an important agent of sediment transport. Generally, material deposited by wind consists of fine sand (usually made of quartz). These sand deposits are often very thick. Sandstone resulting from wind deposition can be distinguished from sandstone resulting from flowing water by the associated internal structures (*see* 'Sedimentary structures' on page 86).

CHEMICAL SEDIMENTS

These sediments result from precipitation of substances dissolved in water, usually as a result of evaporation of the water. Depending on the substance precipitating, the resulting rock has different names; often the name of the mineral that precipitated is used to describe the rock. The general term for such a rock is an **evaporite**. One of the most common evaporites is sodium chloride (table salt, termed **halite**), formed during the evaporation of sea water and can occur as layers hundreds of metres thick.

Occasionally, precipitation occurs as a result of a chemical reaction. An example is the precipitation of iron oxide from sea water, which occurred frequently during the early evolution of the Earth (*see* Chapter 4). The high iron content results in a red- to black-coloured rock that consists mainly of iron oxide, known as **iron formation**. Sometimes the precipitation of iron was accompanied by silica precipitation. Silica is white, so the resulting rock consists of white, red and black layers and is known as **banded iron formation** (*see* **E** below).

Conglomerate

Mudstone

Sandstone

Siltstone

Banded iron formation

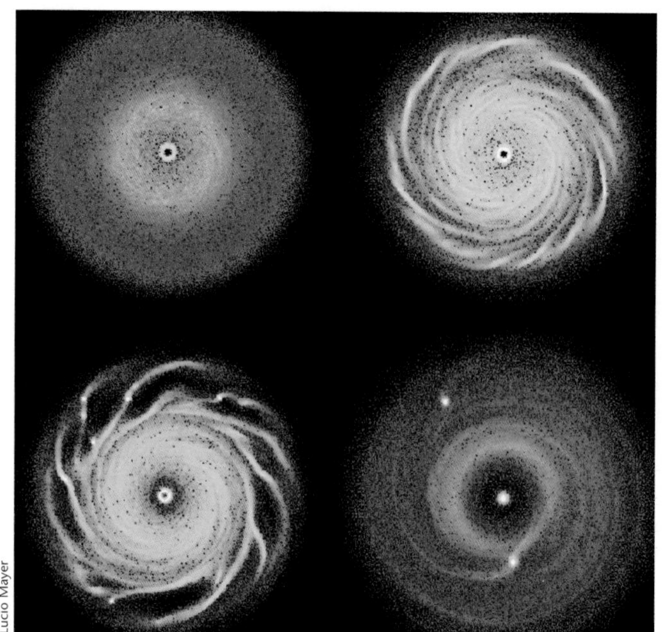

Lucio Mayer

Figure 3.3 ***About 4 600 million years ago, a cloud of gas and dust in the outer reaches of the Milky Way galaxy began to collapse under the influence of gravity. As it shrank it began to rotate ever faster, flattening into a disc shape. Its interior also became hotter until the centre reached several million degrees. Nuclear fusion of hydrogen atoms commenced, producing the Sun. Computer simulation suggests how planets may have formed: material in the disc (top left) began to clump to produce smaller bodies (planetesimals, top right) that grew with time (bottom left) and finally coalesced to form planets (bottom right). The inner planets – Mercury, Venus, Earth and Mars – formed in the hotter, inner region of the disc and are made mainly of rocky material, while the outer planets formed in the cooler, outer disc and consist mainly of gases.***

made up of dense silicate, or rocky, material. The next two planets out from the Sun – Jupiter and Saturn – are gas giants that have rock-ice cores enveloped by hydrogen, helium and other gases. Further out are the two smaller gas giants, Uranus and Neptune, which are so far from the Sun that very little of its radiation reaches them and they are in a permanent state of deep freeze. The outermost planet is Pluto with its giant moon, Charon, whose orbit is off the plane (or ecliptic) of the other planets.

Pluto lies at the inner edge of a vast region of the Solar System that extends to a distance of about 50 Earth-Sun radii from the Sun, known as the Kuiper Belt. In this region, Pluto-like objects seem to abound and at least 700 have so far been identified. Large amounts of unaccreted material from the solar nebula, dominated by ice laced with silicate and carbonaceous dust, lie dispersed in the Oort Cloud, far beyond the Kuiper Belt. Occasionally, lumps of material from these distant regions enter the Solar System as comets.

It is indeed fortunate that the Earth has large planetary neighbours in more distant orbits, and especially Jupiter, for their strong gravitational fields shield the inner planets from impacts by comets. The majority of comets approaching the inner portion of the Solar System are captured by the gravitational field of Jupiter and flung back into deep space. Some cosmologists believe that advanced life would never have been able to evolve on Earth without the protection provided by Jupiter.

THE MOON

The process of planetary growth was probably fairly rapid in geological terms, lasting perhaps less than 60 million years. About 10 million years after the start of planetary formation the Earth had grown to some 65% of its current size and had largely segregated into mantle and core. At this time (about 4.53 billion years ago) it was struck a glancing blow by a Mars-sized object that vaporised part of the mantle. The vapour collected in orbit around the growing Earth, where it condensed and accumulated under gravity to form the Moon. The Earth and Moon continued to grow after this collision to reach their final sizes. The Moon-forming event was so severe that it tilted the Earth's axis of rotation and caused it to precess, like a spinning top knocked off balance, so that the axis of rotation sweeps through a conical path, completing a full revolution once every 26 000 years (the precession of the equinoxes).

The collision left Earth with a sizeable Moon, which is the cause of ocean tides. Even more important was the tilt in the axis of rotation that the collision caused, for it is this tilt that is responsible for the seasons.

But the consequences of the Moon's presence run far deeper. All planets are subject to gravitational forces exerted by the Sun, and more irregularly by neighbouring planets, each in their own orbit. J Laskar and P Robutel of the Bureau des Longitudes in Paris investigated the gravitational effects that planets (and moons) exert on one another, making some startling findings.

Their calculations revealed that gravitational forces have a profound effect not only on rotational frequency (for example, we always see the same face of the Moon), but more importantly on the inclination of the axis of rotation relative to the plane of the orbit. The tilt in the axis of rotation of Mars, for example, varies from 0° to as much as 60° over a period of tens of millions of years, and in a completely chaotic manner. Gravitational perturbations in the inclination of the axis of rotation of Venus were evidently so severe that the planet actually turned upside down and now rotates in the opposite direction to the other inner planets. Uranus lies on its side, with an inclination of 97°, possibly for the same reason.

The Moon saved the Earth from these chaotic oscillations. It is sufficiently massive to have anchored the orientation of the Earth's axis of rotation (**figure 3.4**), so that it deviates by only 1.3° from the average of 23.3°. In the absence of the Moon, Lasker and Robutel calculated that the Earth's axis of rotation would vary from 0° to about 85°, and in a chaotic way. Such variations would have had a profound effect on the Earth's climate, its seasons and the length of its days, and it is unlikely that advanced forms of life could have evolved on the planet.

The formation of the Moon and the large outer planets, especially Jupiter, were propitious events in the early history of the Solar System, for they created conditions favourable for the subsequent appearance and evolution of life, and advanced life in particular. The Earth would probably be a very different place without the Moon or Jupiter.

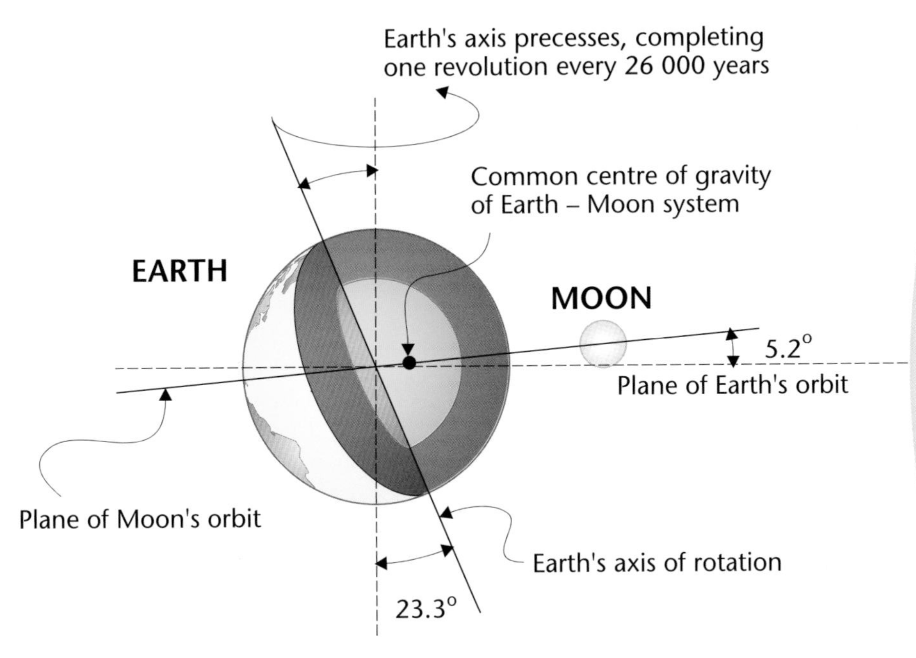

Figure 3.4 *The Earth and Moon rotate around a common centre of gravity that lies within the Earth but displaced from its centre. This motion has stabilised the inclination of the Earth's axis and protected it from gravitational perturbations induced by other planets.*

MEASURING THE AGE OF A ROCK

Many different, naturally occurring, radioactive chemical elements are employed to measure the ages of rocks. The most common method used to date old rocks employs the radioactive element uranium. There are two kinds of uranium atoms, which are chemically the same, but one is slightly heavier than the other: these are known as isotopes of uranium. Both are radioactive, which means that the atoms are unstable and change or decay into atoms of another chemical element, called the daughter element, over time.

The isotopes of uranium decay to produce different isotopes of lead. The rate of decay is constant, and is called the **half-life**; this is the time it takes for half the atoms in a sample of the material to decay to the daughter element. The half-lives of the uranium isotopes are particularly well known, largely because these isotopes are used for power generation in nuclear reactors and for nuclear weapons manufacture.

Certain minerals incorporate uranium in their structure when they form, but exclude lead. This occurs because of the different chemical properties of uranium and lead. One such mineral is zircon, which is a zirconium silicate. As time passes, uranium atoms in the zircon will decay to lead atoms. These lead atoms are trapped in the crystal structure, even though chemically unsuited to their environment, because the surrounding atoms hold them in place. The longer the time elapsed, the more lead there will be and the less uranium. The time that has elapsed since the zircon formed can thus be calculated from the measured concentrations of the two isotopes of uranium and their respective daughter isotopes present. This gives the age of formation of the zircon.

This method of dating can only be used to date rocks in which the zircon formed at the same time as the rock itself, and is therefore restricted mainly to igneous rocks. Sedimentary rocks cannot be dated in this way, because the zircons they contain come from older rocks.

BROKEN PLANETS AND METEORITES

The region between Mars and Jupiter is characterised by a huge number of orbiting rock fragments, the **asteroids**, representing a planet that failed to form, together with the debris of several planetesimals that were broken up in mutual collisions. The largest of the asteroids is Ceres, 1 000 km in diameter.

Although Jupiter's strong gravitational field in the main protects the inner planets from cometary collisions, it also has a malevolent aspect. Asteroids are sometimes forced off course by the gravitational influence of Jupiter and fly through space, occasionally intersecting the orbit of Earth. When they hit the Earth, the repercussions can be devastating. Some of the consequences of past collisions are discussed later in this book (*see* Chapters 4, 9 and 11).

While hits by large extraterrestrial bodies are fortunately rare, there is nevertheless a constant rain of smaller bodies into the Earth's atmosphere from the Asteroid Belt, as well as cometary debris, which amounts to about 30 000 tonnes per year. Most of these are very small particles that burn up in the atmosphere (forming meteors or shooting stars), but some fragments from the Asteroid Belt are large enough to penetrate the atmosphere and reach the surface. They are known as **meteorites**. Other rare meteorites are derived from material ejected from the surfaces of Mars and the Moon, which themselves have been hit by meteorites and comets large enough to throw material into space.

Thousands of meteorites have been found on Earth and studied in detail, and their age has been measured to be 4 600 million years. A proportion of them are fragments from the cores of broken planets, made of iron and nickel. One example is the famous Hoba meteorite in northern Namibia, which is the largest meteorite known (**figure 3.5**). The majority, however, are fragments of small asteroids consisting

Uranium dating is unsuitable for very young rocks, because the half-lives of uranium isotopes are very long (thousands of millions of years) and uranium is a rare element, so very little of the daughter isotopes will be present in young zircons. For such rocks, a commonly used method is based on a radioactive isotope of the more common element potassium, which decays to argon. Argon is a chemically unreactive gas, and when a rock forms it contains no argon. Argon accumulates in the rock from the decay of potassium as time passes. The age of formation of the rock is obtained from measurements of the amount of argon and the radioactive isotope of potassium present in the rock.

University of the Witwatersrand

The late Prof Hugh Allsopp was a South African pioneer whose discoveries revolutionised the science of rock dating.

Another well-known method of dating, the Carbon 14 method, is not used in rock dating. The reason for this is that the half-life of the radioactive isotope of carbon is very short, and it is not possible to measure the age of materials older than about 40 000 years.

The methods used to date rocks are simple in principle; in practice they involve complex and expensive equipment and usually special, ultra-clean laboratories for preparation of the samples. There are also many potential sources of error, and great care has to be exercised in selecting samples for dating and in the interpretation of the results. Modern equipment makes it possible to obtain extremely precise ages, with errors of measurement typically better than one part in a thousand.

Figure 3.5 *Tens of thousands of bodies, varying in diameter up to 1 000 km, orbit the Sun between Mars and Jupiter. They are known as asteroids. Occasionally small asteroids intersect the Earth's orbit and crash through the atmosphere. We refer to these as meteorites. The asteroids represent a planet that failed to form, as well as the debris of planetesimals that were broken up in mutual collisions. Some of these smashed planetesimals contained nickel-iron cores. The nickel-iron Hoba meteorite near Grootfontein in northern Namibia, shown here, is a fragment of the core of such a body and is the largest known meteorite. The parent bodies of most meteorites formed 4 600 million years ago. Meteorites therefore provide information about conditions prevailing during the formation of the Solar System.*

Carl Anhaeusser

mainly of magnesium and iron silicates similar in composition to Earth's mantle, but containing tiny flecks of iron-nickel metal. Unlike the rocks of Earth, meteorites have remained essentially unchanged since they formed about 4 600 million years ago, and have given us a glimpse of the processes that took place in the Solar System as the planets were forming, as well as the nature of the raw material from which our own planet was built.

COOLING AND DIFFERENTIATION OF THE EARTH

In its earliest state the Earth was probably entirely molten, heated by the incessant bombardment of planetesimals as it grew by amalgamating asteroid-like debris from the solar disc under the force of gravity. During its first few tens of millions of years the molten Earth cooled and differentiated into a number of concentric layers – a core, mantle and crust, stratified according to density, with dense iron-nickel metal forming the core. The next few hundred million years of Earth's history saw continued sinking of dense material and the solidification of a thin (less than 30 km thick), lower-density, outer crust.

Throughout this time, though, the Earth (together with the other planets of the Solar System) continued to be bombarded by an intense flux of asteroids and comets, as the Earth's gravitational field swept up solid material in the disc within its orbital region. This ongoing bombardment was responsible for the almost complete destruction of the earliest vestiges of the crust. Those parts of the early crust that were not destroyed by impact events were probably later recycled into the mantle by plate tectonic processes, in particular subduction.

Consequently, there is very little remaining of the rocks that formed in the first few hundred million years of Earth history. This chaotic period of time is referred to as the Hadean Era, derived from *Hades*, the underworld of Greek mythology (i.e. Hell). The oldest terrestrial material that has been dated so far are tiny zircon (zirconium silicate) crystals found in a sedimentary rock from Western Australia, one of which is 4 400 million years old. Although the sediments within which the tiny zircon crystals are found are themselves younger, the existence and age of these grains clearly attests to the presence of Hadean rocks that have subsequently been eroded and obliterated.

In contrast to the Earth, the Moon still preserves much of its very early crust. Rocks brought back by the Apollo and Luna missions to the Moon that have been dated are up to 4 500 million years old. Heavily pock-marked by meteorite impact scars (**figure 3.6**), the Moon is geologically inactive and its early crust remains relatively intact. The Earth would have suffered even more cosmic blows than the Moon as it has a stronger gravitational field, but no trace of this period remains. Only the Moon can provide us with insight into the turbulent events that characterised the very early period of Earth's history (*see* 'Geology of the Moon' on page 74).

The Hadean Earth would not have possessed oceans as it was too hot. As the bombardment waned and the Earth cooled and solidified, water vapour, carbon dioxide and other gases began to form the early atmosphere. These were degassed from the molten interior and possibly augmented by cometary collisions with the Earth. With further cooling, it began to rain as water vapour condensed out of the atmosphere, and the oceans formed. When this happened is not clear. Surface water was definitely present by 3 900 million years ago but may have appeared 300 to 400 million years earlier.

NASA

Figure 3.6 ***The surface of the Moon is very old and still carries the scars of the impacts that characterised the period during which the planets formed.***

THE GEOLOGICAL TIME SCALE

Starting in the 18th century, early geologists in Europe began to produce geological maps. They realised that rocks, particularly sedimentary rocks, could be arranged in order of deposition, i.e. in a relative age sequence. This ordering was based on the stacking of layers, younger on top of older, and on the types of fossils contained in the rocks. They noted that the types of fossils present, which were mainly marine organisms, periodically underwent sudden changes in the varieties of species present. They assigned names to the broad intervals that had similar fossil types, often choosing names from places where rocks of that type were well developed. These names, such as Jurassic and Cretaceous, are still in use today.

The significance of the changes in fossil types was not fully appreciated until Charles Darwin and Alfred Wallace published their theory of evolution. It then soon became generally accepted that the changes in fossils were the result of evolutionary processes. It was realised that long periods of time must have been required to produce these changes, but until the discovery of radioactivity and the development of absolute dating techniques, no one knew how much time was involved.

Dating of rocks using radioactive isotopes has now placed the periods identified by means of fossil types into an absolute time framework. Moreover, dating has also allowed rocks that do not contain fossils to be placed in their correct time slot. Information on both fossil types and age determination has been compiled into a **standard geological column**, or standard geological time scale (*see* diagram, right). The column is divided hierarchically into varying time intervals: **Eons** represent very long periods, **Eras** somewhat shorter periods, and **Periods** and **Epochs** are still shorter.

There are three eons: the **Archaean, Proterozoic** and **Phanerozoic**. Organisms with hard body parts that gave rise to fossils occur only in the Phanerozoic, whereas older rocks contain no fossils, or at best traces of very primitive single-celled organisms. The Archaean and Proterozoic are collectively known as the **Precambrian**. The Phanerozoic is divided into three Eras, the **Palaeozoic** (ancient life), **Mesozoic** (middle life) and **Cenozoic** (recent life), based on fossil types. The subdivision of the Precambrian into Eras is arbitrary. The Phanerozoic Eras are further subdivided into Periods, also based on fossil types.

Sudden changes in fossil types that mark the Era boundaries are now known to be mainly due to sudden mass extinction events, of which the most severe occurred at the end of the Permian, when about 96% of species became extinct. In the more famous end-Cretaceous extinction, which marked the end of the dinosaurs, about 70% of species became extinct.

GEOLOGICAL TIME SCALE

	EPOCH	PERIOD		ERA	EON
	Pleistocene (Holocene)	Quaternary		Cenozoic	PHANEROZOIC
2	Pliocene	Neogene	Tertiary		
5	Miocene				
24	Oligocene	Palaeogene			
34	Eocene				
55	Palaeocene				
65		Cretaceous		Meso-zoic	
142		Jurassic			
206		Triassic			
248		Permian		Palaeozoic	
290		Carboniferous			
354		Devonian			
417		Silurian			
443		Ordovician			
495		Cambrian			
545				Neo-Proterozoic	PROTEROZOIC
1000				Meso-Proterozoic	
1600				Palaeo-Proterozoic	
2500					ARCHAEAN

STRATIGRAPHY

The Earth's crust responds to forces generated in the mantle below, continents moving both laterally and vertically as a result of the action of these forces. At times, continents may be submerged below sea level, and at other times raised and exposed to agents of weathering and erosion. During periods of submersion, sediments will accumulate, resulting in sedimentary rock. Different types of sedimentary rock accumulate, depending, for example, on the rate of subsidence and the rate of sediment supply. There may also be occasional volcanic activity associated with subsidence of the crust. The net result is that periods of subsidence are generally associated with layered accumulations of various types of sedimentary and igneous rocks. Depending on the extent of the area affected by subsidence, these accumulations may be large, extending over millions of square kilometres, or they may be small, covering only tens of square kilometres.

During periods of uplift, weathering and erosion occur, and a previously accumulated pile of sedimentary and volcanic rocks may be partially or completely removed. An uplift event may be followed by later subsidence, during which a new period of sediment accumulation will occur, often depositing sediment on a floor of older, partly eroded sedimentary rock.

By careful mapping of the distribution of sedimentary and associated volcanic rocks, it is possible to identify those formed within the same period of subsidence, and to separate them from accumulations formed during other periods of subsidence. All those rocks deposited during the same period of subsidence belong to a single event, and for convenience are grouped together and referred to as a **Supergroup**.

Supergroups are given names, usually taken from regions where they are particularly well exposed – for example, the Karoo Supergroup, which underlies the Karoo region, and the Witwatersrand Supergroup, whose rocks form the Witwatersrand ridges.

Supergroups thus consist of layers of many different rock types formed during one period of accumulation, each layer representing different conditions of accumulation. The oldest will lie at the bottom of the pile and the youngest at the top in a vertical profile through the pile. During a single period of accumulation, there may be pronounced changes in the prevailing conditions – for example, from persistent deep-water conditions to a prolonged period of shallow-water conditions, followed perhaps by a period of volcanic activity.

Such changes result in the accumulation of very different types of rocks. It is therefore possible to subdivide the layers making up a Supergroup into smaller entities, based on the types of rocks present. These subdivisions are, in hierarchical order, termed **Groups, Subgroups, Formations, Members** and **Beds**, the last-mentioned representing individual layers. Each of these smaller subdivisions is also given a name, usually the name of the locality where those rocks are best represented. The study of such groups of rocks and their interrelationships is referred to as **stratigraphy**.

John Hancox

Layered sedimentary rocks of the Karoo Supergroup.

The variety of rocks forming a supergroup is conveniently portrayed in diagrammatic form as a vertical column, with the various rock types indicated using different symbols. The oldest layer is placed at the bottom. Vertical heights of symbols on the column are proportional to the thicknesses of the various layers they represent, while the subdivisions are indicated adjacent to the column (*see* diagram, right). Such a diagrammatic representation is known as a **stratigraphic column**.

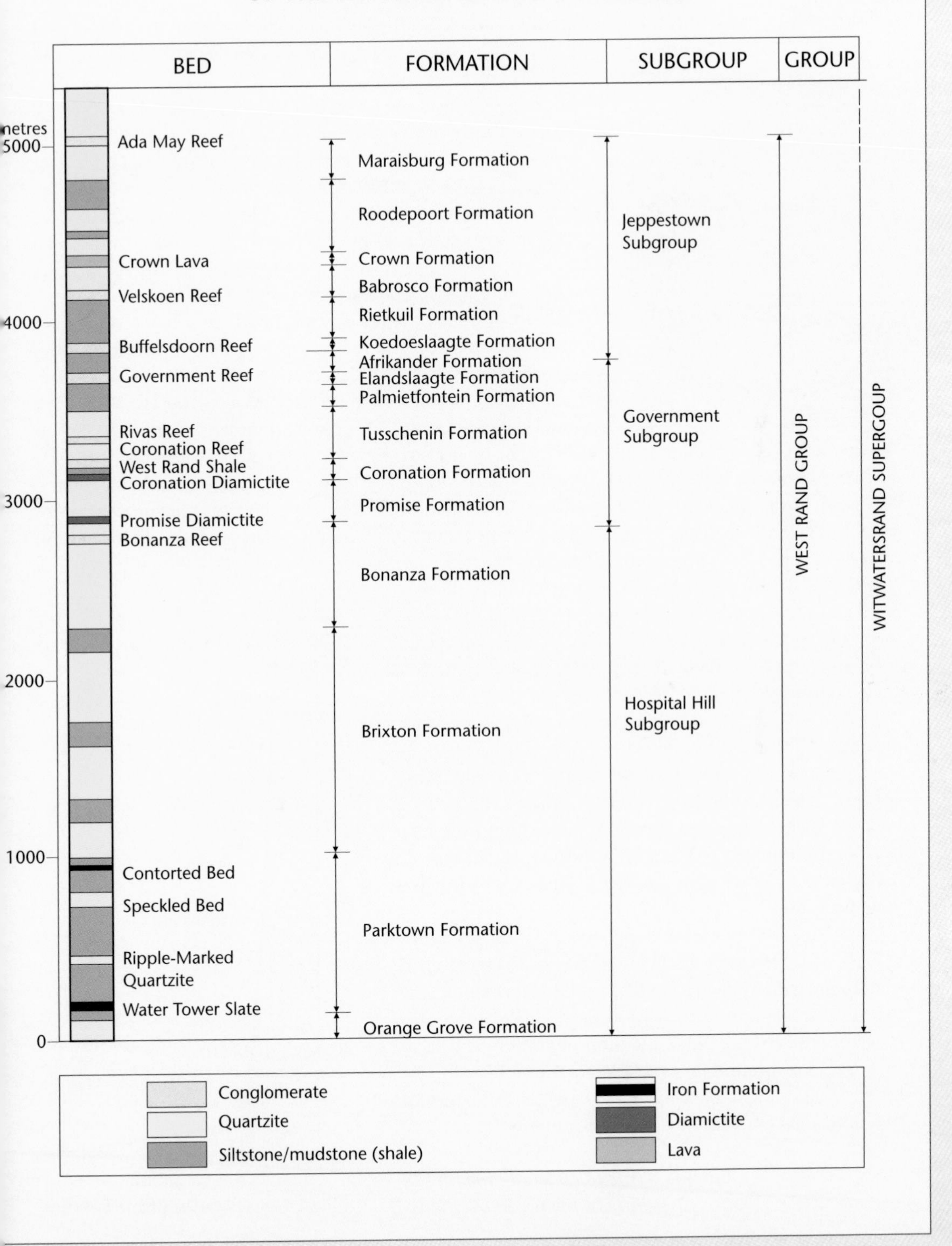
STRATIGRAPHIC COLUMN FOR PORTION OF THE WITWATERSRAND SUPERGROUP
BED
FORMATION
SUBGROUP
GROUP
metres
5000
4000
3000
2000
1000
0
Ada May Reef
Crown Lava
Velskoen Reef
Buffelsdoorn Reef
Government Reef
Rivas Reef
Coronation Reef
West Rand Shale
Coronation Diamictite
Promise Diamictite
Bonanza Reef
Contorted Bed
Speckled Bed
Ripple-Marked Quartzite
Water Tower Slate
Maraisburg Formation
Roodepoort Formation
Crown Formation
Babrosco Formation
Rietkuil Formation
Koedoeslaagte Formation
Afrikander Formation
Elandslaagte Formation
Palmietfontein Formation
Tusschenin Formation
Coronation Formation
Promise Formation
Bonanza Formation
Brixton Formation
Parktown Formation
Orange Grove Formation
Jeppestown Subgroup
Government Subgroup
Hospital Hill Subgroup
WEST RAND GROUP
WITWATERSRAND SUPERGOUP
Conglomerate
Quartzite
Siltstone/mudstone (shale)
Iron Formation
Diamictite
Lava

GEOLOGY OF THE MOON

The geological evolution of the Moon has been pieced together from the study of photographs of its surface, various geophysical studies carried out both from orbiting satellites and on its surface, and from the study of samples brought back by the Apollo astronauts and the Russian Luna missions.

The ancient Lunar Highlands are heavily cratered, while the younger Maria (sing. mare) have relatively few craters.

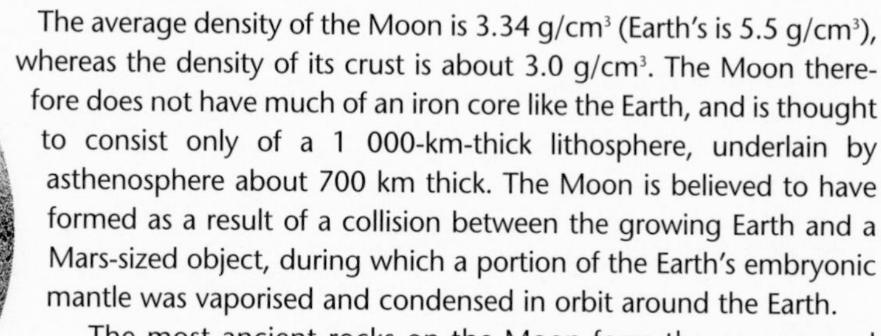

The average density of the Moon is 3.34 g/cm^3 (Earth's is 5.5 g/cm^3), whereas the density of its crust is about 3.0 g/cm^3. The Moon therefore does not have much of an iron core like the Earth, and is thought to consist only of a 1 000-km-thick lithosphere, underlain by asthenosphere about 700 km thick. The Moon is believed to have formed as a result of a collision between the growing Earth and a Mars-sized object, during which a portion of the Earth's embryonic mantle was vaporised and condensed in orbit around the Earth.

The most ancient rocks on the Moon form the very rugged Lunar Highlands and are made of an igneous rock containing abundant calcium feldspar (called anorthosite). These rocks form the paler areas visible on the lunar surface. The Highlands are extremely pock-marked by craters. Highland rocks are up to 4 500 million years old and are believed to represent the early crust that formed during solidification of a once completely molten Moon. The very large number of craters on the Highlands indicates that at this stage of planet formation the rain of space debris was an important geological process.

Around 3 900 million years ago, a number of very large bodies hit the lunar surface, forming gigantic circular basins. Between 3 900 million and 3 100 million years ago, large volumes of basalt erupted and filled these circular depressions, which are visible as the dark areas on the lunar surface. These are known as the Lunar Maria, meaning seas, because early astronomers thought they were seas. Since 3 100 million years ago, relatively fewer impacts have disturbed the Moon's surface. The rate of in-fall of space debris was therefore huge during the first few hundred million years of lunar history, but tailed off quite markedly after about 3 800 million years ago (*see* diagram, right).

It is likely that the Earth experienced a similar history of early crust formation that was accompanied by intense bombardment. The period predating crust formation was probably characterised by a process of aggregation and in-fall of cosmic fragments, the intensity of which was so severe that it resulted in a completely molten Earth. The Moon became geologically inactive soon after its formation and so its ancient initial crust is preserved.

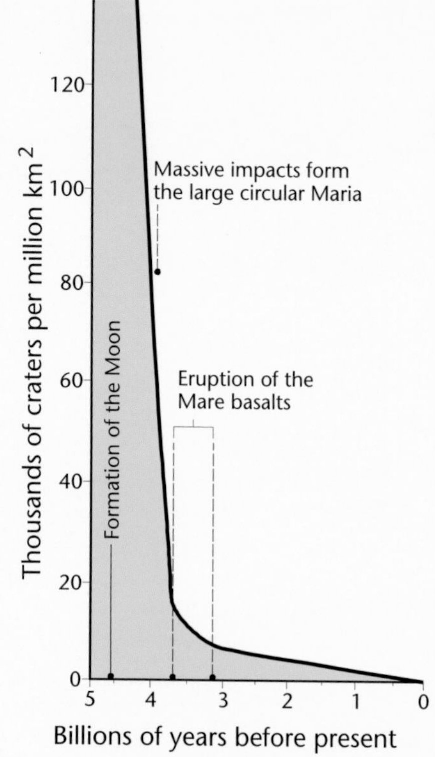

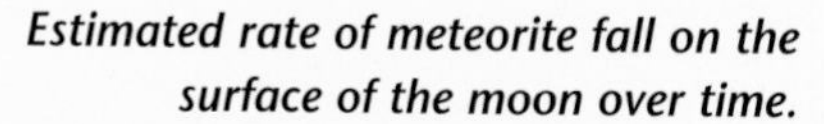

Estimated rate of meteorite fall on the surface of the moon over time.

Figure 3.7 *The granodiorite (grey) in this outcrop is the oldest rock in southern Africa, and formed 3 644 million years ago. It is cut by younger granites (pink).*

AGES OF ROCKS

The Earth formed about 4 600 million years ago, and that is the age of the Earth as a whole. The rocks we find on the surface of the Earth are much younger. This may appear paradoxical – a paradox best explained by way of an analogy. We measure our age from the day we were born, but the atoms making up our bodies are much older. Virtually all were present on the Earth when it formed. Many of these atoms were synthesised in stars that existed long before the Earth formed, and most – the hydrogen atoms – date back to the Big Bang itself. So the materials of which we are made are very old. Rocks are just the same. When we measure the age of a rock, we record the time of its formation, although the material from which it is made is much older (*see* 'Measuring the age of a rock' on page 68).

THE ANCIENT CRUST

The oldest known rock on Earth, a granodiorite (a plutonic igneous rock intermediate between a granite and a diorite: *see* 'Classification of igneous rocks' on page 34) from Canada, is 4 100 million years old. The oldest reasonably well preserved crust that clearly represents an identifiable primitive continental landmass is the 3 900- to 3 800-million-year-old Amitsoq terrane of southwest Greenland, which also includes sedimentary rocks. However, this ancient terrane (a region with similar geology and geological history) is highly fragmented and metamorphosed, and does not provide a clear record of events in the early stages of Earth history. Far better preserved are the rocks of southern Africa and Western Australia, where the geological record is remarkably complete.

Barberton Mountain Land

The region in the Mpumalanga Lowveld known as the Barberton Mountain Land has emerged as one of the classic terranes for the study of the Earth's ancient crust, including some of the earliest forms of life. In the mid-1960s the late Al Engel of the University of California at La Jolla wrote:

> *The Barberton Mountain Land offers the geologist a unique opportunity to study the early stages in the evolution of the Earth. There, remnants of the oldest upper mantle, oceanic crust, and an overlying island arc-like rock complex are fossilized in a sea of granite and granitic gneiss ... Studies of these rocks offer deep insight into many aspects of terrestrial differentiation, especially the early evolution of oceanic and continental crusts, the seas, and the atmosphere.*

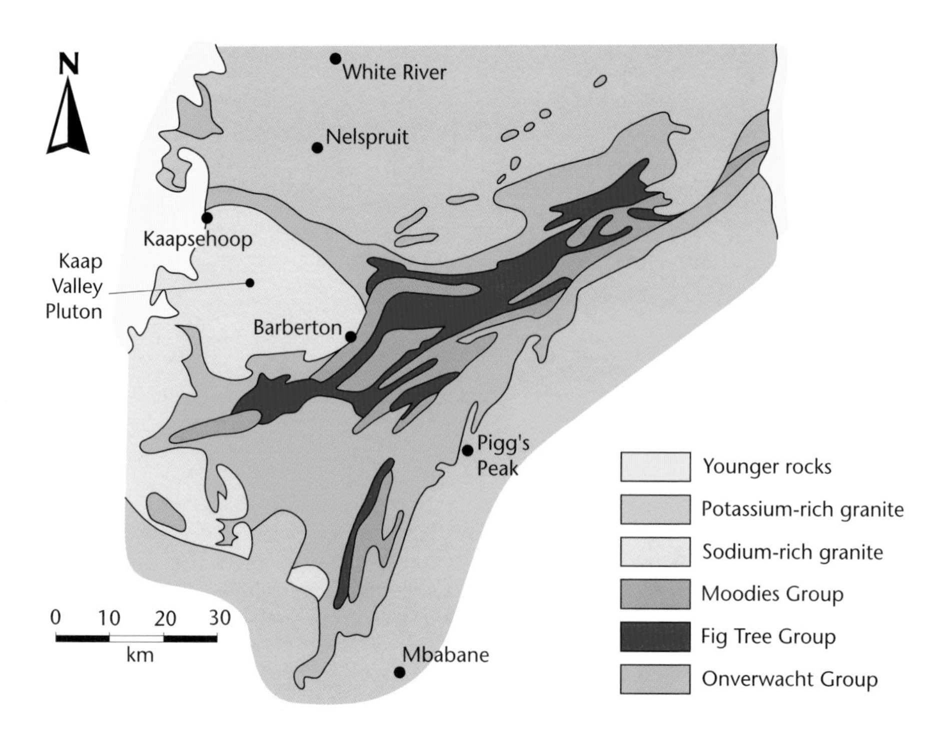

Figure 3.8 *The rocks of the Barberton region represent the best-preserved example of the Earth's ancient oceanic and continental crust. This crust formed between 3 600 and 3 000 million years ago. Shown here is a geological map of the region. Rocks forming the early oceanic crust have a distinctive green colour. The linear fragments of these rocks, as typified by the Barberton region, are known as greenstone belts. Although originally horizontal, the layers have been folded on themselves, and the oldest now form the outer parts of the belt, while the younger rocks lie in the core. These rocks have been intruded by large bodies of granodiorite (e.g. the Kaap Valley pluton) and granite (the Nelspruit granite batholith).*

These prophetic words are even more appropriate today as numerous studies continue to unravel the secrets of the oldest crustal remnants on Earth. What has emerged is an amazing story, not only of the rocks themselves, but also of the geological environments that prevailed over 3 000 million years ago. This early period of Earth history extending from 4 600 to 2 500 million years ago is known as the Archaean Eon (*see* 'The geological time scale' on page 71).

The rocks of the Barberton Mountain Land consist of two main components: a layered pile of volcanic and sedimentary rocks perhaps 20 km thick, known as the **Barberton Supergroup** (*see* 'Stratigraphy' on page 72). This is enveloped in a sea of granodiorite (**figure 3.7**) and granite batholiths, as illustrated in **figure 3.8** (*see* also 'Geological maps', right). The originally horizontal volcanic and sedimentary rock pile has been folded on itself in the form of a trough and the layers now stand on edge. A stroll across the upturned layers is equivalent to a walk through time as the oldest layers were originally at the bottom of the pile and the youngest at the top. The volcanic and many of the sedimentary rocks are dark green and are therefore commonly referred to as **greenstones** and the entity as a whole as a **greenstone belt** because of its linear form (**figure 3.8**). The granites that surround these rocks are grey to pink in colour and contrast spectacularly with the greenstones.

GEOLOGICAL MAPS

(Note: It is recommended that you read 'Stratigraphy' on page 72 before reading this box.)

Maps are widely used to convey information about the distribution of some feature over a particular region. The scale of a map is usually stated in relative distances: for example, a scale of 1:50 000 means that a distance of 50 000 cm (i.e. 500 m) on the ground is represented by 1cm on the map. The scale of a map is also often indicated by a scale bar.

Geological maps are used to portray the distribution of rocks beneath the soil. Detailed maps show the distribution of individual rock types, denoted on the map by different symbols or colours. This cannot be done if the map is to cover a large area, such as an entire country, for example. For such maps, we need to group rock types together. This grouping is done using the principles of stratigraphy (*see* page 72). Detailed maps show each individual bed (i.e. rock type), whereas less detailed maps show only Groups or perhaps only Supergroups. Geological maps also have a **legend** that provides information as to the meanings of colours or symbols used on the map. The rock types on the legend are usually arranged vertically in order of age, the youngest at the top. The stratigraphic affiliation of the various rocks, if known, is normally also given in the legend.

Rocks often are deposited in horizontal layers (e.g. sedimentary rocks) or have some kind of layering associated with them. Earth movements may result in disturbance of the layers, which may vary in intensity from simple tilting to intricate folding. Where tilting of layers has occurred, information about the attitude of the layers is also usually included on geological maps. This takes the form of an arrow denoting the direction in which the layers **dip**, together with the angle, measured from the horizontal, at which they dip, and a line perpendicular to the direction of dip, known as the **strike**.

Knowledge of dip and strike allows one to interpret what happens to layers below the Earth's surface. For example, knowledge of the dip and strike of a gold-bearing rock layer will allow an estimate to be made as to how deep below surface the layer will occur some distance from the place where it crops out on surface. Thus, a geological map together with dip and strike information, enables the construction of geological cross-sections (*see* diagrams, right).

Rock layers may also be disturbed by **faults**. As a result, rock layers become displaced so that they are no longer continuous on a map. Displacements can be very large, often exceeding many kilometres. Faults are therefore also included on geological maps.

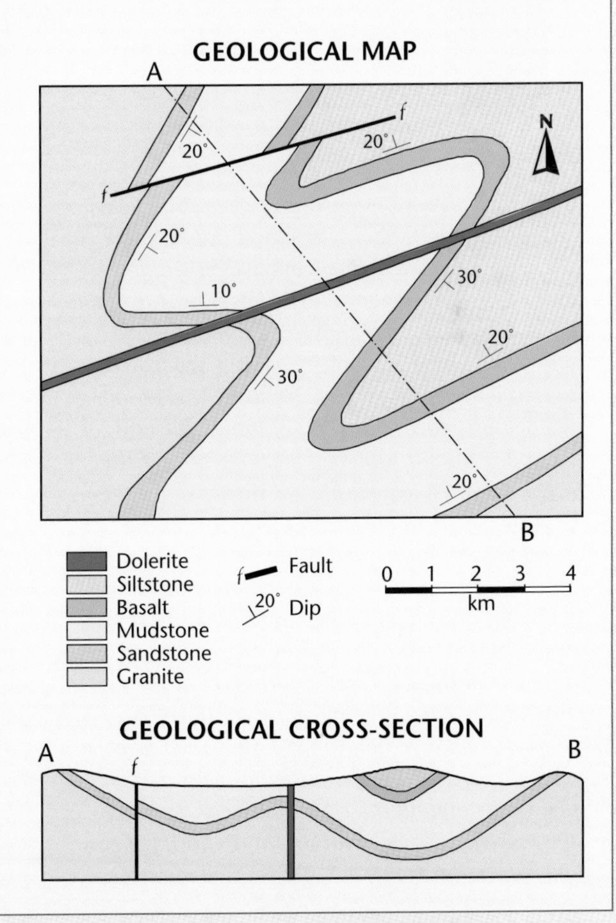

An example of a geological map and an interpretative geological cross-section.

The oldest rocks in the Barberton Supergroup consist of a stack of volcanic rocks approximately 7 km thick. This forms the lower part of the **Onverwacht Group**, one of the three major divisions of the Barberton Supergroup, the other two being the largely sedimentary **Fig Tree** and **Moodies Groups** (**figure 3.9**)

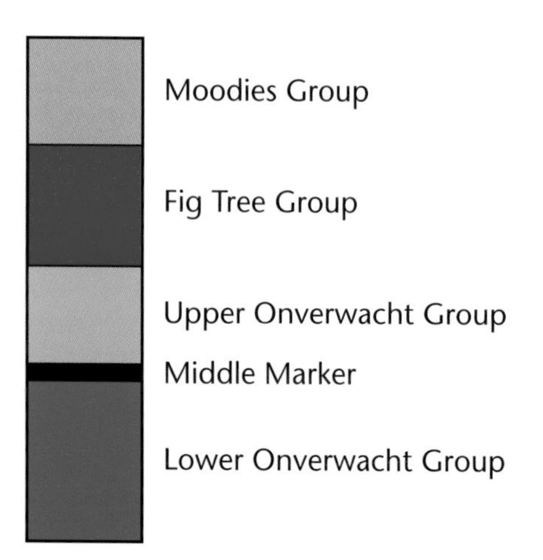

Figure 3.9 *The layered rocks forming the Barberton greenstone belt have been subdivided into discrete packages. The oldest (lowermost) package is known as the Onverwacht Group, and consists of two parts. The lower part is made up of a volcanic rock known as komatiite and represents former oceanic crust. It is separated from the overlying volcanic rocks by a layer of sedimentary rock consisting of iron oxide and silica, the Middle Marker, which formed on the ocean floor. The upper part of the Onverwacht Group consists of basalt and dacite. It formed at a subduction zone when rising magma erupted on the sea floor, burying the sediment layer. Sediment also accumulated in deep-water trenches associated with the subduction zone to produce the distinctive layered turbidites of the Fig Tree Group. Finally, as the early continent began to grow, and rose significantly above sea level, eroded material was deposited in rivers and shallow water around the early continent, forming the Moodies Group of rocks. These consist of conglomerates (gravels), sandstones and mudstones.*

The volcanic rocks found in the lower Onverwacht Group are related to basalt but have a distinctive chemical composition. They were recognised as a new class of rock in 1969 by twins Morris and Richard Viljoen, then students at the University of the Witwatersrand. These rocks, which were first discovered in the Komati River valley east of Badplaas in the southern part of the Barberton greenstone belt, have appropriately been named **komatiites** (**figure 3.10**) and are now recognised around the world in most localities where ancient greenstone sequences are preserved.

The komatiites poured out as lavas in Archaean oceans some 3 600 to 3 200 million years ago, forming primitive oceanic crust. They commonly show pillow structures (**figure 3.10**) like those on modern ocean floors. Moreover, the komatiitic flows were locally buried from time to time by ocean floor sediments consisting of iron- and silica-rich mud. The komatiites have a unique chemical composition rarely found in younger volcanic regions. What makes them unique is their high content of magnesium. Melting experiments have shown that rocks of this type must have formed at temperatures ranging from about 1 300°C to 1 650°C. In today's world, the highest temperature attained by basaltic lavas erupting on ocean floors seldom exceeds about 1 200°C. This temperature difference suggests that the Archaean volcanic rocks developed from a mantle decidedly hotter than that existing today, and that the Earth has cooled significantly over the past 3 000 million years.

The Middle Marker

The sea-floor sediments that accumulated after the komatiite eruptions had ceased consolidated to form a distinctive, widespread layer of sedimentary rock referred to as the **Middle Marker**. This rock consists of iron oxide and carbon- and silica-rich chert layers (*see* 'The formation of sedimentary rocks' on page 64) together with layers of calcium carbonate. These ancient rocks tell us that in spite of its hotter interior, the Earth already possessed oceans 3 500 million years ago, and like the oceans of today they contained abundant calcium carbonate. It is not possible to determine the precise environment in which the komatiite lavas erupted; all we can deduce is that they formed by volcanic eruptions under the sea. This activity persisted for some time, building up a thick pile of pillow lavas. Then it ceased and a layer of sediment buried the lavas.

It is tempting to interpret these events from what we know about modern-day plate tectonic processes (*see* Chapter 2). Komatiites resemble modern ocean-floor basalts and were possibly erupted at a mid-ocean ridge where new oceanic crust was being formed. At that time, komatiite rather than basalt would have been the main rock type of the oceanic crust because the Earth was hotter. As newly formed crust moved away from the ridge on the ocean floor conveyor, volcanic activity ceased and sediment began to accumulate on the sea floor, burying the komatiite pillow lava. This sediment became the Middle Marker. The package of rocks was carried laterally, conveyor belt-style, until it encountered a subduction zone.

This interpretation of the sequence of events that produced the rocks of the lower Onverwacht Group is plausible, but by no means proven, and there are other interpretations. Research continues on this interesting period of Earth's history and no doubt our understanding of the story these rocks have to tell will improve with time.

The rock pile overlying the Middle Marker, also approximately 7 km thick, forms the upper part of the Onverwacht Group. It consists of a variety of volcanic rocks, including basalt and dacite, somewhat similar to those found in modern-day island arcs. It seems that the early oceanic crust was subducted into the upper mantle, driven by rapidly convecting, small-scale convection cells in the hot upper mantle. Subduction of the oceanic crust led to partial melting and the production of basalt and granodiorite magmas. Some of this erupted onto the ocean floor, burying and preserving it (**figure 3.11**), while most of the granodiorite crystallised in the crust to form batholiths and plutons (*see* 'Igneous intrusions' on page 46). These volcanic and plutonic rocks constituted the Earth's early island arcs. There is rather more basalt in the upper Onverwacht Group than in modern-day island arcs, possibly also a consequence of the hotter mantle in the past.

Associated with these ancient island arcs of the Archaean oceans were trenches that formed as oceanic crust cascaded downwards into the mantle. As the volcanic islands began to emerge from the seas adjacent to the trenches, they underwent rapid erosion. The eroded sediment was dumped into the trenches (**figure 3.11**), where it accumulated as very distinctive, fine-grained and layered sedimentary deposits known as **turbidites** (*see* 'Sedimentation in ocean trenches' on page 82).

Carl Anhaeusser

Figure 3.10 *Although the Earth's surface temperature 3 500 million years ago was much the same as today, its interior was hotter when the rocks of the Barberton region were forming. A very distinctive rock type related to basalt, known as komatiite, erupted to form the early oceanic crust. Eruption processes produced bulbous forms known as pillow lava, shown here, just as undersea eruptions do today.*

In the Barberton greenstone belt these sedimentary rocks began forming more than 3 400 million years ago, leading to the development of a 2.5-km-thick, predominantly sedimentary pile known as the **Fig Tree Group**. Most of the Fig Tree sedimentary rocks consist of sandstones, siltstones and mudstones, the mudstones very often black in colour due to the presence of abundant carbon (in the form of graphite). These carbonaceous sediments formed in an environment where virtually no free oxygen was present. Nevertheless, there are indications that

BARBERTON REGION: FORMATION OF EARLY OCEANIC AND CONTINENTAL CRUST

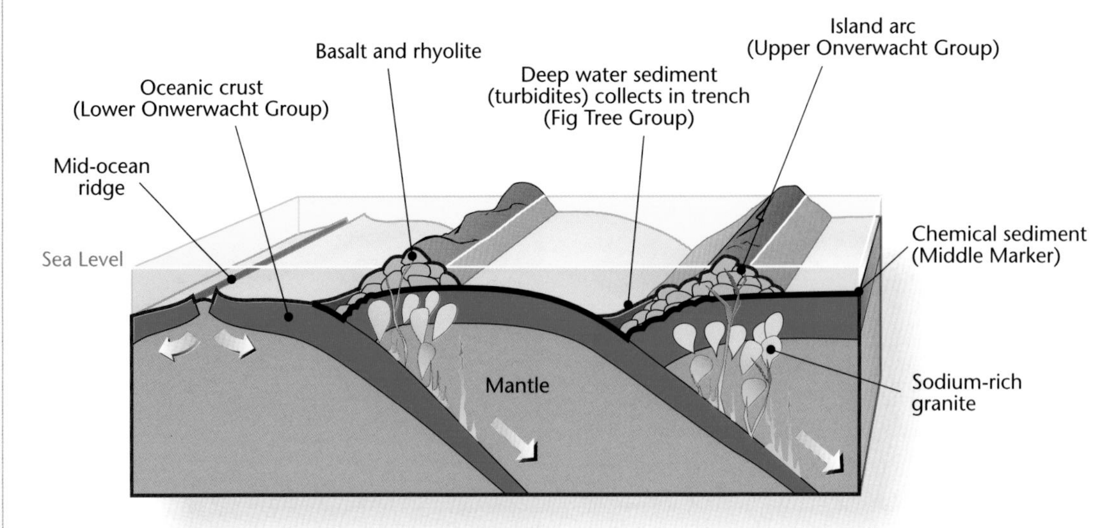

Figure 3.11A *Komatiite magma intruded into the axis of a mid-ocean ridge (left side of figure), forming pillow lavas and new oceanic crust (Lower Onverwacht Group). As this crust moved away from the ridge, the pillow lavas became buried by a layer of chemically precipitated iron oxide and silica (Middle Marker). Oceanic crust was simultaneously being subducted, and water released caused local melting, producing basalt and granodiorite magmas, some of which erupted on the ocean floor, burying the Middle Marker (Upper Onverwacht Group). Granodiorite magma also crystallised in the crust, forming plutons (e.g. Kaap Valley pluton). The volcanoes emerged from the sea, forming ancient island arcs. Sediment eroded from these islands was deposited in adjacent ocean trenches to form the Fig Tree Group.*

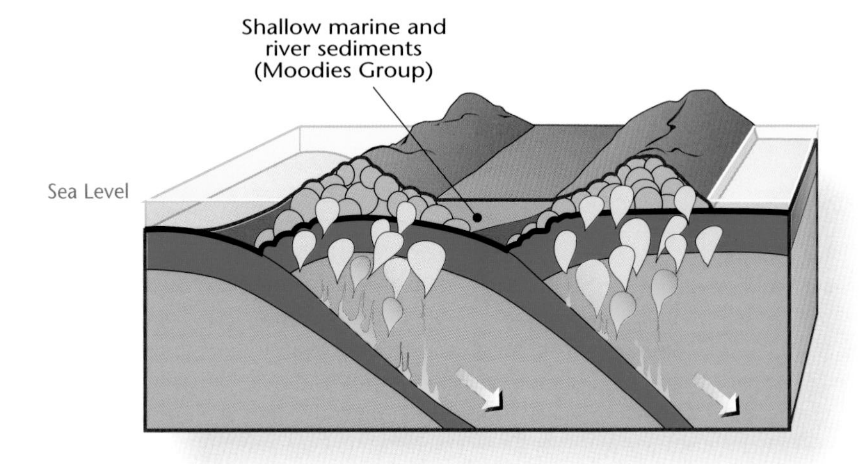

Figure 3.11B *The granodiorite plutons rendered the island arcs buoyant, and they could not be subducted. Collision of arcs resulted in amalgamation and crustal thickening, forming the first micro-continents. Sediment shed by the amalgamated and now deeply eroded island arcs included pebbles of granodiorite and abundant sand that was deposited in rivers and shallow seas on and around these growing micro-continents. These sediments are represented by the Moodies Group.*

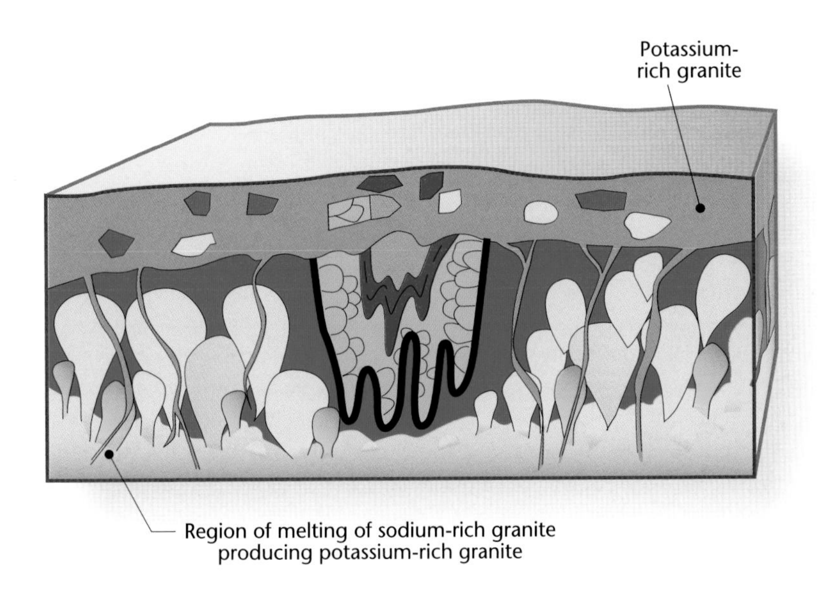

Figure 3.11c *Thickening of the crust caused melting of the more deeply seated granodiorites, producing granite magmas. These intruded upwards and collected as extensive, sheet-like batholiths.*

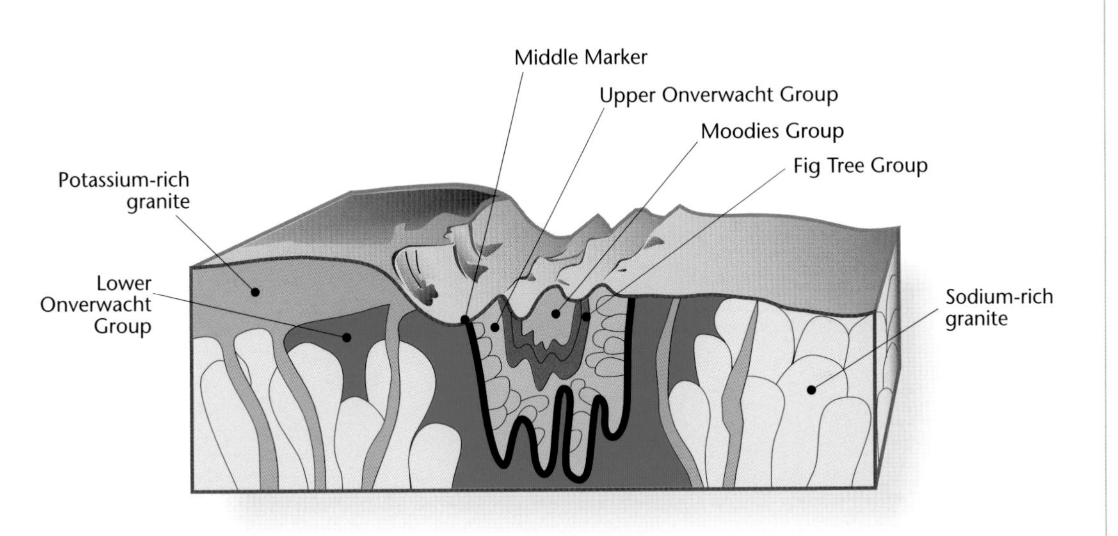

Figure 3.11d *Recent erosion of portions of the granite batholiths has exposed the underlying greenstone belt and its associated granodiorite plutons.*

SEDIMENTATION IN OCEAN TRENCHES

Sediment is delivered to the oceans by rivers, and it generally accumulates on the continental shelf (*see* 'Sedimentation on continental shelves' on page 100). Sand collects in the near-shore environment, silt and mud further out to sea. Sediment can accumulate to great thicknesses and may become unstable; it can collapse in an undersea avalanche down the continental slope. Instability is often triggered by earthquakes.

Undersea avalanches are termed **turbidity currents** and consist of a mixture of sediment and water. Initially, they accelerate down the slope, but slow down at the base of the slope, where they deposit the contained sediment. Coarser sediment deposits first, followed by increasingly finer sediment. The result is a layer of sediment at the bottom of and beyond the slope that is relatively coarse at its base (usually fine sand) and fine (mud) at its top. The thickness depends on the volume of material in the turbidity current, but is seldom more than a metre, usually tens of centimetres.

Each turbidity current produces a separate layer. Sedimentary deposits formed at the base of the continental slope therefore consist of a stack of these size-graded layers, called **turbidites**. The frequency with which turbidity currents occur depends on many factors, such as the width of the continental shelf, the rate of sediment supply to the shelf, and earthquake activity. Trenches that form at subduction zones experience frequent turbidity currents. Landward of the trench is a mountain belt or island arc, where erosion is rapid, so large quantities of sediment are supplied to the shelf. The shelf is typically narrow, so there is not much space to accommodate the sediment and these areas experience frequent earthquakes. The net result is frequent turbidity currents, which produce a very distinctive pile of size-graded layers of sediment in the trench.

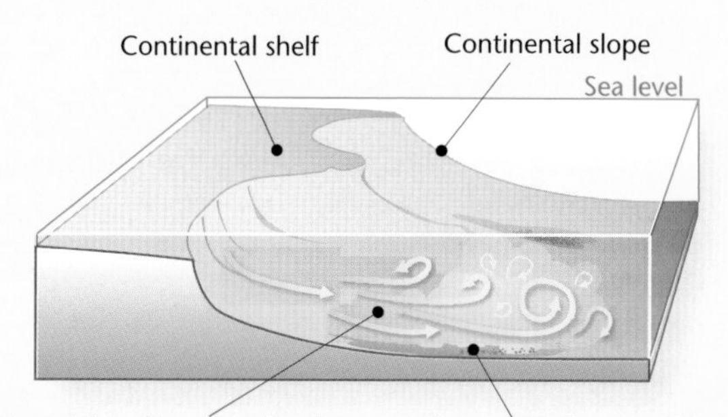

Sediments deposited by turbidity currents formed these well-layered rocks in the southern Karoo.

Uwe Reimold

Figure 3.12 ***Several layers consisting of small spherules, such as illustrated above, have been found in the Fig Tree Group. The spherules originally consisted of glass and were produced by melting of the crust during very severe meteorite impacts. Five such layers have so far been found, indicating that major meteorite impacts were still relatively common 3 300 million years ago.***

life was already flourishing at that time. Primitive organisms, possibly including photosynthesising cyanobacteria, were growing in the shallow water around the volcanic islands. Their growth produced domical structures known as **stromatolites** (*see* 'The formation of sedimentary rocks' on page 64) and they left scattered replicas of their cells in the form of microfossils. This is discussed further in Chapter 6.

By Fig Tree times, the rate of meteorite in-fall had declined markedly (*see* 'The geology of the Moon' on page 74), but there is evidence to suggest that impacts were still common. South African mineralogist Sybrand de Waal has identified strange nickel-rich, iron minerals in the Barberton area that may be oxidised fragments of a nickel-iron meteorite. In addition, geologists from Stanford University, led by Don Lowe, have found several (at least five) sedimentary layers in the Onverwacht and Fig Tree Groups made up entirely of small spheres (**figure 3.12**). The spheres appear very similar to the droplets of molten rock produced by large meteorite impacts, an observation confirmed by chemical evidence.

In all cases, the meteorites impacted on komatiitic or basaltic crust. Lowe is of the opinion that all of the layers located thus far reflect impacts more severe than the one that wiped out the dinosaurs 65 million years ago and could have been made by objects ranging in diameter from 20 to 50 km. A spherule layer of identical age to one in the Onverwacht Group has recently been found in similar rocks in the Pilbara region of Western Australia, and Lowe believes the layers were formed by the same impact, which occurred 3 465 million years ago. There is no evidence of craters, however, but the rocks have been so folded and buckled that we would be unlikely to recognise a crater even if it was present.

THE BEGINNINGS OF A SOUTHERN AFRICAN CONTINENT

As is the case with modern plate tectonic activity, granodiorite magma was produced at sites of Archaean subduction and island-arc formation. Some erupted as lavas on the ocean floor, while the remainder crystallised in the crust to form batholiths (**figure 3.11**). Once this type of primitive crust formed it was not easily destroyed, as the rocks are of lower density and thus more buoyant than the volcanic rocks of oceanic floor and mantle. This prevents their subduction.

Instead, plate tectonic processes began to cause amalgamation of island arcs. Fusion of island arcs created micro-continents made predominantly of granodiorite, which rose above sea level. Weathering and erosion of exposed rock produced sediment that was transported by rivers to accumulate in shallow water around the edges of the micro-continents (**figure 3.11**). Here it formed conglomerates (gravels) and sandstones (sand), while mudstones (mud) and banded iron formations (a chemical precipitate of iron oxide and silica) accumulated in deeper water (*see* 'The formation of sedimentary rocks', page 64).

These types of sedimentary rocks form the uppermost portion of the Barberton Supergroup in another 2.5 km-thick pile of rocks known as the **Moodies Group**. Still preserved in these 3 200-million-year-old rocks are sedimentary structures such as cross-bedding and ripple marks identical to those that are seen forming today, as illustrated in **figure 3.13** (*see* 'Sedimentary structures' on page 86). Some of these structures indicate that the sediments were deposited in basins influenced by tidal activity. This affirms that the Moon, from time immemorial, has exerted a significant gravitational influence on our planet and that the water bodies of the Archaean were not lakes but true oceans. Studies of tide-generated layering in sedimentary rocks of the Moodies Group by Ken Eriksson from Virginia and his students have revealed that the spring tide-neap tide interval was only nine days long, indicating an 18-day lunar month. This implies that the Earth was rotating more rapidly, so day length was shorter than today and there were more days per year.

At that time there were no terrestrial plants. The atmosphere was very different from today: there was little or no free oxygen, and water vapour, carbon dioxide and nitrogen made up the bulk of the atmosphere. Atmospheric pressure was considerably higher (*see* Chapter 6). Carbon monoxide, ammonia and even hydrogen cyanide and hydrogen sulphide may also have been present.

This atmosphere was probably far more chemically aggressive than today, rapidly decomposing rocks with which it came into contact. Without the stabilising effect of vegetation, no soils could have formed and decomposing rock would have been quickly eroded by rain, streams and rivers. Rocky outcrops would undoubtedly have dominated the continental landscape.

CONSOLIDATION OF THE EARLY CONTINENT

As the earliest micro-continents amalgamated they were subjected to continuing magma addition, with more granodiorite being added. The earliest granodiorites, like those found in the Barberton-Swaziland region (**figure 3.7**), range in age from about 3 600

Carl Anhaeusser

Figure 3.13 ***Ripple marks preserved in 3 200-million-year-old sedimentary rocks of the Moodies Group. The layers were originally horizontal but are now vertical due to folding.***

Figure 3.14 ***Mixed rocks called migmatites formed during the intrusion of granodiorite magma into older volcanic rocks.***

to 3 200 million years, and they are chemically distinctive, having a high sodium content. By contrast, the later granitic rocks in this region are potassium-enriched and range from about 3 100 to 2 700 million years in age. These rocks were added incrementally to the growing continent.

The granodiorites generally appear on geological maps as circular or oval-shaped plutons, which invaded the greenstone belts from below like rising hot-air balloons, probably while partly solid. This upward emplacement style (or **diapirism** – analogous to the rising globular blobs in novelty lava lamps) caused considerable structural disturbance to all the rocks in the region immediately surrounding them.

The greenstones became folded on themselves as they sank into the cusps between the rising plutons. Rocks along the margins of the plutons became intensely folded, forming **gneisses**. Pieces of the surrounding rocks torn off by the intrusions were heated and plastically transformed into complex, mixed rocks known as **migmatites** (**figure 3.14**). The largest intrusion of this type, the Kaap Valley pluton, is responsible for the circular, approximately 30-km-diameter De Kaap valley near the historical gold mining town of Barberton (**figure 3.8**).

The later granitic rocks occur in the form of massive, sheet-like batholiths. Some of these bodies – such as the Nelspruit batholith north of Barberton, the Mpuluzi batholith southeast of Badplaas and the Pigg's Peak batholith in Swaziland – exceed 60 km in diameter. The magmas appear to have formed by remelting of the earlier granodiorite as the crust thickened. They intruded through the older rocks and spread out as extensive, horizontal sheets (**figure 3.11**). These various additions of granite, which began to develop approximately 3 200 million years ago, progressively stabilised the continental crust. The stabilisation was aided by the formation of a thick keel in the mantle beneath the early continental crust.

At a still later stage, between about 3 000 and 2 700 million years ago, the Archaean crust was again intruded by a few smaller granite plutons that represented the final stages of continent development

SEDIMENTARY STRUCTURES

Most sedimentary rocks are formed by deposition of sediment from flowing or standing water. At the time of deposition, certain features may form in the sediment that are preserved in the resulting rock. These are known as sedimentary structures, and often provide useful information about the conditions under which deposition occurred.

STRATIFICATION

Sedimentary rocks are often stratified or layered, known as **bedding**. Each layer or bed usually represents an increment of deposition. For example, in glacial lakes that freeze over in winter, the bottom sediment shows pronounced layering: during the summer thaw, melt-water streams deposit silt in the lake, while during winter very fine sediment slowly settles from the water column. These seasonal changes produce alternating mud and silt layers, known as **varves**.

Pronounced layering also develops in the sedimentary deposits on the beds of lakes fed by rivers which experience periodic flooding. Under normal conditions, fine sediment rich in organic matter accumulates, but during a flood, a layer of coarser silt will be deposited on the lake bed. As the flood wanes, finer and finer material is deposited, so the layers formed by a flood become finer upwards through the layer, called **graded bedding**. Graded bedding also forms in sediment layers deposited by turbidity currents (*see* **A**, right, and 'Sedimentation in ocean trenches' on page 82).

Sand is usually transported along the bed of a stream or across a beach by water or wind, and often moves as a series of parallel ripple structures. Particles are eroded from behind each ripple, are carried up its back, and then avalanche over the crest to form a relatively steep face to the ripple. Gradually the ripple advances while others follow behind (**B**). The surfaces that were eroded (that is, behind the ripple) form flat layers or **beds**.

Within each of these beds the lower part of the sloping avalanche face is usually preserved. These sloping surfaces form **cross bedding** (**C**), which is useful because it gives information on the direction of current flow during sediment deposition. In sand deposited under the influence of tides, the cross beds will slope in opposite directions in alternate beds. The thickness of a cross-bedded layer can range from a few centimetres for small ripples, to a metre or more for dunes that form in large rivers, to tens of metres for wind-blown sand dunes such as those in the Namib Desert (**D**).

SURFACE FEATURES

Ripple-covered bedding surfaces are often preserved in sedimentary deposits. When the rock is exposed, it may break apart along such surfaces, exposing the fossil **ripple marks** (**E**). Muddy sediment deposited on a riverbank usually dries in the sun and cracks due to shrinkage. A flash flood may deposit sand over the mud-cracked surface, filling the cracks with sand and depositing a layer of sand over the mud. This too can be exposed during subsequent erosion, producing fossil **mud cracks** (**F**).

Coastal mudflats form in areas protected from wave action where there is a supply of fine sediment. At low tide, the muddy surface is exposed, but this is submerged as the tide comes in. In hot, arid climates, sea water in the mud evaporates and salt crystals form little cubes at the sediment surface. The crystal shape is occasionally preserved in the sedimentary rock, forming **salt casts** (**G**).

Another feature occasionally observed on bedding surfaces are tracks and burrows left by animals of various kinds (including humans). These are known as **trace fossils** (**H**). Isolated raindrops falling on soft mud usually form small indentations, known as **raindrop impressions**. These too can become buried and preserved (**I**).

By mapping out the types of sedimentary rocks present in a particular area and examining the sedimentary structures they contain, it is possible to reconstruct the geographical environment that existed at the time the sediment layers were laid down.

Graded bed, with conglomerate at the base grading up to mudstone at the top

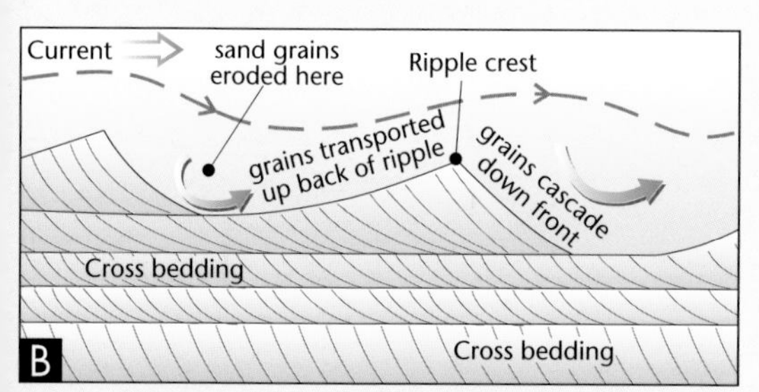

Diagram showing how cross bedding forms

Cross bedded sandstone

Very large cross bedding in wind-blown sand in the Namib Desert

Fossilised ripples in a sandstone

Fossilised mud cracks

Salt casts

Fossilised animal tracks

Fossilised raindrop impressions (right) and modern-day equivalents (left)

Figure 3.15 *Granites, which intruded into the older rocks between 3 000 and 2 700 million years ago, are more resistant to erosion. They give rise to magnificent scenery in the Barberton Mountain Land.*

Carl Anhaeusser

in the region. These plutons – such as the Mpangeni pluton in the Crocodile River gorge east of Nelspruit and the Mbabane pluton in Swaziland – today give rise to some spectacular scenery (**figure 3.15**).

In the Barberton Mountain Land we see fragments of the oldest known oceanic and continental crust. As the micro-continent grew, more island arc material was added on the northern side, along with other fragments of ancient ocean floor. Today these later additions can be seen as northeasterly trending linear greenstone belts, such as the Murchison Range, Pietersburg and Sutherland-Giyani belts in the Limpopo Province (**figure 3.16**). Crustal material was also added on the western side of the continent, producing linear belts such as those at Amalia and Kraaipan in the North West Province. Remnants of ancient oceanic crust in a sea of granodiorites, granites and gneisses can also be seen between the Witwatersrand and Pretoria, especially in the Muldersdrif area. These linear greenstone belts possibly mark the boundaries or sutures between amalgamated micro-continental fragments (**figure 3.17**).

A remarkable aspect of this early continent is that once formed, its rocks have remained essentially unchanged for 3 000 million years. Such blocks of ancient crust, with their distinctive greenstone-granodiorite-granite association, have been termed **cratons** to distinguish them from later-formed continental crust. Although most of this ancient craton, which is known as the **Kaapvaal Craton**, is buried beneath younger rocks, it forms the core of southern Africa (**figure 3.17**). This ancient continent was originally probably much larger – we have no idea how large – but has lost pieces by later rifting. The Pilbara region of Western Australia, as well as portions of Madagascar and India, probably once formed part of the Kaapvaal Craton (see Chapter 5). Cratons similar to Kaapvaal formed at about the same time or slightly later elsewhere on the globe, and terranes like those of the Barberton and Pilbara regions occur in other parts of Africa, Greenland, Antarctica, Canada, Finland, Brazil and Siberia.

THE SIGNIFICANCE OF THE BARBERTON MOUNTAIN LAND

The rocks of the Barberton area preserve a record of the formation of the Earth's first sizeable continent. However, there is a long gap in Earth's early geological record from the formation of the Earth 4 600 million years ago to the creation of the early crust in the

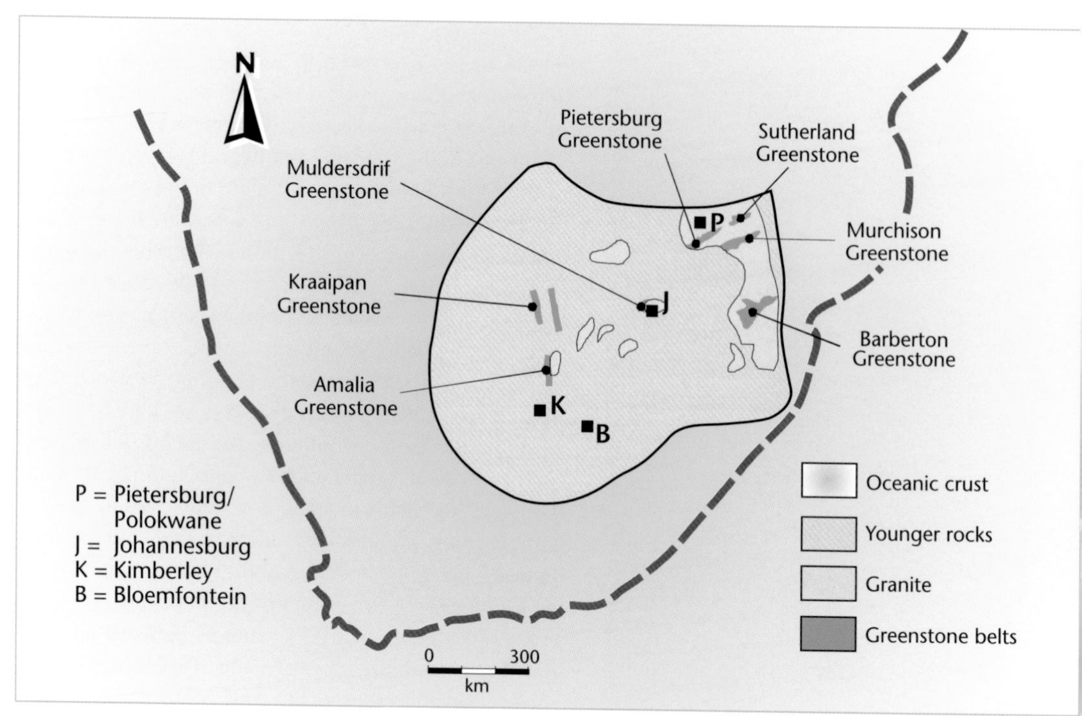

Figure 3.16 *The micro-continent represented by the rocks of the Barberton area continued to grow by the accretion of similar terranes on the northern and western flanks. They are marked by rocks similar to those in the Barberton area. Accretion produced a small continent – the Earth's oldest – known as the Kaapvaal Craton. Much of it is today buried under younger rocks, as illustrated. It was originally much larger, but the eastern section has been separated by later rifting. (The coastline of South Africa and the locations of cities are shown for reference.)*

Barberton region. Events in the earlier portion of this time gap have been pieced together from the study of the Moon and indicate an extremely violent and hot beginning to Earth history. Initially, bombardment from space was the dominant process, but decreased in intensity over the first 500 million years. No clear evidence of this violent period remains on Earth, but as on the Moon, which still carries the scars of bombardment, it is likely that a crust formed during this period.

Recycling of the remnants of this ancient crust, probably by plate tectonic processes, seems to have started more than 4 000 million years ago. Evidence for this is provided by 4 100-million-year-old granitic rocks, commonly produced at subduction zones, that have been found in Canada, as well as the 4 000- to 4400-million-year-old zircons found in Western Australia. These have chemical characteristics indicating that they were derived from continental crust.

Rocks of the 3 900-million-year-old Amitsoq region of west Greenland contain banded iron formation, indicating that surface water, possibly even oceans, existed on Earth by that time. But because of their superb preservation, the rocks of the Barberton region provide the first glimpse of the large-scale geological processes involved. When these rocks were forming, the Earth differed in important ways from today. Its interior was hotter, while its surface temperature was much the same as today, sustaining oceans. The temperature gradient in the Earth would therefore have been greater, perhaps leading to a more vigorous form of plate tectonics, with far smaller plates.

Island arcs may have been more abundant, and rapidly amalgamated, forming the first micro-continents. The variety of granite types is greater in these ancient terranes compared to today, and the early period of granodiorite, followed by later granite, is

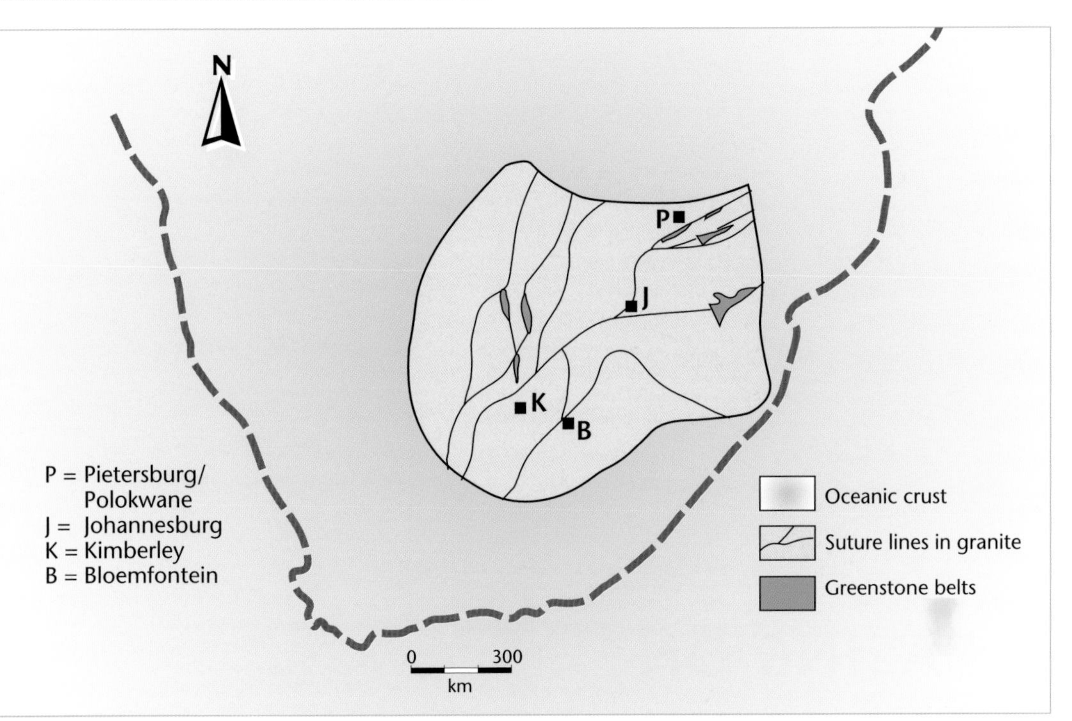

Figure 3.17 ***The Kaapvaal Craton is believed to have been assembled through the amalgamation of many micro-continents and the greenstone belts are believed to mark the suture lines between the fragments. This map (compiled by Maarten de Wit of the University of Cape Town and his colleagues) shows the possible locations of individual micro-continents. (The coastline of South Africa and the locations of cities are shown for reference.)***

unique to this period. Also different in these ancient terranes is the absence of significant metamorphic belts along zones of island arc collision, possibly also a consequence of a hotter Earth with smaller plates and a more rapidly recycling crust.

In other respects the Earth was similar to today. Oceans were present and subject to tidal influences. Oceanic crust was being created by underwater volcanic eruptions and recycled at subduction zones. Continents had formed and drainage networks had developed on these early continents. In addition, eroded material was being deposited in adjacent oceans to form early particulate (clastic) sedimentary rocks. Life was flourishing in the oceans and photosynthesising organisms had probably already appeared.

These primitive forms of life were extracting carbon dioxide from the water, causing the mineral calcite to precipitate in distinctive stromatolite structures. Moreover, sizeable continents, which included the Kaapvaal Craton, had formed. This particular continent provided a basement on which a remarkable variety of sedimentary and other formations were deposited over the subsequent 3 000 million years, producing southern Africa's rich geological legacy.

SUGGESTED FURTHER READING

Bowler, S. 1996. 'Formation of the Earth.' *New Scientist*, Inside Science Section, vol. 152, issue 2060, 14 December.

Gribbon, J. 1987. *In Search of the Big Bang*. Corgi Books, London.

Hawkins, S. 1988. *A Brief History of Time*. Bantam Books, London.

Johnson, MR, Anhaeusser, CR and Thomas, RJ. 2005. *The Geology of South Africa*. Council for Geoscience, Pretoria.

THE STORY OF
EARTH & LIFE

4

BASINS ON THE EARLY CONTINENT

Glen McGavigan / Sishen Mining

Rocks deposited on the Kaapvaal Craton between 3 074 and 1 900 million years ago contain much of South Africa's mineral wealth, such as the vast iron ore deposits at Sishen.

ROUTE MAP TO CHAPTER 4

Age (years before present)	Event
3 100 million	• The core of the Kaapvaal Craton, portion of the world's oldest continent, had stabilised. Subduction zones were still active along its western and northern boundaries, whereas its eastern boundary was probably attached to similar continental crust (now parts of Madagascar, India and Australia).
3 074 million	• The young continent experienced rifting, causing thinning of the crust and the deposition of sedimentary and volcanic rocks of the Dominion Group and the lower Pongola Supergroup (Nsuze Group).
2 970 million	• Cooling caused subsidence of the rifts. The Kaapvaal Craton submerged below sea level, and sedimentary rocks of the Witwatersrand Supergroup (West Rand Group) and upper Pongola Supergroup (Mozaan Group) began accumulating.
2 900 million	• Compressional forces generated along the subduction zone on the northern margin of the Kaapvaal Craton began to transmit southwards through the crust, activating large faults. Uplift of the crust along these faults created mountains and fragmented the lower Witwatersrand Supergroup strata (West Rand Group) into a number of small depressions. Rivers bringing sediment from the uplands formed extensive alluvial plains in local regions of subsidence. Gold and other dense minerals became concentrated in these rivers, forming the spectacular Witwatersrand Supergroup gold deposits.
2 714 million	• The Zimbabwe Craton, a micro-continent similar to the Kaapvaal Craton, collided with the latter continent, rupturing the crust. Vast amounts of lava erupted to form the Klipriviersberg Group of the Ventersdorp Supergroup, ending the deposition of the Witwatersrand Supergroup.
2 708 million	• The force of continental collision continued to rupture the continent. Movement along large faults created elongated mountains and valleys. Sediment eroded from the mountains collected in the valleys, forming the Platberg Group of the Ventersdorp Supergroup. The crust finally stabilised, the mountainous terrane was eroded, and sedimentary and volcanic rocks of the Bothaville and Pniel Groups were deposited.
2 650 million	• Rifting of the Kaapvaal Craton occurred again, taking place between large, ancient fractures in the crust. The resulting trough filled with sediment and some lava, forming the Wolkberg Group. Continued subsidence resulted in invasion of the sea across the Kaapvaal Craton. River systems were drowned (Black Reef Formation) and an extensive shallow sea formed in which primitive life (bacteria) thrived. Their prolific growth resulted in the accumulation

	of more than a kilometre of dolomite, as well as vast quantities of iron and manganese, which were precipitated by oxygen released by the cyanobacteria (Chuniespoort and Ghaap Groups of the Transvaal Supergroup).
2 350 million	• Renewed rifting and subsidence caused local uplift and partial erosion of the iron and dolomite deposits. An increase in sediment supplied to the shallow sea swamped the cyanobacterial colonies, and shallow water marine deposits of the Pretoria Group were deposited.
2 061 million	• Vast amounts of rhyolite erupted across the centre of the Kaapvaal Craton, initiating the formation of the Bushveld Complex. This was followed by the intrusion of basalt magma beneath the rhyolite, which collected to form a sill about 7 km thick. Slow crystallisation and repeated introduction of new magma resulted in segregation of different minerals, producing the largest chromium, platinum and vanadium deposits on Earth. Finally, granite magma intruded beneath the rhyolite cover, forming granites and completing the Bushveld Complex.
2 049 million	• Intrusion of an unusual alkali-rich magma into the crust gave rise to the Phalaborwa Complex, which is a source of copper, phosphorus and vermiculite.
2 023 million	• A large asteroid struck the Kaapvaal Craton near the present-day town of Vredefort in Free State, forming a crater about 300 km in diameter. Rebound after the impact brought rocks up from depth to form a dome-like structure, the Vredefort Dome, in the centre of the crater. The crater itself was subsequently eroded, exposing on surface the rocks which once lay below the original crater floor.
2 000 million	• Deep-water sedimentation on the western margin of the Kaapvaal-Zimbabwe Craton gradually gave way to shallow water conditions and eventually river deposits began to accumulate, forming the Olifantshoek Supergroup.
1 900 million	• Sedimentary deposits on the western margin of the Kaapvaal Craton became folded and were thrust towards the east. At the same time they were intruded by granite magma. This may have been caused by the collision of the Congo Craton with the Kaapvaal-Zimbabwe Craton. This collision appears to have ruptured the Kaapvaal Craton and rifts formed in the interior in which the Waterberg and Soutpansberg Groups of rocks were deposited. The sedimentary rocks deposited at this time are stained red by iron oxide, and provide the earliest evidence for the presence of significant oxygen in the Earth's atmosphere.
1 100 million	• Widespread intrusion of dykes occurred across the Kaapvaal and Zimbabwe Cratons, producing the Timbavati gabbros and the Umkondo dolerites.

A CORNUCOPIA OF VALUABLE MINERALS

The Kaapvaal Craton is part of the world's oldest known continent, but this is not its major claim to fame. What has really made the craton famous is the diversity and wealth of its mineral deposits and the long record of Earth history it preserves. Over a period of approximately 3 000 million years the craton provided a basement on which some of the world's most spectacular mineral deposits formed. These include the world's largest gold, chromium, platinum, vanadium and andalusite deposits. In addition, the craton hosts major iron, manganese, phosphate, fluorspar, diamond, asbestos, coal and copper deposits. This chapter looks at the geological events during the late Archaean and Proterozoic Periods (*see* 'The geological time scale' on page 71), which led to this cornucopia of minerals, as well as the insight they provide into the slowly evolving Earth.

THE CONTINENTAL BASEMENT

The core region of the Kaapvaal Craton had formed and stabilised by about 3 100 million years ago, although growth of the craton may have continued for much longer by the addition of island arcs and their associated igneous intrusions (granite-greenstone terranes) around its margins. In particular, the northern and western margins seem to have been sites of continuing growth, suggesting that subduction zones probably bordered these margins of the Kaapvaal craton (**figure 4.1**). Meanwhile, the stable interior of the continent experienced erosion, and granites that had crystallised at depth became exposed on surface.

This marked the start of a completely new and different style of geological process across the continent. The early period of oceanic crust formation and recycling that produced island arcs and associated granite and granodiorite intrusions,

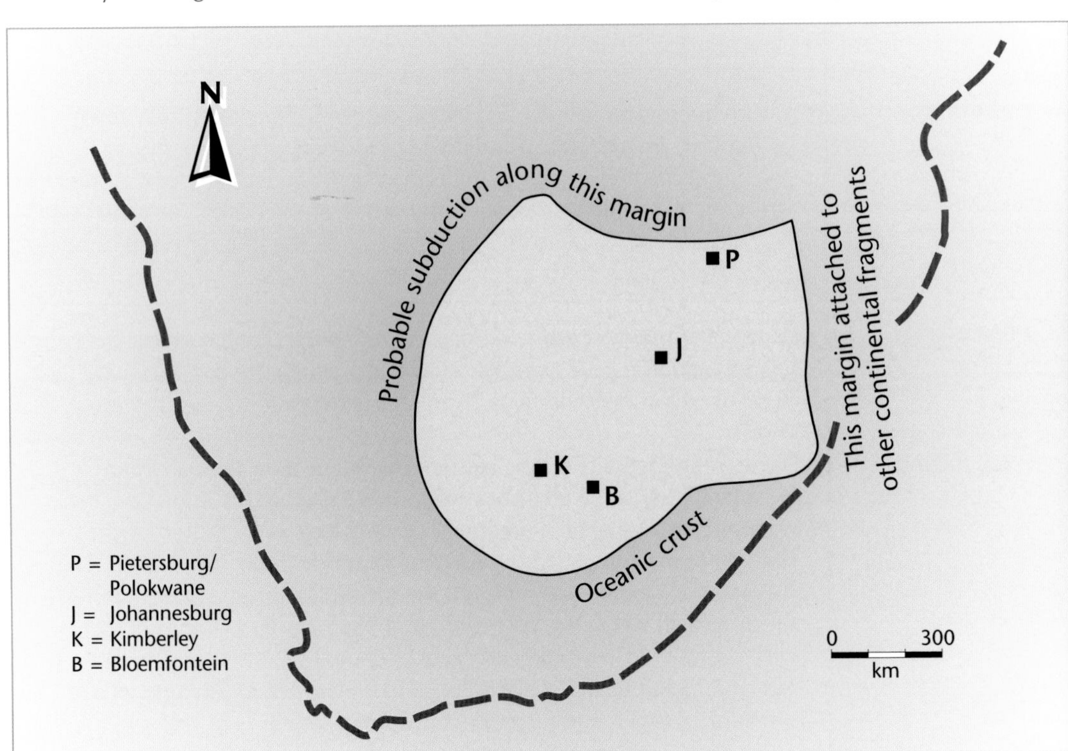

Figure 4.1 *The world's oldest continent, a portion of which is preserved as the Kaapvaal Craton, had stabilised by about 3 100 million years ago. Subduction zones probably lay along its western and northern margins, while the eastern margin was possibly connected to similar continental material that today forms parts of Madagascar, India and Australia. What lay to the south is not known. (The coastline of South Africa and locations of major cities are shown for reference.)*

such as in the Barberton area, gave way to a history dominated by the rise and fall of the continent relative to sea level, interspersed with major igneous events. We no longer see any evidence in the geological record on the Kaapvaal Craton of ocean floors and their recycling. Nevertheless, the events in the continental interior suggest that plate tectonic processes were still very active, but maturing in form to resemble those currently operating on Earth.

From time to time depressions formed on the continent in which sediment and igneous rocks accumulated. These form the great basins of the Kaapvaal Craton. Remarkably, in spite of the continent being stretched and buffeted by neighbouring plates, these accumulations on the continent were relatively unaffected and remain to this day in a pristine state. They provide a record of events on the continent spanning 3 000 million years. These rocks not only preserve the geological history, but also provide insight into the Earth's changing atmosphere and the evolution of its life.

Bruce Cairncross

Figure 4.2 ***Rifting of the young craton about 3 000 million years ago led to the accumulation of volcanic and sedimentary rocks of the Dominion Group and Nsuze Group, shown here.***

RIFTING OF THE YOUNG CONTINENT

The earliest accumulation of rocks on the basement provided by the Kaapvaal Craton is preserved sporadically in the North West Province, and is known as the **Dominion Group**. The rocks lie directly on granite basement and consist of quartzites with minor conglomerate layers overlain by volcanic rocks. The conglomerates contain gold, and were mined from 1888 onwards at the Dominion Gold Mine about 25 km west of Klerksdorp. These conglomerates are remarkably similar to those of the Witwatersrand Basin, described later. The quartzites are overlain by volcanic rocks, including both basalts and rhyolites. The sedimentary component is relatively thin, seldom exceeding 100 m, but the total thickness of the pile is large, reaching 2.6 km in the Klerksdorp area.

The rhyolites erupted 3 074 million years ago. Although the age of the underlying sedimentary rocks is not known, they are unlikely to be older than 3 100 million years – the time of stabilisation of the craton. The Dominion Group has been severely fragmented by faulting and folding, and is largely obscured by overlying younger rocks and by thick soils in the region. Its full extent is therefore unknown. The events leading to the deposition of the Dominion Group are uncertain, but the association of basalts and rhyolites on continental crust is one of the hallmarks of continental rifting (*see* 'Sedimentation in continental rifts' on page 98). It is therefore likely that these rocks record a very early rifting event, probably the earliest experienced by the young continent. The Dominion rift failed to develop into an ocean – rifts commonly fail in this way – and the newly born Kaapvaal continent remained intact.

These events had a parallel in northern KwaZulu-Natal and Swaziland where a suite of rocks similar to the Dominion Group, known as the **Nsuze Group** (**figure 4.2**), developed. This forms part of the **Pongola Supergroup**. The Nsuze rocks consist mainly of volcanic rocks, interlayered with minor sedimentary rocks. Several of the conglomerates were mined on a small scale for gold, but yields were generally low and mining was short-lived.

SEDIMENTATION IN CONTINENTAL RIFTS

Continents are frequently subjected to lateral stretching, which may lead to rifting. The process initially causes thinning of the hot, plastic lower crust and underlying lithospheric mantle, while the cool, brittle upper crust responds by fracturing (**A**). This thinning occurs in a linear belt, often along some pre-existing weak zone in the crust. As the lithosphere thins, hot asthenosphere rises from below, causing the crust to bulge upwards. Further fracturing of the upper crust occurs as the bulge rises and the crust thins, creating a depression along the axis of the surface bulge – a rift valley (**B**).

Rivers flow into the depression and lakes may form along its length, such as in the East African Rift Valley. Sediment therefore accumulates in the rift valley. As the hot asthenosphere rises below the rift, pressure is reduced and the asthenosphere begins to melt. The melt, or magma, slowly segregates and moves upwards because of its lower density compared to the surrounding rock. Heat from the rising magma, as well as heating by the asthenosphere, may also cause some melting of the lower crust (**C**).

The mantle melt erupts on the surface to produce basalt, while the melted crust produces rhyolite volcanoes. Rifts therefore often have both rhyolite and basalt volcanoes associated with them. The material filling the rift valley consists of sedimentary deposits, which may include both lake and river sediments and even shallow marine deposits if the rift valley subsided below sea level. Rhyolite and basalt lavas and volcanic ash are interlayered with the sediment deposits. Continued rifting will result in the development of basaltic ocean crust in the centre of the valley.

The valley may continue to widen and become flooded by the ocean (*see* Chapter 2). If, however, the rift fails, as often happens, cooling of the asthenosphere causes it to sink (**thermal subsidence**), dragging the lithosphere and thinned crust downward (**D**). Because the crust is thinned by rifting, it may ultimately sink below sea level, creating a shallow depression or basin floored by continental crust, in which sediments accumulate. Such depressions are very much wider than the original rift. When viewed in cross-section (**E**), the resulting sedimentary deposits resemble a steer's head, with the sediments filling the large basin representing the horns and the underlying, more limited rift sediments representing the head.

THE AGE OF GOLD

Failed rifts on continents follow a characteristic evolutionary path. The upwelled, hot asthenosphere beneath the thinned crust and lithosphere eventually cools and subsides. This is referred to as **thermal subsidence** because it is driven by cooling. The land surface is dragged downward (*see* 'Sedimentation in continental rifts' above). The relatively narrow rift zone thereby evolves into a very wide, shallow depression that often subsides below sea level.

Drowning of the Craton

Subsidence of this type appears to have followed the failure of the Dominion-Nsuze Group rifting, and much of the Kaapvaal continent started to subside below sea level. Invasion of sea onto the continent marked the start of a major period of sediment accumulation, producing the **Witwatersrand Supergroup**, which hosts the world's most important gold accumulation.

Initially sand was deposited in the shallow sea, commencing sometime around 2 970 million years ago. As the sea deepened, however, finer-grained sediment accumulated and the sand gave way to mud and finally to iron oxide- and silica-rich sediments (banded iron formation) produced by chemical precipitation from the sea water (*see* 'Sedimentation on continental shelves' on page 100). Interruptions in subsidence or an increase in sediment supply to the subsiding basin resulted in

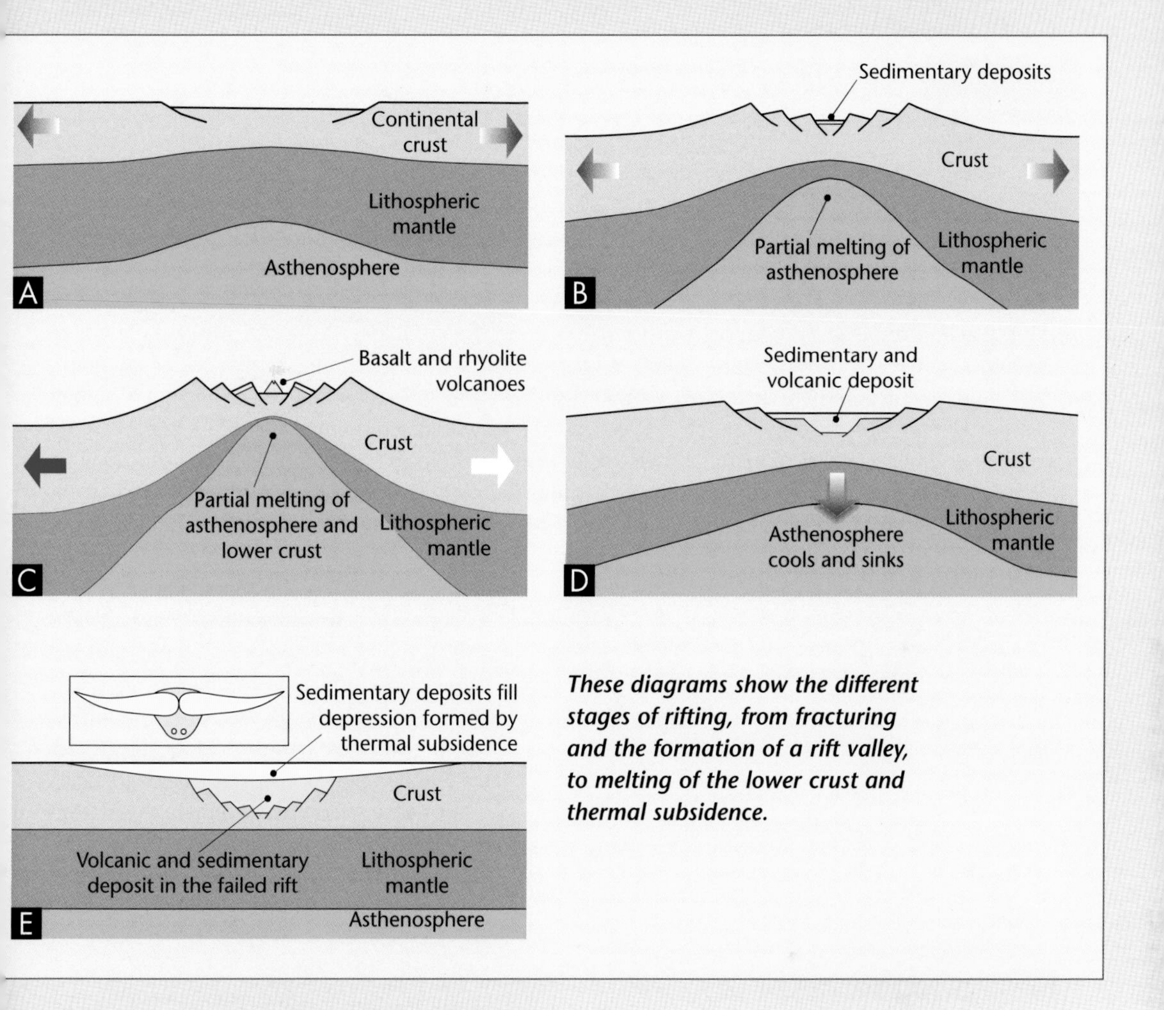

These diagrams show the different stages of rifting, from fracturing and the formation of a rift valley, to melting of the lower crust and thermal subsidence.

shallowing of the ocean as more sediment accumulated, and chemical sediment gave way to mud and then to sand. The sediment was derived from the elevated continental areas to the north and west, where accretion of island arcs was possibly still continuing. The resulting group of rocks forms the lowermost division of the Witwatersrand Supergroup, termed the **Hospital Hill Subgroup** (**figure 4.3**).

Then followed a period of fluctuating sea level. At times of low sea level, rivers flowed across the depression, depositing gravel and sand. At other times deep-water conditions prevailed over much of the continent, and mud and even iron-rich sediments accumulated. Also deposited during this period was an unusual rock type called **diamictite**. This consists of rounded boulders up to tens of centimetres in diameter, set in a very fine, almost muddy matrix.

The most common way this poorly sorted (i.e. big and small particles mixed together) sedimentary rock forms, although not the only way, is as a result of glacial activity, when moving ice is the agent of erosion and deposition (*see* 'The formation of sedimentary rocks' on page 64). Sea level falls when the Earth's ice caps expand and rises when they melt. The association of diamictite with fluctuating sea level at this time suggests that glacial activity may indeed have occurred, marking the earliest known glaciation on Earth, about 2 950 million years ago.

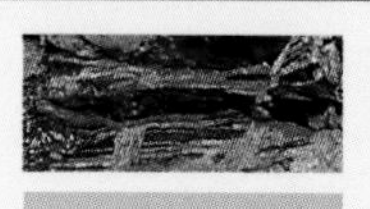

SEDIMENTATION ON CONTINENTAL SHELVES

Sea level constantly changes over millions of years due to the rise and fall of continents, changes in the size of the polar ice caps and a variety of other causes. Areas above sea level are subjected to weathering and erosion, and fragmental material is transported by rivers to the coast where the sediment is deposited. The processes involved can produce layering of different types of sediment.

Diagram **A** illustrates a typical coastline with a river discharging sediment into the ocean. Currents and wave action disperse the sediment and move it along the coastline. In the near-shore area, where the water is relatively shallow and the effects of wave action and currents are strong, coarser sediment, usually sand, will accumulate. Mud will remain dispersed and will only accumulate in much deeper water, where currents and wave action are less severe. Very little sediment is carried to areas remote from land, and here biological material such as skeletons or shells of free-swimming organisms will be the only material to accumulate. Sometimes, chemical sediment may accumulate in these deep-water environments if precipitation is taking place. Prior to about 1 800 million years ago, iron oxides accumulated in this region.

If the land is sinking slowly (or sea level is rising), the ocean becomes deeper, and the zones of accumulation of different sediment types shift accordingly, as shown in **B** and **C**. Where previously mud had accumulated, skeletal material is now deposited (iron-rich sediment in the past); where sand was deposited, mud accumulates. This causes the stacking of layers of different types of sediment: sand at the bottom, then mud and finally skeletal material. These layers will eventually be lithified and become rock: sandstone, mudstone and limestone (or iron formation in ancient rocks). The thickness of the layers depends on the rate of subsidence compared to the rate of sediment supply: slow subsidence and a large supply of sediment will usually produce thicker layers.

If subsidence ceases, but sediment continues to be supplied to the shoreline, the water gradually becomes shallower as a result of sediment accumulation, the shoreline builds outward, and the zones of sand, mud and skeletal material move seaward. This results in stacking of layers in the reverse order to that encountered previously (**D**), and limestone will give way upward to mudstone and finally to sandstone. This is one of the ways in which layering in sedimentary rocks arises, and is common in deposits formed along continental shelves.

Original shoreline
A
Position of original shoreline
B
Position of original shoreline
Sand
Mud
Lime
C
D

Diagrams showing the stacking of different kinds of sedimentary rocks during rising (A to C) and static (D) sea level.

The group of rocks formed during this period of fluctuating sea level is known as the **Government Subgroup** (**figure 4.3**). Thereafter, subsidence and sediment accumulation rates were more or less equal, and the basin remained largely inundated by the sea, with mud and sand accumulating to form the **Jeppestown Subgroup** sequence of rocks. Deposition of these rocks ended approximately 2 910 million years ago.

The resulting three groups of rock layers, which attain a maximum thickness of about 5 000 m, are collectively known as the **West Rand Group** (**figure 4.3**) and they form the lower portion of the Witwatersrand Supergroup. The rocks of the West Rand Group have been traced over much of the Kaapvaal Craton, extending from Swaziland in the east to Ottosdal in the west, and from Thabazimbi in the north to beyond Bloemfontein in the south (**figure 4.4**). This indicates the wide extent of incursion of the sea over the continent.

Today the outlying portions occur as scattered fragments, separated as a result of subsequent erosion, but they once formed continuous layers over the entire area. Very little of the West Rand Group of rocks actually outcrops on the surface, and most of the fragments shown in **figure 4.4** are buried beneath younger rocks; their presence has only been revealed by deep drilling and studies of the Earth's magnetic field. They include rocks of the **Mozaan Group**, part of the Pongola Supergroup, which are developed in northern KwaZulu-Natal and Swaziland.

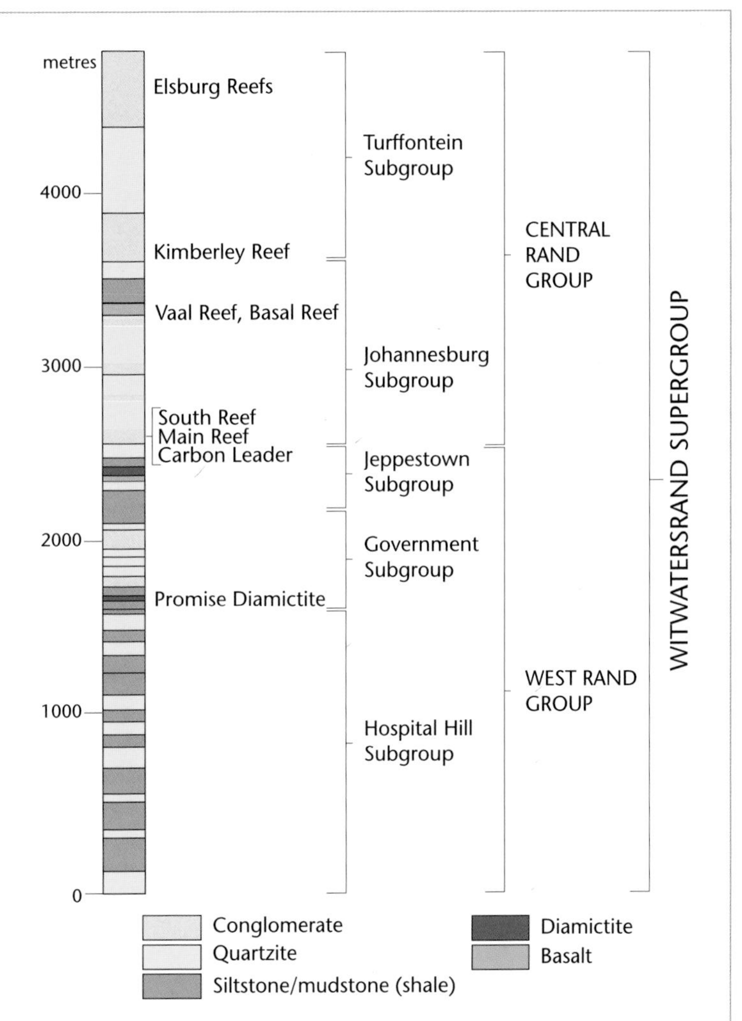

Figure 4.3 *Following a period of rifting of the Kaapvaal Craton that thinned the crust, the continent subsided below sea level and sediments started accumulating about 2 970 million years ago. The resulting pile of sedimentary rocks, up to 7 km thick, is known as the Witwatersrand Supergroup. The rocks comprising the supergroup are shown here in the form of a composite stratigraphic column. The lower West Rand Group was deposited largely in a shallow sea, whereas the upper Central Rand Group is an accumulation of river deposits and is mainly terrestrial. Certain of the conglomerate layers in the Central Rand Group, known as reefs, contain rich concentrations of gold and uranium.*

Although outcrops of the West Rand Group rocks are rare today, they nevertheless are responsible for some significant topographic features, the most important being the Witwatersrand ridge that extends from Randfontein to Germiston. The quartzites and the banded iron formations are very resistant rock types and form ridges, while the intervening, softer mudstones underlie lower

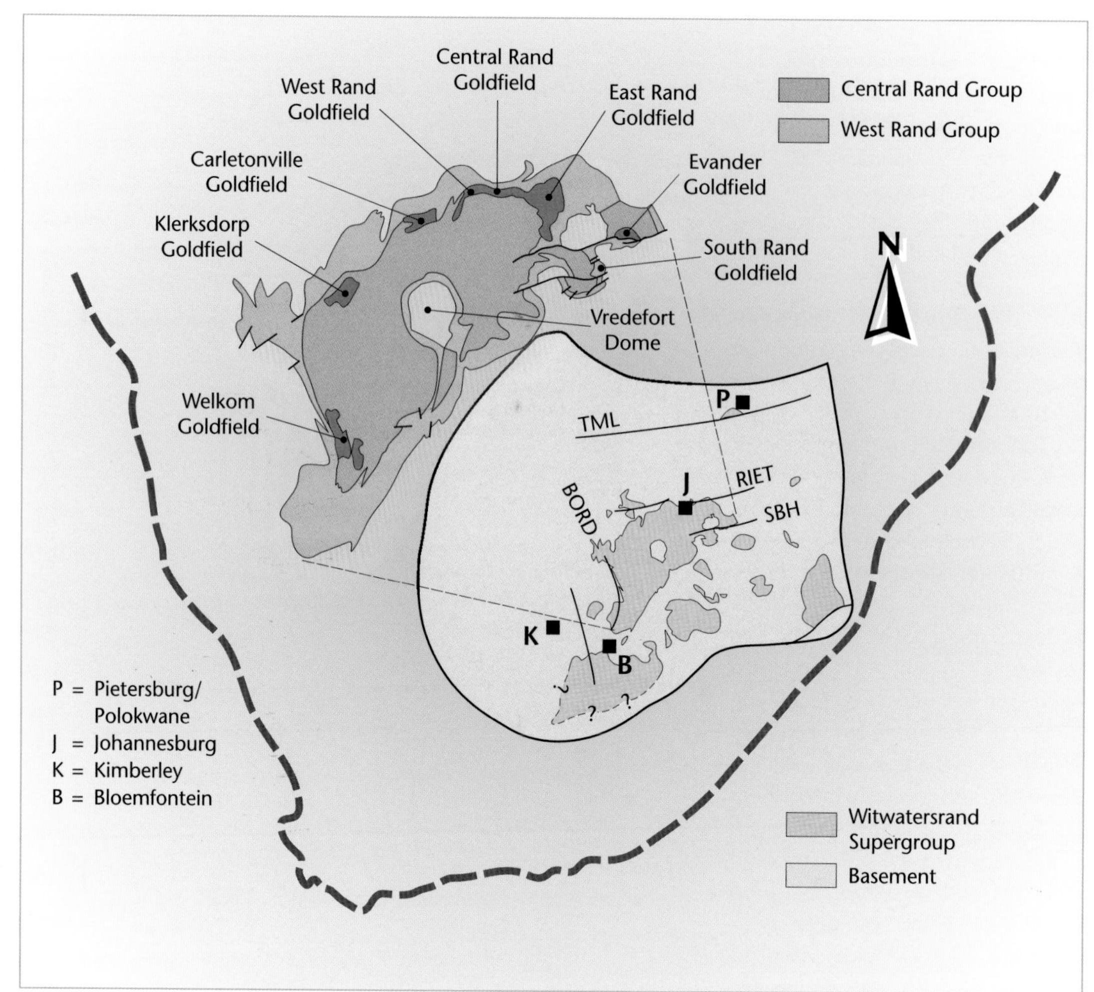

Figure 4.4 *The rocks of the Witwatersrand Supergroup were originally widely distributed over the Kaapvaal Craton, but much has been removed by erosion, leaving only the scattered remnants shown here. The enlarged inset shows the main area of preservation of the Witwatersrand Supergroup basin. The major goldfields occur in an arc around the western and northern sides of the basin. The locations of these goldfields were determined by earth movements along faults such as the Thabazimbi-Murchison Line (TML), the Rietfontein Fault (RIET), the Sugarbush Fault (SBH) and the Border Fault (BORD).*

ground. The lowermost quartzite forms the Witwatersrand escarpment, which rises steeply above the more easily weathered granite-greenstone basement to the north.

Rise of the Craton

Conditions surrounding the Witwatersrand depression changed radically at the end of deposition of the Jeppestown Subgroup, probably due to changes in the nature of the collisional margin along the continent's northern edge. Subduction was probably oblique and the angle shallow, with little volcanic activity above the subsiding slab (*see* Chapter 2). The interior of the continent began to experience severe compressional stresses and the crust ruptured, leading to development of large-scale fractures or faults (*see* 'Folding and fracturing of rocks' on page 41) across the craton. These included the Thabazimbi-Murchison fracture, the Rietfontein and Sugarbush fractures, and the Border structure (**figure 4.4**).

Many of these fractures developed along the old suture lines, where island arcs had amalgamated during growth of the Kaapvaal continent. These sutures were somewhat weaker than the average cratonic crust and remain so today. Movement on the fractures involved lateral sliding, as well as vertical slip. The overall effect of these movements was to cause some sections of the crust to rise relative to others, producing mountainous terrain within and around the formerly extensive West Rand Group depression (**figure 4.5**). The depression became fragmented into a number of sub-basins, separated by uplands.

River systems eroded these rising uplands. Erosion was probably very effective because of a more corrosive atmosphere and the total absence of vegetation. Sediment was carried down to the subsiding regions by the rivers, where it was deposited on wide, gravel-covered erosion surfaces called pediments and further downstream on extensive alluvial plains. On these vast plains the rivers divided and rejoined in complex patterns of channels and sand and gravel bars. West Rand Group rocks were stripped from the uplifted blocks by erosion, exposing underlying basement granites, while West Rand Group strata in the subsiding blocks adjacent to the fractures became folded as a result of block movement. The sea was pushed back (**figure 4.5**) by uplift and sedimentation, and the sedimentary material accumulating in the depressions became dominated by the river deposits, consisting mainly of sand (now quartzite), interlayered with gravel (now conglomerate).

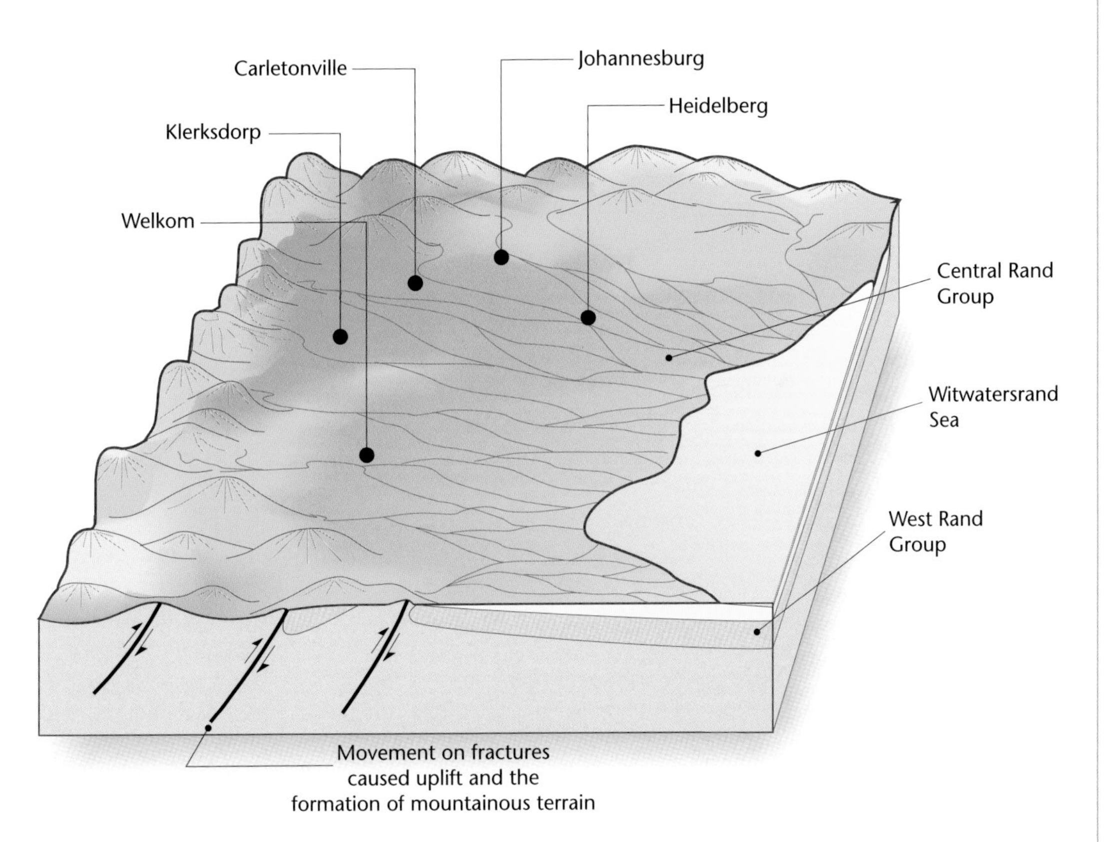

Figure 4.5 *During deposition of the Central Rand Group of the Witwatersrand Supergroup between 2 900 and 2 700 million years ago, crustal movement along faults caused mountains to develop, separated by regions of subsidence. Rivers eroding material from the upland areas coalesced downstream to form vast riverine plains. Gold and other dense minerals became concentrated in the stream beds on these plains as a result of the winnowing action of the flowing water.*

Continued uplift of the mountains and subsidence of the lower lying areas resulted in thick accumulations of river deposits in the sub-basins. The resulting rocks are collectively known as the **Central Rand Group** (**figure 4.3**) and attain a maximum thickness of about 2 000 m. The West Rand and Central Rand Groups together are termed the **Witwatersrand Supergroup**. The sub-basins containing Central Rand Group rocks were distributed over a large area, but it is not known how many there were. The oval-shaped basin shown in the enlarged inset in **figure 4.4** is the best known, and is referred to as the **Witwatersrand Basin**. Smaller sub-basins occur at Evander and the South Rand, and others are known to exist in northern KwaZulu-Natal and in the Free State around Bethlehem, extending into northern Lesotho.

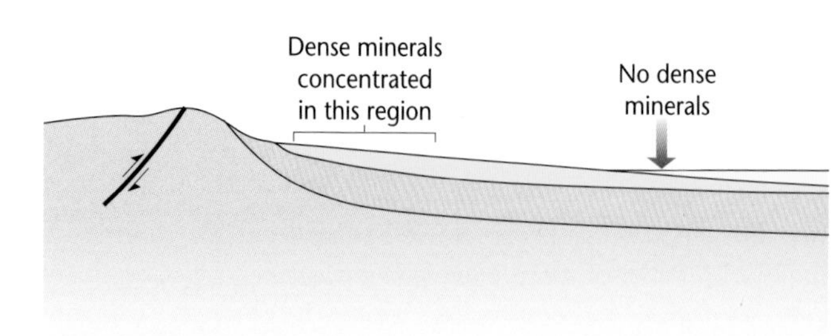

Figure 4.6a ***Dense minerals such as gold were concentrated in sediments close to the mountain source, largely to the north and west, whereas those deposited further away have little gold, as illustrated in the upper cross-section.***

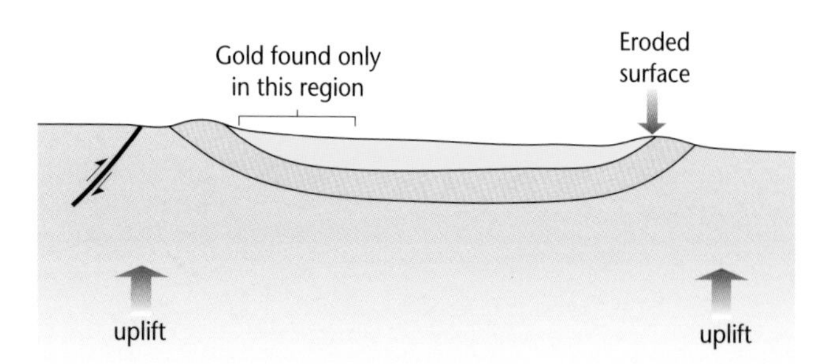

Figure 4.6b ***The deposit was subsequently uplifted around its margins whereas its centre was depressed, producing the basin shape. Gold mineralisation is therefore confined to one side of the basin (see also* figure 4.4*).***

Gold accumulates

Flowing water has the ability to sort particles according to both size and density. Larger or denser particles (the latter referred to as heavy minerals) are less easily moved than less dense or smaller particles. Under conditions of sustained flow, with a slow but steady supply of sediment, heavy minerals accumulate along with larger particles, while less dense material is washed away. Heavy minerals only become concentrated over a limited length of flow, and downstream the concentration will decrease (**figure 4.6**). In this way, certain conglomerate (gravel) layers of the Central Rand Group became enriched in heavy minerals, among which were particles of gold, uraninite (uranium oxide), pyrite (iron sulphide) and, very occasionally, diamonds.

It is possible that additional gold was added to the conglomerates long after deposition (*see* 'The gold controversy' on page 106). Not all conglomerates are enriched in these heavy minerals: only those deposited under suitable conditions became enriched. These conditions require sustained flow in a slowly eroding setting with a limited sediment supply and, of course, a suitable source of heavy minerals. Conglomerates deposited under conditions of rapid sediment accumulation are not enriched in heavy minerals, so they contain little gold or none at all.

Economic concentrations of gold are developed mainly on the northern and western margins of the Witwatersrand Basin (**figure 4.4**) whereas the rocks in the southeast contain little or no gold, having formed too far from the source (**figure 4.6**). Conditions suitable for the concentration of gold and other dense minerals occurred at different times in different regions where Central Rand Group strata were accumulating, and produced economically mineralised conglomerate layers at different levels in the Central Rand Group of rocks. These mineralised conglomerates are termed reefs by miners. None of these reefs is uniformly mineralised, as gold content generally decreases with distance away from the source regions of the sediment.

Within a given reef, higher concentrations of gold often occur in linear zones (so-called pay shoots), probably marking former river channels.

Towards the end of this period of sedimentation, uplift along the fractures became progressively more severe, and very coarse gravels were shed into the subsiding depressions. In certain areas, notably around Klerksdorp and Carletonville, erosion of lower gold-bearing reefs occurred during this late phase of uplift and the eroded gold became concentrated in a layer at the very top of the Central Rand Group, known as the Ventersdorp Contact Reef.

The rocks of the Witwatersrand Supergroup are not in themselves unusual, but the quantity of gold they contain is truly spectacular – without equal anywhere else on Earth (*see* 'The gold controversy' on page 106). Time may have been an important factor in creating the golden treasure house of the Witwatersrand Supergroup. The contrasting sediment accumulation rates of the West Rand and Central Rand Groups are quite remarkable: the 5 000 m of West Rand sediment accumulated in 60 million years, whereas the 2 000 m of Central Rand Group accumulated in 200 million years – the latter under the influence of river deposition, in which sedimentation is generally more rapid.

It is likely, therefore, that hidden within the Central Rand Group are long periods when no sediment accumulated, and even periods when erosion of sediment layers predominated. The long period of sedimentation, and especially the intervening periods of erosion, may well have been important. They allowed for slow and therefore efficient winnowing; this concentrated heavy minerals in certain of the conglomerate layers, which produced the unique gold deposits of the Central Rand Group. In contrast, the conglomerates of the more rapidly deposited West Rand Group contain very little gold. The ultimate source of the gold, however, remains one of the great enigmas of South African geology.

Minerals present in the conglomerate layers of the Witwatersrand Supergroup tell us something about the nature of the Earth's atmosphere 2 900 million years ago. The conglomerates formed in streams and rivers, and the water would have equilibrated with the atmosphere. Under conditions prevailing today, neither uraninite nor pyrite are stable in streams, and are destroyed by reaction with dissolved oxygen. Uraninite actually dissolves in this process, while pyrite forms sulphuric acid and an extremely insoluble red or orange iron oxide. We can conclude from the presence of uraninite and pyrite in the conglomerates that the atmosphere at that time did not contain free oxygen.

The presence of pyrite in the conglomerates has given rise to one of the more serious negative aspects of gold mining. After the gold has been extracted the waste material, which includes the pyrite, is pumped to large dumps. Oxygenated rain water reacts with the pyrite, releasing sulphuric acid into the ground water. Acid mine drainage, as the phenomenon is known, has become a worrying problem in many areas where gold has been mined.

The gold mines are today very deep. Mining is taking place on some mines at depths of almost 4 km below surface. The pressure exerted by the overlying rocks at such depths is huge and occasionally the rock surrounding the open workings disintegrates under the stress in events known as rock bursts. Sadly, lives of miners are sometimes lost as a consequence. A very active research programme is underway to develop technology to prevent this loss of life.

A VOLCANIC DELUGE

The intense uplift in the final stages of Witwatersrand sedimentation culminated in the rupturing of the Kaapvaal Craton. Huge fractures developed, up which basalt magma from the mantle flowed, erupting on surface in extensive fissure eruptions (*see* 'Igneous intrusions' on page 46). Witwatersrand sedimentation stopped abruptly as the very fluid basaltic lava flowed across the land surface, burying the river systems. This volcanic event commenced 2 714 million years ago, and indications are that it was short-lived, lasting perhaps less than six million years. Nevertheless, in this relatively brief period, the pile of lava attained a maximum thickness of 2 000 m and covered an area in excess of 100 000 km^2.

Today, these lavas form the Klipriviersberg hills south of Johannesburg and have been named the **Klipriviersberg Group**. They also form prominent topography in the Heidelberg area. This volcanic event marks the beginning of a new period of accumulation on the Kaapvaal Craton, termed the

THE GOLD CONTROVERSY

The Witwatersrand Supergroup is host to the most remarkable gold resource on Earth. Since first discovered in the Witwatersrand conglomerates in 1886, about 50 000 tonnes of gold have been produced from these rocks. This amounts to around 31% of all gold ever mined throughout history. Huge quantities still remain, but mostly at concentrations that are not economic to mine at the present gold price. How this immense quantity of gold came to be concentrated in one small region remains a controversial subject.

Gold occurs in the conglomerate rocks as gold metal. Not all conglomerates contain gold: only those that lie on former erosion surfaces carry the metal. The conglomerate layers represent former river gravels and the gold is usually most concentrated at the bottom of the layer, although there may be small amounts dispersed through the entire layer. The thickness of the conglomerate layers varies widely, from a single layer of pebbles a few centimetres thick to layers tens of metres in thickness. The gold-bearing conglomerates also contain several other minerals, all characterised by above-average density, notably minerals containing uranium (uraninite and brannerite), iron sulphide (usually pyrite), and various others. In addition, a carbon-rich material known as kerogen, or bitumen, is often present as well, though it has a low density compared to most other minerals.

The amount of gold in conglomerates that are currently being mined is typically very small, usually 4–10 g of gold per ton of rock. Iron sulphide is usually the most abundant of the dense minerals, making up about 3% of the conglomerate. It occurs as rounded pebbles (so-called buckshot pyrite) and as small crystals, usually cubes. It has a gold colour and is often referred to as fool's gold. The bitumen occurs as small rounded balls less than 1 mm in size (called flyspeck carbon), or sometimes as a continuous layer about 1–2 cm thick (carbon or bitumen seams), usually at the bottom of a conglomerate layer. Invariably, the bitumen is intimately mixed with uranium-bearing minerals, and is usually very rich in gold.

There is a fierce debate raging about how the conglomerates came to contain so much gold. There are two extreme views. The **placerists** believe that the gold came from remote source rocks, now almost completely eroded away, and was brought to the Witwatersrand Basin by rivers. Flowing water has the ability to separate minerals according to their densities – a process made vividly evident by the black-coloured sands of the KwaZulu-Natal coast, where dense, titanium-rich black minerals have been concentrated by wave action. Such concentrations of dense minerals formed by the action of flowing water (or wind) are known as **placer deposits**. Gold is much denser than most other minerals and readily becomes concentrated in river gravels. Placerists cite the fact that gold is always associated with other dense minerals, as would be expected if the gold had been concentrated by the action of flowing water. The presence of bitumen has been ascribed to life processes. The bitumen seams are interpreted as remnants of primitive life forms (algal or bacterial mats), which acted as traps for fine gold particles.

In contrast, the **hydrothermalists** argue that whereas there may have been some gold concentrated by flowing water, most of the gold was introduced long after the conglomerate had formed and became lithified. They argue that the fine, filigree shapes of many of the grains of gold are so delicate that they could never have been transported by water. They also point to the absence of magnetic iron oxide (magnetite), a very common, dense iron mineral (for example, in banded iron formation) from the conglomerates. The hydrothermalists maintain that this must have been an abundant mineral in the original gravels, where it was concentrated by flowing water along with other dense minerals like uraninite. Later, the magnetite became converted to iron sulphide by interaction with sulphur- and gold-bearing hot water (so-called **hydrothermal fluid**). Consumption of the dissolved sulphur as a result of the formation of iron sulphide reduced the solubility of gold, which also precipitated.

To the hydrothermalists, the spherical buckshot pyrite is no more than former rounded grains and pebbles of magnetite, now changed to iron sulphide. They also point to the presence of other, minor associated minerals that suggest interaction with hot water. In their view, the bitumen is not of biological origin, but also formed by precipitation from fluids, in this case fluids containing oil-like substances.

Energetic radiation arising from radioactive decay of uranium interacted with the oil molecules, causing them to combine to form larger molecules that precipitated as solids. In this way, carbon became fixed around the grains of uranium-bearing minerals.

The debate continues, but some have adopted the middle ground, the so-called **modified placerists**. In their view, gold was initially concentrated in the conglomerates by flowing water, but later some of the gold was dissolved and re-precipitated by hot water passing through the conglomerates. Some additional gold may have been added during this process.

Recent technological advances in rock dating techniques have made it possible to measure the age of formation of gold and iron sulphide particles. Only a few such grains from Witwatersrand conglomerates have thus far been dated, but the results indicate that the gold and iron sulphide from mines in the Klerksdorp area formed 3 030 million years ago. The exact time of deposition of the conglomerate from which the gold grains were extracted is imprecisely known, but must have been sometime after 2 900 million years ago. Thus, the gold is older than the rock in which it occurs, indicating a placer origin for the gold and iron sulphide grains dated. However, many more grains will have to be dated before the placer origin is confirmed.

Whatever caused the concentration of gold in the conglomerates, it is clear that the Witwatersrand contains an unusually large amount of gold, and appealing to one or other process does not really answer the question as to why this region of the Earth's crust contains so much gold. Perhaps the answer is lost in the mists of time, and we will never discover it.

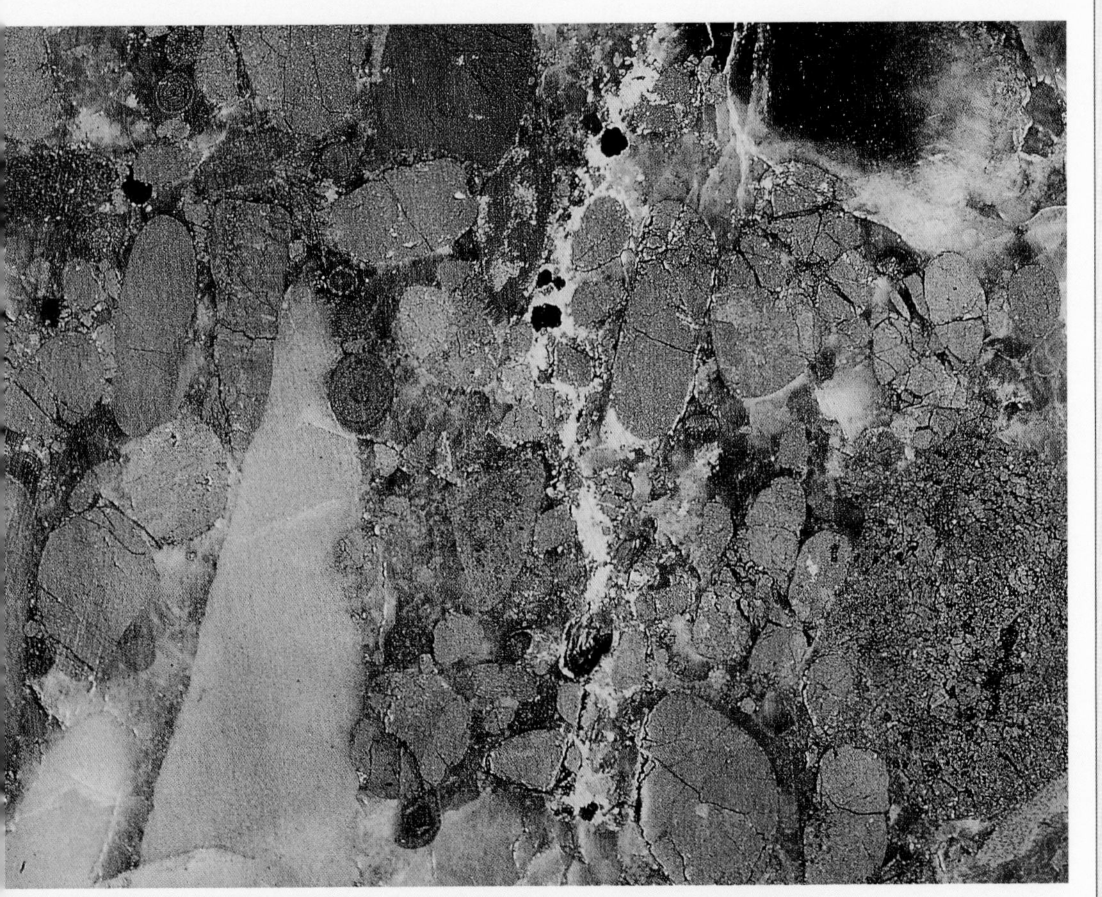

Gold and carbon nodules in buckshot pyrite conglomerate (twice actual size).

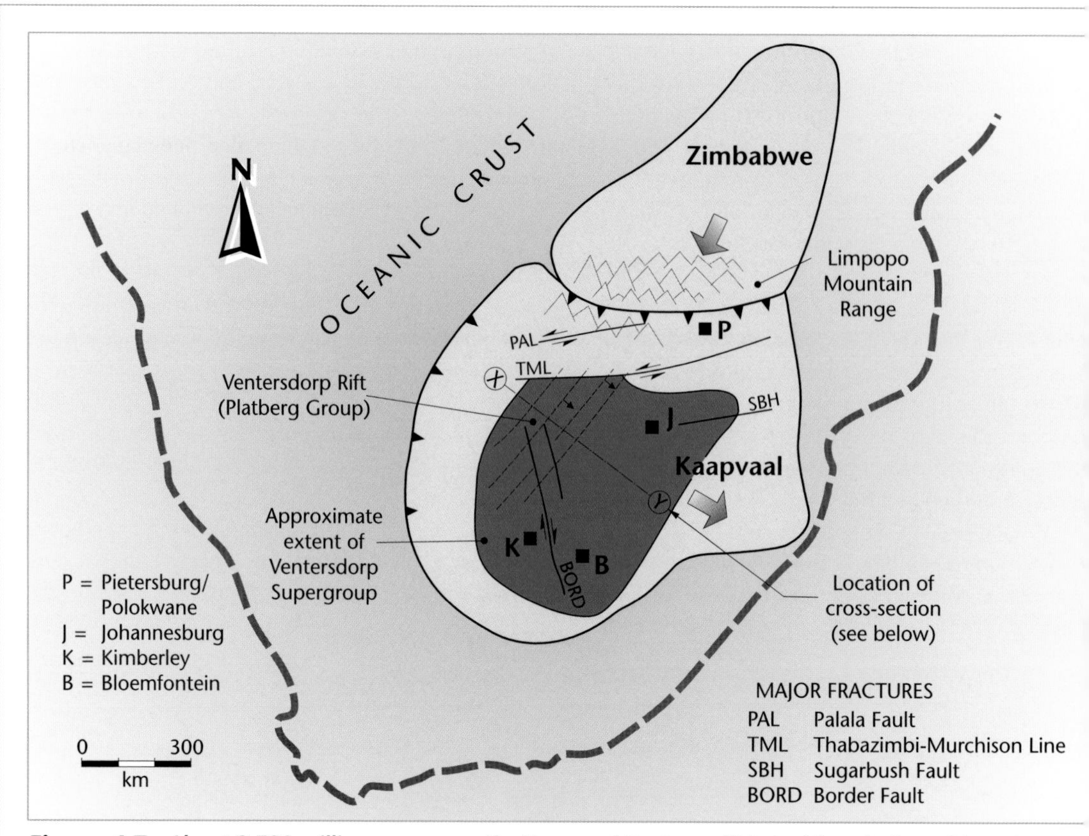

Figure 4.7A *About 2 700 million years ago, the Kaapvaal Craton collided with a similar, although slightly younger, ancient continent known as the Zimbabwe Craton. The collision, similar to that currently taking place between India and Asia, produced a Himalayan-type mountain range along their common boundary, the Limpopo Mountain Range. The impact of the two continents caused the interior of the Kaapvaal Craton to rupture and huge quantities of lava poured through the cracks, burying the Witwatersrand Supergroup strata beneath more than a kilometre of lava (the Klipriversberg Group of the Ventersdorp Supergroup). These eruptions began about 2 714 million years ago and lasted for no more than six million years. Pressure from the north continued and sections of the continent began to slide sideways to accommodate the stress. Movement took place along old fault zones, but many new faults formed at this time as well.*

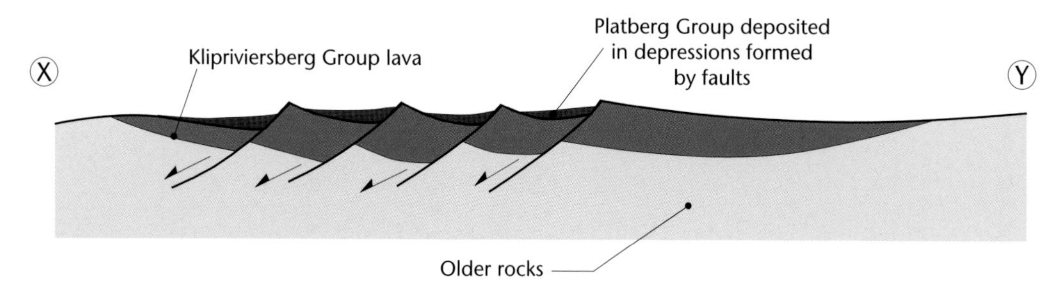

Figure 4.7B *The lateral movement of the crust along the faults produced arrays of long valleys separated by ridges, as illustrated by this cross-section.*

Ventersdorp Supergroup. The cause of rupturing of the continent that gave rise to this new period of accumulation is uncertain, but may be related to the collision of the Zimbabwe Craton with the Kaapvaal Craton.

KAAPVAAL-ZIMBABWE COLLISION AND CONTINENT RUPTURE

Zimbabwe, which is largely underlain by granite-greenstone basement, had an early geological history that in many respects was similar to that of the Kaapvaal Craton. The Zimbabwe Craton formed separately and slightly later than the Kaapvaal Craton, but like the Kaapvaal, was drifting as a micro-continent on the young Earth. Subduction along the northern margin of the Kaapvaal brought the two continents into contact about 2 700 million years ago. Details of the collision event are not fully understood. Moreover, there appears to have been some renewed activity along the boundary about 2 000 million years ago, making reconstruction of earlier events more difficult. We will skip the details and press on with our journey.

As the Zimbabwe continent ploughed into the Kaapvaal Craton, the latter began to break apart, releasing lava through fissures. Portions of the Kaapvaal Craton were expelled sideways (**figure 4.7**), releasing the compressive stresses within the continent. More fractures developed and the uplifted blocks that had formed high ground during Central Rand Group times collapsed, becoming instead great, elongated valleys. Sediment was rapidly shed into these valleys from their margins, forming fan-shaped aprons, often passing down-slope into small lakes (**figure 4.8**). Coarse boulders dominated the upper slopes of these aprons, becoming finer down-slope, and passing ultimately into muddy material on the lake beds.

The lakes also supported primitive life forms that deposited calcium carbonate in stromatolitic forms. Sedimentation was periodically interrupted by volcanic eruptions of basalt and especially rhyolite, the latter often violent, producing copious quantities of volcanic ash. The rocks deposited during this period are collectively known as the **Platberg Group**, named after a small hill west of

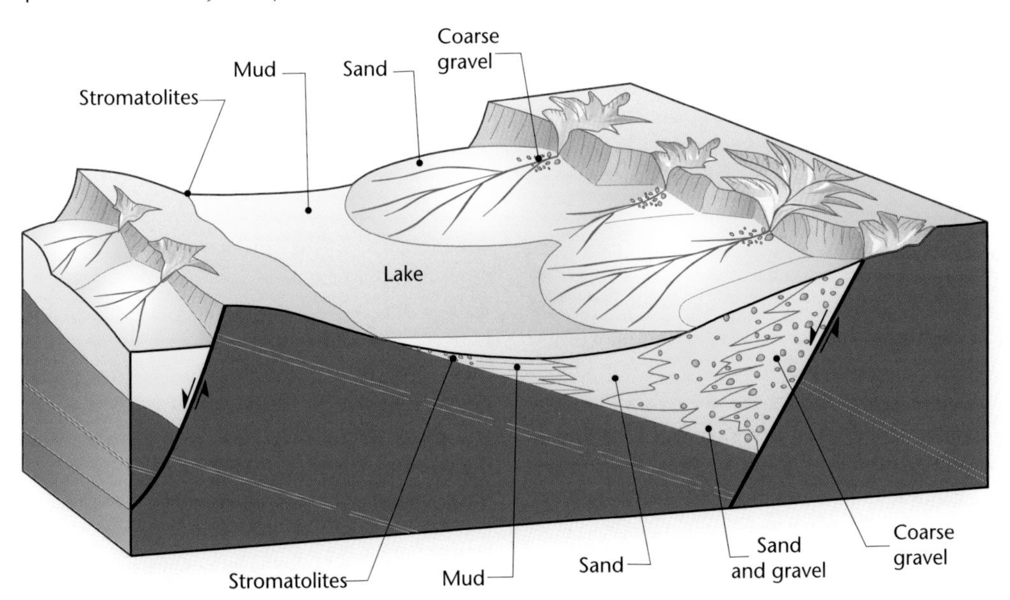

Figure 4.8 *As sections of the Kaapvaal Craton slid sideways in response to continuing pressure from the north, alternating valleys and ridges formed. The valleys deepened and lakes formed. Sediment shed from the adjacent ridges began to build fan-shaped aprons along the valley margins, as shown here. Cyanobacteria growing in the lakes gave rise to stromatolites. These sedimentary deposits form the Platberg Group of the Ventersdorp Supergroup. Scattered volcanic rocks within this group are 2 708 million years old.*

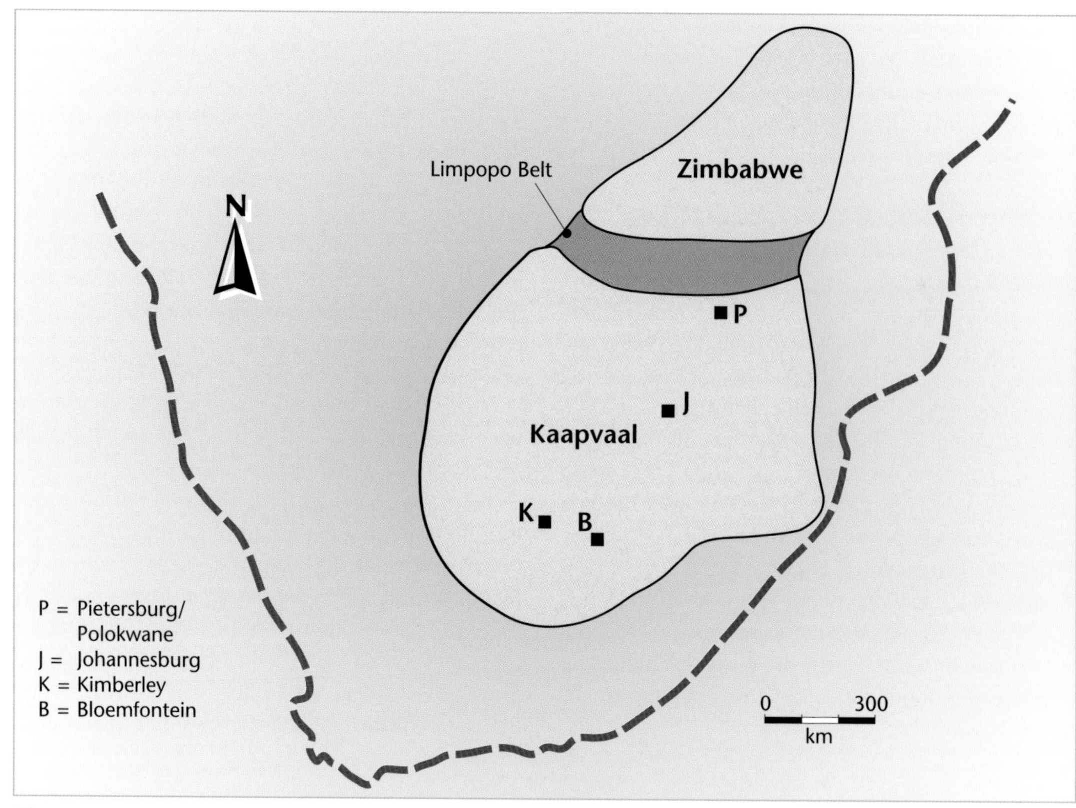

Figure 4.9 ***Erosion of the mountain range along the collision zone between the Kaapvaal and Zimbabwe Cratons brought to surface rocks that had formerly been deeply buried in the crust, where they had experienced great heat and pressure. These metamorphosed rocks today form the Limpopo Belt. The two cratons have remained fused together to the present day. (The coastline of South Africa is shown in this diagram for reference.)***

Potchefstroom where they are well exposed. The volcanic rocks indicate that eruption occurred 2 708 million years ago. These were indeed turbulent times, particularly compared to the preceding 200-million-year accumulation of Central Rand Group sediments.

This fragmentation of the Kaapvaal Craton was rapid, lasting probably about eight million years, and it culminated in a major rift depression across the continent extending from a position south of Kimberley northwards towards Mokopane/ Potgietersrus (**figure 4.7**). Slow subsidence of the rift captured rivers, and sediments began to accumulate. These were buried by periodic eruptions of basalt through the thinned crust. The resulting rocks are known as the **Bothaville** and **Pniel Groups**.

The collision of the Kaapvaal and Zimbabwe continents caused thickening of the crust along their common boundary as sedimentary rocks on the floor of the intervening ocean became compressed between the converging land masses. These folded rocks were heated and the crust melted, producing granite magmas. The thickened crust formed a mountain range with a deep root, in a manner similar to the modern Himalaya, which formed as a result of the collision of India with Asia (*see* Chapter 2).

The collision of the Zimbabwe and Kaapvaal Cratons heralded a change to a more modern plate tectonic style, as earlier collisions of micro-continents during the assembly of the Kaapvaal Craton did not produce such massively thickened crust. The mountain range resulting from the collision of the two continents was rapidly eroded, causing the rise of its root zone, so rocks previously buried to a depth of at least 30 km in the crust became exposed on the surface.

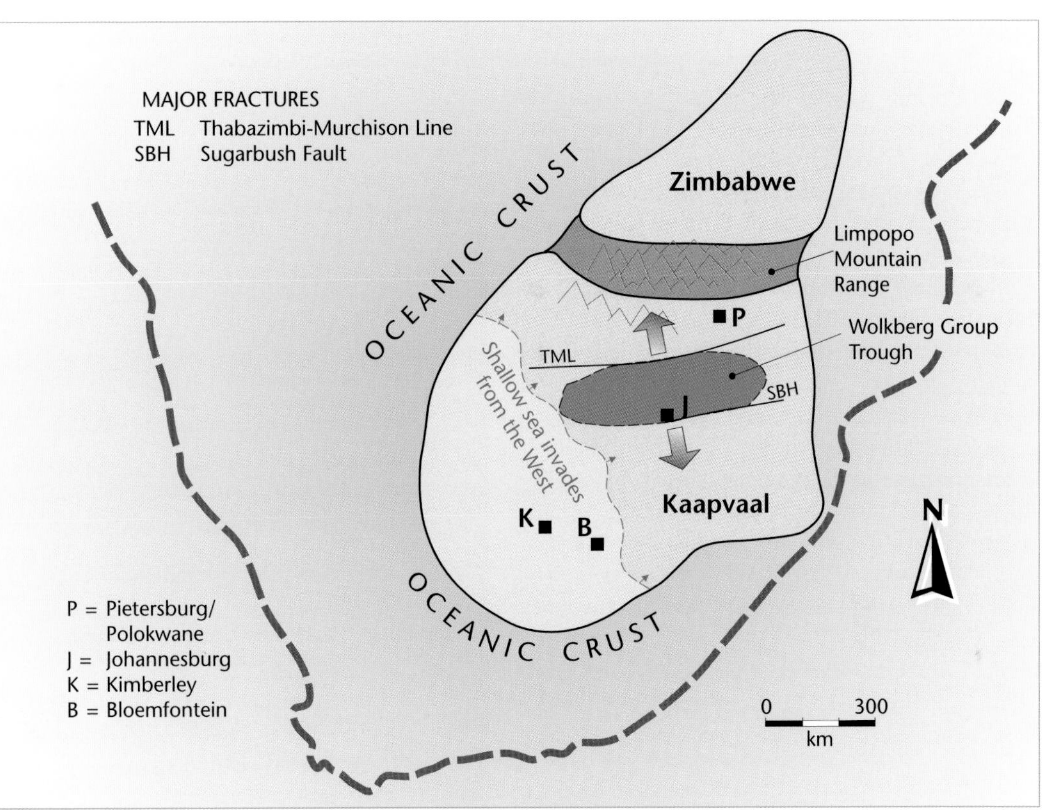

Figure 4.10 ***Stretching of the crust of the Kaapvaal Craton started about 2 650 million years ago, forming a rift between the Thabazimbi-Murchison Line and the Sugarbush Fault, in which sedimentary rocks of the Wolkberg Group accumulated. As this trough became deeper, the whole Kaapvaal Craton began to subside below sea level and sediments of the Transvaal Supergroup began to accumulate.***

Today, minerals characteristically formed at very high temperatures and pressures (metamorphic rocks) occur in rocks on the surface along the Limpopo River. These rocks form a linear zone, known as the **Limpopo Belt**, along the boundary between the two formerly separate continents, and mark the suture between the two continents (**figure 4.9**). It is difficult to imagine that the hot, flat and monotonous country north of the Soutpansberg was once the site of snow-covered peaks like the Himalaya, but the geological evidence indicates that such a range certainly did exist.

RENEWED RIFTING OF THE CRATON

Although copious quantities of lava poured onto the crust during the Ventersdorp rifting event and the crust of the continent had been thinned by stretching, it failed to rupture completely. Instead, it stabilised. Around 2 650 million years ago, a new period of stretching and rifting of the continent began. Ancient sutures in the continental crust played an important part in localising the rift that formed in the region between the Thabazimbi-Murchison and Sugarbush fractures (**figure 4.10**). Sand and mud were carried by rivers into the rift trough. There they accumulated along with occasional basaltic lava flows to form the **Wolkberg Group**. These rocks today form part of the eastern escarpment in the Blyde River Canyon area (**figure 4.11**). The sands have been metamorphosed to produce very hard quartzites that form the prominent cliff faces of the canyon and of the escarpment. Equivalent rocks occur to the west and extend by way of Thabazimbi as far as eastern Botswana.

This rifting was followed by a period of thermal subsidence, when almost the entire Kaapvaal Craton subsided below sea level to form a large, shallow continental shelf on which sedimentary rocks of the **Transvaal Supergroup** began accumulating. Uplands persisted in the north, remnants of the mountain range between the Zimbabwe and Kaapvaal Cratons. As the Kaapvaal Craton subsided, river systems draining its surface drowned and were buried by beach and shallow-water marine deposits. This gave rise to the conglomerate, sandstone and mudstone deposits of the **Black Reef Formation**. These rocks cap the high ridge of the eastern escarpment (**figure 4.11**) and are widespread across the craton, extending as far to the west as the Northern Cape and Botswana.

Features preserved in the rocks of the Black Reef Formation suggest that at times during deposition the region must have been characterised by extensive mud flats. Ripple marks produced by the ebb and flow of tides are preserved in the rocks, as are mud cracks formed as the mud dried at low tide. Even casts of cubic salt crystals about 1 cm across abound in the mudstones of the upper Black Reef Formation (**figure 4.12**), suggesting that the mud flats must have developed in restricted bays that periodically dried out.

THE BACTERIAL AGE

As the sea encroached onto the Kaapvaal Craton, a new phenomenon appeared – a product of life on a scale never seen before. Indications of life in the oceans are present in rocks as old as those at Barberton, but occurrences are localised and fragmentary. In the rocks overlying the Black Reef Formation there is evidence for life on an abundant scale as cyanobacteria came to dominate the shallow sea. The environment was probably similar to a modern-day coral reef such as the Great Barrier Reef of northeastern Australia. In those days, however, coral had not yet evolved and its environmental niche was occupied by cyanobacteria, which photosynthesise, extracting carbon dioxide from the water in which they live. This causes an imbalance in the water chemistry, leading to precipitation of calcium carbonate that adheres to the slimy bacterial layer.

The bacterial layer reforms over the calcium carbonate deposit. Gradually, layer upon layer of calcium carbonate is added, forming stromatolites. Such deposits formed across most of the continent at this time and produced some of the world's most

Walter Knirr / IOA

Figure 4.11 *The very resistant quartzites of the Wolkberg Group, which represent sand deposited in an ancient rift, form many of the cliffs in the Blyde River Canyon area of the eastern escarpment.*

spectacular examples of stromatolites. The shape assumed by stromatolites depends on water depth, tidal range, and wave and current activity (**figure 4.13**). In very shallow water between the low- and high-tide marks (the intertidal zone), they form mat-like structures or occur as tiny spherical bodies called oncoliths, which are formed by the growth of bacteria under the influence of the to-and-fro movement of water (**figure 4.14**).

In slightly deeper water they form interlinked finger-like forms that enlarge to domes (**figure 4.15**) up to a metre high and half a metre in diameter as the water deepens. In deep water (but less than 100 m), stromatolites take the form of elongated whale-back structures tens of metres long and metres high (**figure 4.16**), as well as large cone-shaped mounds. Essential requirements for the development of stromatolites are a water depth of less than 100 m (the bacteria need light to grow) and a very low input of sediment from the land. Stromatolites grow very slowly, and they are swamped and overwhelmed if rivers supply large quantities of sediment to the shallow seas where they are forming.

Cyanobacteria and their associated stromatolites still exist today, living fossils apparently unchanged since the Archaean Eon. However, today they only

Figure 4.12 *Casts of salt crystals (each about 1 cm across) in mudstone from Bourkes Luck in Limpopo Province. These salt crystals grew on muddy tidal flats that formed as the Kaapvaal Craton slowly subsided below sea level some 2 600 million years ago.*

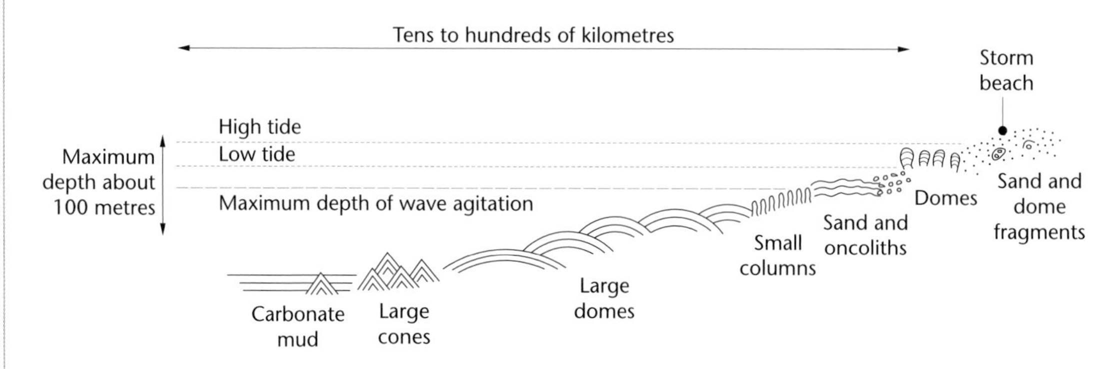

Figure 4.13 *Subsidence of the Kaapvaal Craton produced a shallow sea in which cyanobacteria thrived. These photosynthesising bacteria consumed carbon dioxide and caused the precipitation of calcium carbonate from the sea water, which stuck to the slimy bacterial colonies. Continued precipitation of calcium carbonate and growth of the cyanobacteria gave rise to reef-like structures tens to hundreds of kilometres wide. The shape assumed by the bacterial colonies varied with water depth, tidal range and wave activity. This diagram illustrates the variety of forms assumed by the bacterial colonies, known as stromatolites. Precipitation of carbonate eventually produced a layer more than 1 km thick across most of the Kaapvaal Craton. Some time after deposition, the calcium carbonate was converted to calcium-magnesium carbonate, known as dolomite.*

occur in a few isolated areas where conditions are suitable for their growth. The reason for this is that cyanobacteria are a favoured food of sea snails (gastropods), and the wide distribution of these animals restricts stromatolite growth to regions where the sea water is too saline to support them. Shark Bay in Western Australia is a famous locality where stromatolites are currently forming (**figure 4.17**). During the Proterozoic Eon there were no grazing marine animals, and so the cyanobacteria reigned supreme.

The original stromatolitic calcium carbonate deposits were chemically changed to a calcium-magnesium carbonate (known as **dolomite**) some time after deposition, but the

Terence McCarthy

Figure 4.14 *In shallow water gently agitated by currents, calcium carbonate precipitated by the cyanobacteria produced spherical structures known as oncoliths, shown here.*

Figure 4.15 *Between the high- and low-tide marks, bacterial colonies accumulated domical shaped stromatolites, viewed here from above.*

Terence McCarthy

Figure 4.16 *In deep water, below the influence of wave activity, large, elongated stromatolite domes formed, such as can be seen exposed near Boetsap in the North West Province. Here elongated domes about 15 m wide meet in cusp-like forms.*

processes involved are not well understood. In addition, dolomite became partially replaced by silica in the form of chert, which usually occurs as discontinuous layers in the dolomite. Dolomite is slightly soluble in rain water, whereas chert is not. On outcrops, the grey dolomite slowly dissolves to produce the characteristic elephant-skin appearance on weathered surfaces, and the chert layers stand out in sharp relief (**figure 4.18**).

The dolomites are widespread and probably originally covered almost the entire Kaapvaal Craton. Today they outcrop in two areas: in the Northern Cape, North West Province and Botswana, where they form the **Ghaap Group**; and in the North West, Gauteng and

Nic Beukes

Figure 4.17 *Stromatolites still form today, such as here at Shark Bay in Western Australia. They are rare, however, because the bacteria that form them are eaten by marine grazing animals; stromatolites only form where conditions are unsuitable for such animals.*

Mpumalanga provinces, where they are referred to as the **Chuniespoort Group** (**figure 4.19**). The time interval over which the dolomite accumulated is not very precisely known, but probably occurred between 2 600 and 2 400 million years ago.

The cyanobacteria were restricted to water depths of probably less than 100 m. In deeper water, carbonate muds accumulated, whereas in very deep water, iron and manganese oxide deposits formed on a massive scale. The reason iron and manganese oxides precipitated is related to the composition of the Earth's atmosphere at the time. During the early period of Earth history, the atmosphere contained almost no free oxygen. Iron and manganese released by weathering of rocks on the Earth's surface and from undersea hot springs were present in their soluble, reduced form, dissolved in the oceans.

Terence McCarthy

Figure 4.18 ***When exposed to the atmosphere, the dolomite of the Transvaal Supergroup develops a characteristic weathered appearance. Slow dissolution of the grey dolomite results in a surface resembling elephant skin. Insoluble layers of chert in the dolomite stand out as ledges on the weathered surface.***

Cyanobacteria photosynthesise and release oxygen as a waste product. This oxygen dissolved in the water, and the shallow water where they lived became slightly oxygenated while the deep ocean water remained oxygen-free. Diffusion of oxygen from the surface water to the deeper oceans brought the oxygen in contact with the reduced iron and manganese. This caused oxidation of these metals to essentially insoluble forms and they precipitated as oxides, creating iron formations. This process had probably been happening for a long time before the formation of the dolomites, as iron formations occur in rocks even older than those at Barberton – such as the 3 800-million-year-old Isua iron formation in the Amitsoq terrane in Greenland – but these older deposits appear to be far smaller.

Cyanobacteria appear to have been on the increase, and peaked during the time of Transvaal Supergroup dolomite deposition. During this climax period of the cyanobacteria, vast quantities of oxygen were produced and the great reservoir of iron and manganese in the oceans was precipitated, leading to the formation of major iron and manganese deposits, not only in South Africa but world-wide. Once this reservoir of iron and manganese was consumed, significant free oxygen began to appear in the atmosphere for the first time, changing forever the course of the evolution of life (*see* Chapter 6).

The shallow Transvaal sea in which the cyanobacteria lived continued to subside. As the water deepened, iron accumulated in a layer across the submerged craton. Although iron formations are thickest and most prominent in the Northern Cape, small remnants of this once extensive layer are also still preserved in the Thabazimbi area and in the region of Penge in the Limpopo Province.

Within a thin interval in the iron formations of the Northern Cape there is an unusual combination of rock types: diamictite is developed in places overlying striated surfaces, suggesting a glacial origin. A layer of basalt containing pillow structures overlies the diamictite, indicating that it erupted under water. Palaeomagnetic studies of the lava reveal that at the time of eruption the region lay in the tropics. This creates a climatic paradox, which recurs some 1 000 million years later (*see* Chapter 6): how is it possible for marine glaciation to occur at tropical latitudes?

An answer to this problem has recently been proposed by Joe Kirschvink of the California

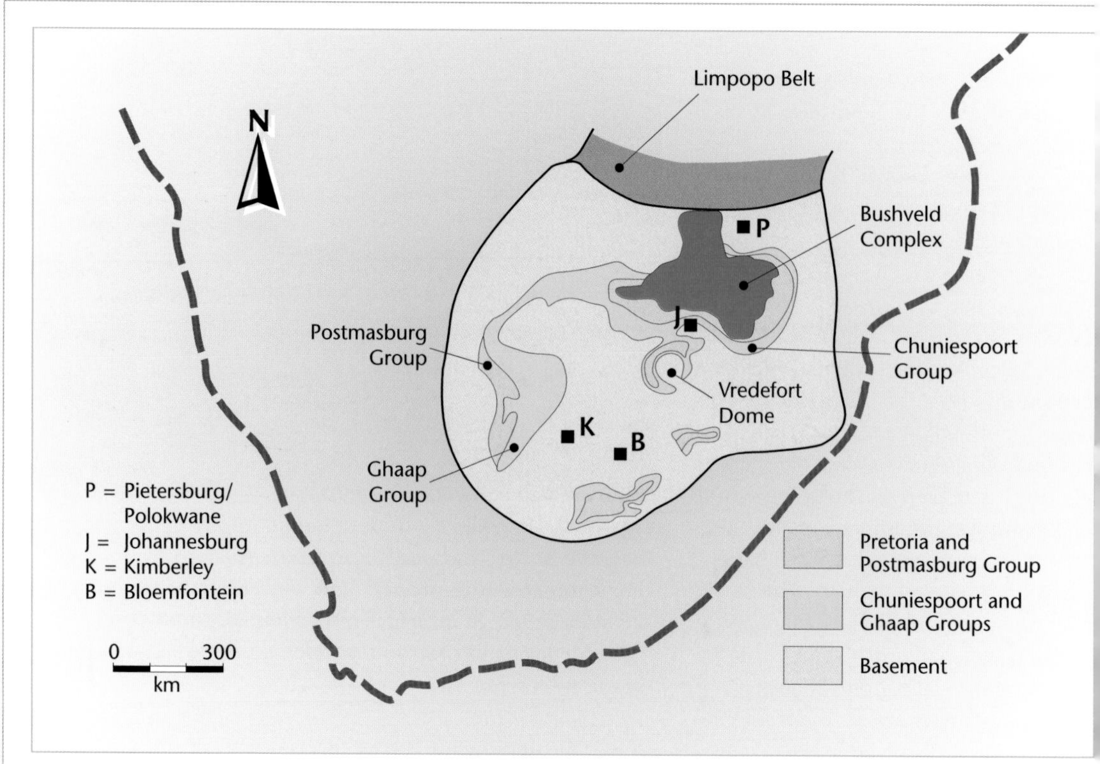

Figure 4.19 *The dolomites of the Transvaal Supergroup were once very widespread across the Kaapvaal Craton, but their extent has been reduced by later erosion. This map shows the present distribution of the dolomite (Chuniespoort and Ghaap Groups), as well as the overlying rocks of the Pretoria and Postmasburg Groups. (The coastline of South Africa and the locations of cities are shown for reference.)*

Institute of Technology. Called the **Snowball Earth Hypothesis**, this remarkable theory postulates that the Earth has undergone several episodes in its history during which the entire ocean, from poles to the equator, was frozen over to form a global ice-sheet several kilometres thick (*see* 'Snowball Earth', right).

DESTRUCTION OF BACTERIAL REEFS

Following deposition of the iron formations of the Transvaal Supergroup, there is a major break in the sedimentary record. The dolomites became locally uplifted and tilted, and much of the iron and manganese deposits and some of the underlying dolomites were eroded from the southern, uplifted portions. The dolomites were removed mainly by dissolution, leaving behind a blanket of insoluble chert fragments, which attains a maximum thickness of 200 m. This marks the beginning of a new stage of sedimentation on the continent – the deposition of the **Pretoria Group**, which formed over the period 2 350 to 2 100 million years ago.

The Pretoria Group consists of a pile of sedimentary rocks, mainly mudstones and quartzites with some basalts, collectively up to 5 km thick. The sediments were deposited almost entirely under marine conditions ranging from muddy tidal flats, today represented by mudstones, to shallow-marine sands that today form very prominent quartzites, such as those that give rise to the Magaliesberg Mountains (**figure 4.20**). The tidal flat environments encouraged stromatolite growth, but nowhere near the scale of the Chuniespoort Group.

The Pretoria Group, in general, is made up of progressively finer sediments (i.e. more mudstones and less quartzite) towards the south, suggesting that most of the sediment was supplied from the north or northwest. Erosion has reduced the area formerly covered by the Pretoria Group and today it is confined mainly to an oval-shaped area

SNOWBALL EARTH

In the 1960s, Cambridge geologist Brian Harland found evidence of glacial deposits (tillites) from widely scattered regions around the world, all about 700 million years in age. What made the find really puzzling was that palaeomagnetic studies (*see* 'Earth's magnetism' on page 36) indicated that some of these deposits formed near the equator. In 1986, rocks of this type were found near Adelaide in Australia, and Joe Kirschvink, an expert in palaeomagnetic studies from the California Institute of Technology, went to investigate. His results confirmed that these rocks must have formed close to the equator during a glacial episode. An iron formation formed by precipitation of iron from sea water was associated with these glacial deposits, which is unusual. Earlier in Earth history, before free oxygen appeared in the atmosphere, the precipitation of iron from sea water to form iron formations was widespread, but the rocks from the Adelaide area were deposited long after the atmosphere became oxygenated. So the iron formations posed a problem.

Kirschvink formulated a novel solution to the twin problems of glacial deposits forming near the equator and their association with iron formations. He proposed that at the time these rocks were deposited, the Earth had almost completely frozen over – hence the name **Snowball Earth**. He reasoned that life on Earth would have been almost completely shut down by the extreme cold, except in areas around active volcanoes and especially those in the deep sea along mid-ocean ridges. The atmosphere and ocean water became separated by a layer of ice, and iron released at undersea hydrothermal vents would have begun to build up in the oceans. The ice cover would have reflected away the Sun's heat, and permanent freezing of the Earth could have resulted, were it not for on-going volcanic activity.

Kirschvink suggested that carbon dioxide released from volcanoes into the atmosphere would have gradually caused its concentration to rise. Without surface biological activity or surface water, there was no way this carbon dioxide could have been removed from the atmosphere. It has been calculated that the carbon dioxide concentration may have reached 350 times its present value, when its greenhouse effect began to raise the surface temperature and melt the ice. Once the ice melted and the atmosphere again came into contact with the oceans, wholesale precipitation of iron occurred, forming the iron formations. Kirschvink's theory thus neatly explained both the glacial deposits and the associated iron formations.

A similar association of glacial deposits and iron formations has been recorded in rocks of this age in Namibia, in the metamorphosed sedimentary rocks of the Damara Belt. Here, in addition to iron formation and tillite, there are very unusual deposits of limestone, the so-called cap carbonates, which have a strange, non-biological look about them. Paul Hoffman and colleagues from Harvard University carried out isotopic studies on the carbon in these limestones, and found that it lacks the characteristic signature of photosynthesis.

Hoffman proposed that the cap carbonates, like the iron formations, also formed in the wake of melting of the global ice sheet. He suggested that as the oceans warmed in the carbon dioxide-rich atmosphere, large amounts of water vapour (also a greenhouse gas) were released into the atmosphere and the temperature rose rapidly, producing a hothouse Earth. The fine glacial debris left by the ice weathered rapidly, releasing a sudden burst of calcium to the oceans, resulting in massive precipitation of calcium carbonate to form the cap carbonate.

Why such a global ice sheet formed at this time is a matter of conjecture, but it has been suggested that it could be related to the distribution of continents on the globe at the time. Many small continents spread around the equator could have disrupted heat transfer from the equator to the poles by restricting ocean circulation, thereby initiating an ice age. As the ice caps expanded, more of the solar radiation would have been reflected from the Earth, lowering the temperature. This may have culminated in a runaway effect, and global freezing. The Snowball Earth hypothesis remains just that, and is still the subject of much controversy and active research.

Figure 4.20 *The Pretoria Group, which overlies the dolomites in the Transvaal Supergroup, was deposited in a shallow sea and consists of quartzites and mudstones. The hard quartzites form prominent mountain ranges such as the Magaliesberg Mountains, shown here at Hartebeespoort Dam west of Pretoria. The layers were originally horizontal, but now dip gently to the north due to subsidence caused later by the Bushveld Complex. Part of the range has been offset by fault activity.*

Grant Cawthorn

extending from Botswana to the eastern escarpment in Mpumalanga, and to areas in the Northern Cape (**figure 4.19**), where it is known as the **Postmasburg Group**.

The deposition of the Pretoria Group marks a radical change in environment: the preceding shallowly submerged continental shelf environment in which the cyanobacteria flourished was completely destroyed. After the brief period of uplift, a shallow sea formed, into which large amounts of sediment were being supplied, largely preventing the growth of stromatolites. The environmental change therefore essentially involved a change in the character of the source area of the sediment rather than of the environment where deposition was taking place. What caused this is still uncertain, but it may have been a response to a renewed phase of rifting similar to that which gave rise to the Wolkberg Group.

THE AGE OF PLATINUM

South Africa generates more than half of all the world's annual production of platinum, chromium, vanadium and refractory minerals. These are derived from a truly remarkable body of igneous rocks – the **Bushveld Complex** – one of the geological wonders of the world (**figure 4.21**). At current mining rates, there are sufficient resources of all these commodities to last for hundreds of years (*see* 'Mineral wealth of the Bushveld Complex' on page 124).

In its broadest sense, the Bushveld Complex comprises three different groups of igneous rocks. The oldest is a series of volcanic rocks, followed by basaltic magma that did not reach surface. The intrusive basalt instead formed a large underground chamber, the largest known in the world, stretching 400 x 300 km across the Limpopo, North West and Mpumulanga provinces, and attaining a maximum thickness of 8 km. Finally, a very different magma was intruded above the older (basaltic) body and crystallised as granite. These three components have been named the **Rooiberg Group**, the **Rustenburg Layered Suite**, and the **Lebowa Granite Suite** respectively, and together make up the Bushveld Complex.

All these events happened 2 061 million years ago in a very short space of time, possibly one to two million years – so short a time that they cannot be distinguished using currently available methods of rock dating. The rocks of the Bushveld Complex were buried under younger sedimentary rocks, but erosion has exhumed them and they now form a huge elliptical structure extending from the Botswana border to the eastern escarpment of Mpumalanga. The origin of the Bushveld Complex is conveniently broken into three stages.

The volcanic event

A great thickness of sedimentary rocks of the Transvaal Supergroup had accumulated on the Kaapvaal continent. On this sediment there erupted a succession of lavas, known as the Rooiberg Group. These lavas are of two types. The first of these were mainly basalt, followed by eruptions predominantly of rhyolites, which are more extensive and

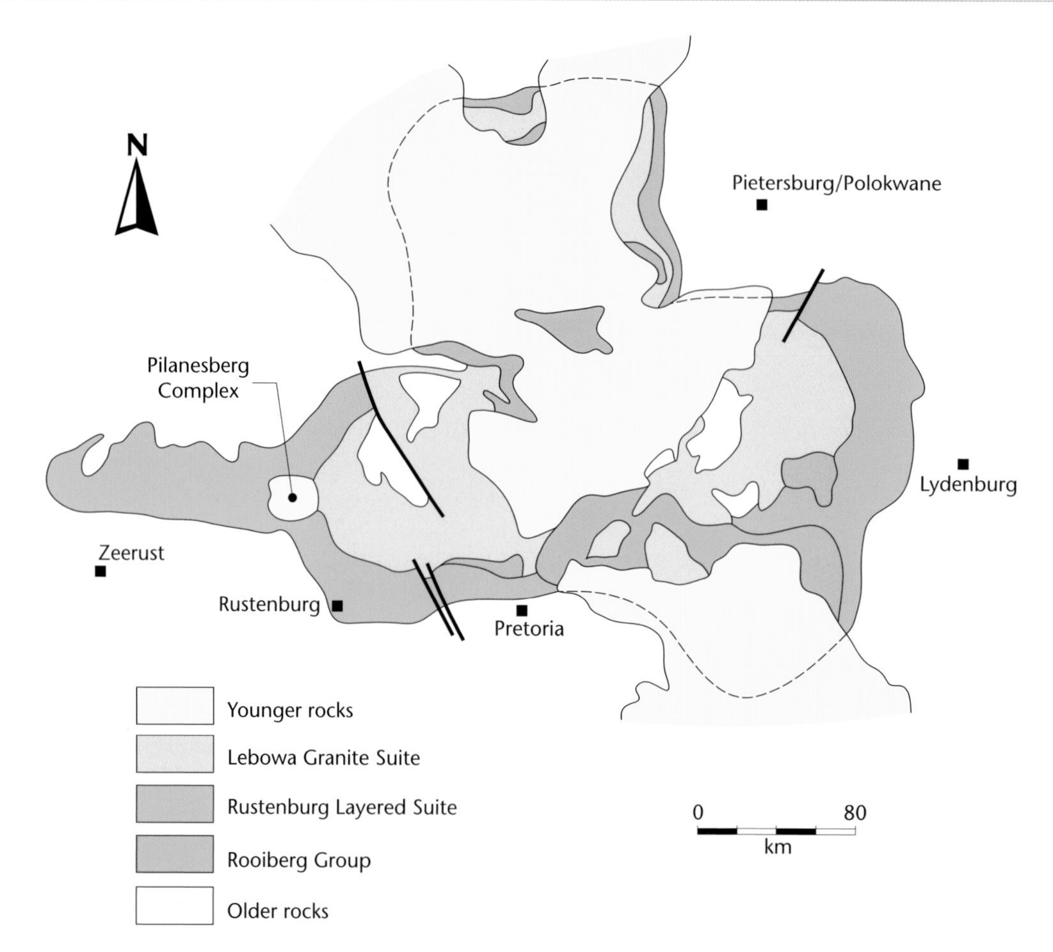

Figure 4.21 ***This geological map shows the surface outcrop of the Bushveld Complex. Dashed lines show probable buried portions. The Complex formed in three stages: rhyolite eruptions on a floor consisting mainly of Transvaal Supergroup sedimentary rocks produced the Rooiberg Group; basaltic magma then intruded below the Rooiberg Group, forming a layer about 8 km thick, which crystallised very slowly, allowing minerals to segregate into layers to form the Rustenburg Layered Suite; finally, granite magma intruded above the Rustenburg Layered Suite but below the Rooiberg Group to form the Lebowa Granite Suite. These various intrusions occurred in a very short space of time 2 061 million years ago.***

thicker than the basalts. On weathering, the lavas produce a rock with a very red surface colouration – hence the name of the hills and village, Rooiberg, near Bela-Bela/Warmbaths, after which these rocks were named.

The closest modern analogue of these eruptions is in the Yellowstone National Park in the United States, where rhyolite eruptions are still ongoing, although none have occurred in living memory. The hot springs and geysers there are current testimony to the continuing volcanic activity. The Bushveld region would have looked similar to Yellowstone 2 000 million years ago, except that there would have been no vegetation.

The first eruptions took place in shallow water, possibly in rivers or lakes, but the eruptions quickly elevated the land surface and later eruptions occurred on land. These later eruptions were not of liquid lava, but were violently explosive, producing huge clouds of hot ash that blanketed the landscape. One intriguing aspect of these lavas is the difficulty in recognising the volcanic craters from

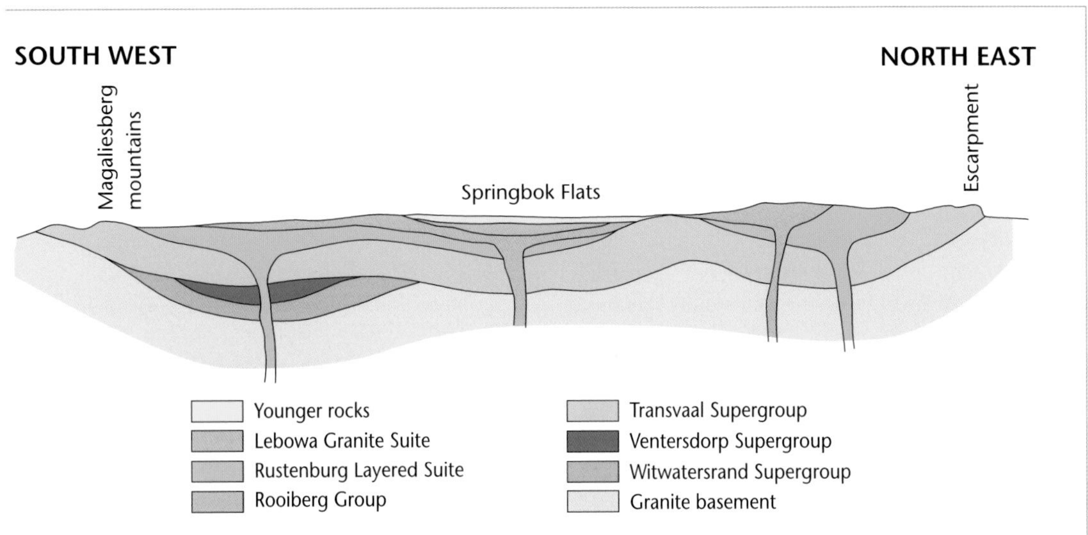

Figure 4.22 ***This schematic geological cross-section through the Bushveld Complex illustrates the relationships between its three component parts.***

which they erupted. In Yellowstone, huge circular craters mark the sites of eruptions, and these can usually be recognised even in ancient volcanic terranes. However, no circular craters have been identified in the Rooiberg volcanic rocks.

The first intrusive event

Immediately after the eruption of the lavas, further igneous activity produced one of the most spectacular suites of rocks in the world. In most of the geological literature it is known as the Bushveld Complex, whereas the more detailed terminology in South Africa defines this as the Rustenburg Layered Suite of the tripartite Bushveld Complex.

Molten rock rising from a depth may sometimes reach the surface of the earth as volcanic eruptions, or it may become trapped at some depth inside the earth. When the latter occurs, it is aided by the presence of layered sedimentary rocks. Between each layer of sediment is a plane of weakness along which the rock can split easily. Magma wedges its way between the layers producing a laterally extensive sheet of magma, called a sill. The most common magma to spread in this way is basalt, which is very fluid and solidifies to produce a very hard rock called dolerite. Many examples of this phenomenon can be seen in the Free State and KwaZulu-Natal, where the sills cap flat-topped hills. These sills range from centimetres to hundreds of metres in thickness, and some may be traced for over 100 km. A sill of magma 1 m thick will cool and solidify within days, whereas a sheet 100 m thick may take 50 years to solidify.

A huge volume of basaltic magma was rapidly generated in the mantle 2 061 million years ago. It began rising into the crust where it injected into the sedimentary rocks of the Transvaal Supergroup, locally inflating the 5 km of sediment to as much as 7.5 km. So rapid was the intrusion rate that new magma was added before the previous material had solidified.

One sill in particular, immediately beneath the rhyolites of the Rooiberg Group, became extremely inflated, reaching a thickness of 8 km. This massive sill or magma chamber extended over an area from eastern Botswana to Lydenburg, and from Pretoria to north of Polokwane/Pietersburg. At least, that is the lateral extent preserved today, but we do not know how much more extensive it may have been originally as its area has been subsequently reduced by erosion (**figure 4.22**). We do know, however, that there are other intrusions of exactly the same age and of very similar rock types distributed over a large part of the Kaapvaal Craton, which are considered to be part of the Bushveld Complex. Whether they were connected with the main body, or whether they were physically separate but contemporary intrusions, is unresolved.

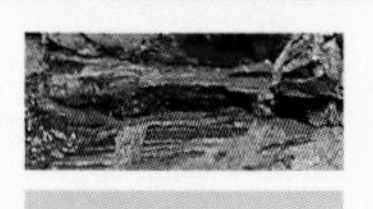

MINERAL WEALTH OF THE BUSHVELD COMPLEX

Most of the minerals and rock types found in the Bushveld Complex are quite common and have no commercial value. But some layers are composed of more unusual minerals and are economically importan The mineral **chromite** is the only economic source of the metal chromium that is used in stainless stee manufacture. The average basalt magma contains about 0.02% chromium. It is not economic to extrac chromium from a rock with so little of the metal. However, the mineral chromite contains 30% chromium

In the Bushveld Complex this mineral accumulated into layers, the thickest of which is about 1 m. Ther are 12 such layers, but only three are of adequate thickness for economic mining at present. Thus, the natural processes of mineral separation that took place in the magma chamber caused an enrichment of abou 1 500 times compared to average basalt magma. Some of the **chromitite** (the name for a rock compose largely of the mineral chromite) layers can be traced for more than 100 km in both the Rustenburg an Steelpoort areas. Equally importantly, they are of uniform composition and thickness down to the maximun depths of drilling, and there is no reason to doubt their continuation to greater depths, so the reserves o chromium are immense.

Very similar to the chromitite layers are the **magnetitite** layers, made of the mineral magnetite. Thes layers occur near the top of the intrusion. **Magnetite** is an iron oxide mineral, but also contains a small proportion of **vanadium**, a metal used for hardening steel. The lowest and first-formed layers contain about 1% vanadium, the subsequent layers contain much less. The fourth layer from the base is 2 m thick and is extensively mined near Roossenekal and Brits. This layer extends for well over 150 km in both the western an eastern sections of the Bushveld Complex.

The third mineral resource is the most important economically. Two layers of rock are mined for thei **platinum** (plus the related metals palladium, rhodium, ruthenium, iridium and osmium, as well as small amount of gold). One is a pyroxene-rich layer, known as the **Merensky Reef**, and the other chromitite, called the **UG2**. Platinum group metals are extremely rare elements in nature, but they hav many important uses – for example, as catalysts in the petrochemical industry, in catalytic converters fo motor vehicle exhausts, in fuel cells that convert oxygen and hydrogen to water (with the release o

Terence McCarthy

The Merensky Reef is a major source of platinum and related metals.

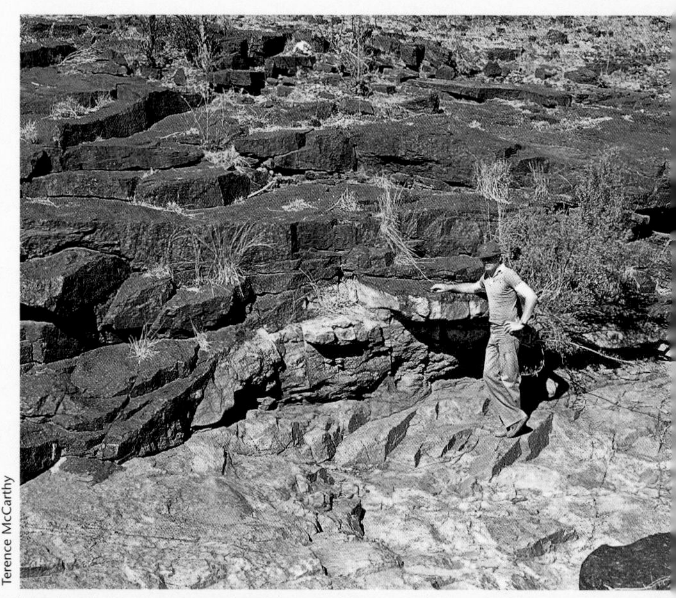

Terence McCarthy

The main magnetite layer is a source of vanadium.

electrical energy), and also in jewellery. The abundance of these elements in a basaltic magma is about 0.000001%. Such small concentrations are usually reported in parts per billion (ppb). The concentration in basalt magmas is thus about 10 ppb.

To put this abundance into perspective, if people were atoms, within a world population of five billion people, only 50 people would be designated platinum. In the two platinum-bearing layers of the Bushveld Complex, the abundance of platinum group metals is up to eight parts per million, or 8 g/tonne. Once again the processes operating naturally in the magma chamber have caused enrichment by a factor of 1 000 in the ore layers.

Even at such concentrations the precious metals are never seen with the naked eye. In the pyroxene-rich ore, there are other distinctive minerals associated with the platinum. These are bright yellow and silver sulphide minerals that are present in small amounts, but because of their colour are readily identified. They contain copper and nickel that are also extracted with the platinum. These two platinum-bearing layers can be traced for in excess of 100 km in both the western and eastern arcs of the Bushveld Complex. Moreover, platinum-bearing rocks of a slightly different kind are also developed in the northern section of the Complex.

The Merensky Reef was discovered near Steelpoort in 1924 by Dr Hans Merensky and was named in his honour, although platinum was known to occur in the rocks of the Bushveld Complex as early as 1908. Small-scale mining began in the eastern limb, but it was quickly realised that there was better infrastructure – especially in terms of water, power and human resources – in the Rustenburg area. By the 1990s mining was taking place along an arc of over 100 km, with a few gaps, from Brits to Thabazimbi. The UG2 was known to contain platinum at the time, but no suitable technology existed to extract it. This technology was developed in the 1970s and exploitation of the UG2 became possible. Several new mines are being developed in the eastern limb (mainly mining the UG2 layer) because of steadily increasing demand for these precious metals.

A totally different type of economic deposit is also exploited in the Bushveld intrusion, namely **dimension stone**. This was used originally for gravestones, but now finds extensive use in facing stone for buildings and for working surfaces in kitchens and bathrooms. Large quarries exist, originally in the area near Belfast but now mainly near Rustenburg, where massive blocks of stone are cut by drilling or cutting with diamond wires. One of the most important requirements of these blocks is that they must have no cracks, joints or internal blemishes. The blocks are cut by huge diamond-edged saws into slabs about 2 cm thick and then polished until they reflect like mirrors. The typical rock type is a gabbro. There is one unusual aspect of the gabbro in the Belfast quarries – the plagioclase is black because each grain contains millions of submicroscopic needles of black magnetite. Thus, the whole rock looks extremely dark, which was a popular attribute in the 1980s, but now paler grey gabbro is more fashionable.

The heat from the cooling Bushveld magma had to escape through the surrounding rocks, which belong mainly to the Transvaal Supergroup. As a result these rocks were heated and underwent a number of changes (metamorphism). The minerals recrystallised to produce extremely hard rocks, such as quartzites, which result from the metamorphism of sandstones. The Magaliesberg and Steenkampsberg mountains consist of this extremely hard rock type, which is why they form such prominent topography. They form the immediate floor to the Bushveld Complex in the eastern and western limbs respectively, so have suffered the highest degrees of metamorphism.

Other sedimentary rocks in the sequence include mudstone. When this rock is metamorphosed it recrystallises to make new minerals. One of these is called **andalusite** and consists of aluminium, silicon and oxygen. This mineral has a very high melting temperature (nearly 2 000°C), so it is used in the manufacture of refractory bricks to line furnaces that have to withstand extremely high temperatures, such as in steelmaking. Mines near Thabazimbi, Zeerust and Steelpoort exploit these metamorphosed shales and the andalusite is separated, cleaned and processed into bricks. At current mining rates, the deposits will last for several hundreds of years.

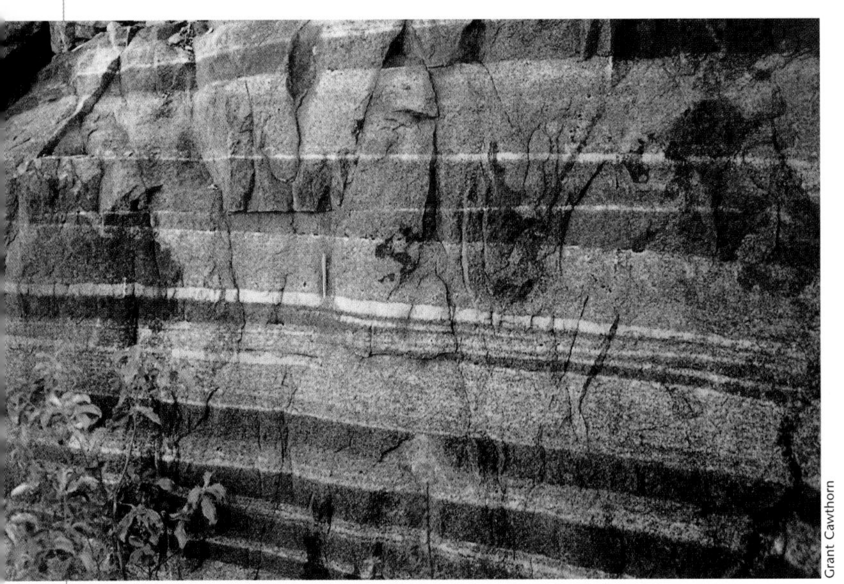

Figure 4.23 ***Very slow crystallisation, as well as periodic influxes of new basaltic magma into the Bushveld Complex, resulted in segregation of minerals of slightly different densities, in this case pyroxene (dark) and plagioclase (light), producing the layering in the Rustenburg Layered Suite.***

Basaltic magma erupts at a temperature of about 1 200°C, and it cools and solidifies very quickly. In fact, at 900°C it is totally solid. In the temperature interval between 1 200°C and 900°C crystals grow from the liquid. They grow very slowly, as the necessary elements or ions need to diffuse through the magma to the site of the growing crystal, and the elements that are not needed have to diffuse away. Magmas are quite viscous so diffusion rates are slow; hence the rate of crystal growth is also slow. A basaltic lava at 1 200°C has a viscosity comparable to that of warm syrup. As a result, in a lava or thin sill, the magma cools and solidifies before crystals of any significant size can form. A piece of lava looks very uninteresting; its crystals are too small to be seen with the naked eye and it looks black or dark grey and featureless.

If the cooling rate is slower – as might, for example, occur in the middle of a thick lava flow or a moderately thick sill – many small crystals start to grow. Fresh pieces of dolerite contain small brown and white needles, perhaps 1–2 mm long, that indicate the presence of crystals. The white mineral is plagioclase, a calcium aluminium silicate; the brown pyroxene, a magnesium iron silicate. Even slower rates of cooling allow bigger crystals to grow, producing gabbro. If the cooling rate is very slow, a different process becomes important: the crystals begin to sink down through the magma, a process with far-reaching consequences.

In a thick sill, pyroxene and plagioclase may sink towards the base of the intrusion. As a result, the layer of rock that forms at the base of the intrusion becomes enriched in plagioclase and pyroxene grains. The remaining liquid has lost these minerals and therefore has a different composition from that which was originally intruded. This process is called **differentiation** or **fractionational crystallisation**, because a fraction of the first-formed minerals separated from the original magma. Elements not incorporated in the minerals that are forming become enriched in the remaining magma.

Eventually different minerals will become stable because of the changing composition of the magma. If those minerals also sink, then a series of layers will form from the bottom upwards, which consist of different minerals. In effect, what differentiation does is to separate the constituent minerals crystallising from the magma into layers. In sills up to 300 m thick, the extent of this sinking and layering of crystals is not generally significant, but becomes more prominent as the thickness of the sheet increases.

Plagioclase and pyroxene have different densities (pyroxene is denser). As crystals sink through the magma the pyroxene may sink more quickly than the plagioclase. The result is that a composite layer may develop that has more of the denser pyroxene towards the bottom and more of the less dense plagioclase near the top. This segregation also results in mineral layering (**figure 4.23**).

Another important process is the intermittent addition of new magma into the chamber. At intervals while the magma is cooling, crystallising and producing layers, new, hotter magma may be added and mixed with the remaining, cooler magma – so the crystallisation sequence goes part way back to the crystallisation of higher temperature minerals. In this way a certain series of layers becomes repeated many times. Such a repeated sequence is called a **cyclic unit**.

Some of the best examples of these cyclic units, which embody the two processes of magma addition and crystal settling, can be seen near the bottom of the Rustenburg Layered Suite. They consist, from the bottom, of the sequence chromite (black), pyroxene (brown), pyroxene and plagioclase together, and plagioclase (white). The different colours of the minerals and rocks make the sequence very obvious (**figures 4.23, 4.24**). Such cyclic units may vary from a few metres to hundreds of metres in thickness. One of these cyclic units has at its base a layer that contains abundant platinum and other rare metals, and is known as the Merensky Reef (*see* 'Mineral wealth of the Bushveld Complex' on page 124).

The combination of cooling and crystallisation, crystal settling and sorting, and magma addition and mixing operated throughout the entire Rustenburg Layered Suite. It produced rocks consisting of a wide range of minerals and mineral proportions arranged in alternating layers. This segregation process led to the concentration of valuable minerals in the Bushveld Complex. Igneous bodies that display such layering are called **layered intrusions**.

The Kaapvaal Craton is composed of a rigid mass of rock about 40 km thick, which has resisted tectonic forces for nearly three billion years. It is made of rocks that have a bulk density of approximately 2.7 g/cm³. Basaltic rocks that formed the Bushveld Complex have a bulk density of about 3.2 g/cm³. The addition of 8 km of this dense material into the crust caused the base of the continent to sag into the hot, somewhat plastic, underlying mantle. This subsidence, called **isostatic adjustment**, caused the centre of the intrusion to sag and the layering to tilt inwards towards the centre of the complex to produce a basin shape. It affected not only the rocks of the Rustenburg Layered Suite, but also the underlying Transvaal Supergroup rocks (**figures 4.20, 4.22**).

Younger rocks subsequently covered the Bushveld Complex, but have been partially eroded and the gentle basin shape is now clearly evident. The complex crops out at surface in three very long arcs, from Thabazimbi to Pretoria in the west, from Mokopane /Potgietersrus to Middelburg in the east, and north of Mokopane/Potgietersrus (**figure 4.21**). This outcrop pattern gives rise to the names western, eastern and northern limbs (or lobes) of the Bushveld Complex.

The second intrusive event

The enormous amount of heat liberated as the basaltic magma flowed through the base of the crust from the mantle caused heating and melting of the deeper parts of the crust. This eventually intruded to higher levels, ponding directly above the

Terence McCarthy

Figure 4.24 ***Slow crystallisation occasionally resulted in complete segregation of certain minerals, which formed layers such as these alternating layers of chromite (a black chromium-iron oxide) and plagioclase (a white calcium aluminium silicate).***

Rustenburg Layered Suite and below a cap of Rooiberg Group rocks, where it crystallised to form granite. The granite is more resistant to erosion than the underlying rocks and so forms the high plateau north of Groblersdal, known as Sekhukhuneland, as well as the prominent ridge of hills running north from Mokopane/Potgietersrus. These granites contain some major mineral deposits, including tin and fluorite. Some quarries also produce dimension stone from these granites, which are sought after for their blood-red colour.

The Bushveld Complex is world famous for its spectacular rocks and valuable mineral resources. There is nothing unusual about the processes that took place inside the magma chamber as the Rustenburg Layered Suite formed, and the same processes are seen in igneous rocks all over the world. Elsewhere there are slight differences in the minerals involved and some details of the layering, but the general mechanisms are the same. What makes the Bushveld Complex unique is its scale. The layers are also much thicker and laterally more continuous than in any other layered intrusion in the world.

While the processes that created the layering are fairly well understood, we do not know what produced this vast quantity of magma in so short a space of time. Some process caused extensive and extremely rapid melting, initially of the crust, forming the extensive Rooiberg Group rhyolite, then of the mantle, and finally of the lower crust. It has been suggested by some that melting may have been caused by a hot mantle plume that rose beneath the continent from the core-mantle boundary. Others have suggested that rapid melting may have been triggered by impacts by large meteorites. In the absence of supporting evidence, however, these proposals remain conjectural.

PHALABORWA

A small hill situated in the lowveld of Mpumalanga, known as Loolekop, was the site of copper mining from as early as 750 AD. Indications of ancient mining attracted the attention of prospectors, including the famous geologist Hans Merensky. It was discovered that Loolekop contained not only copper, but also apatite (a phosphorous-bearing mineral) and vermiculite (widely used in horticulture and insulation). Exploitation of the copper began in earnest in the 1960s with the establishment of Palabora Mining Company (PMC). The State retained rights to the

Figure 4.25 ***A large igneous intrusion known as the Phalaborwa Complex formed 2 049 million years ago in Limpopo Province. This contains diverse rock types and has been mined for copper, phosphate and vermiculite. The open pit mine, 760 m deep, is one of the largest man-made holes in South Africa.***

apatite, as it considered phosphate to be a strategically important commodity (South Africa was dependent on imported phosphate at the time), and created FOSKOR to mine the apatite.

Loolekop has disappeared; in its place is one of the largest man-made holes in South Africa, 760 m deep (**figure 4.25**). Mining is carried out on a massive scale, and at one time PMC held the world record for the mass of rock moved in a 24-hour period – an amazing 521 433 tonnes. Phalaborwa's production meets South Africa's entire copper and phosphate requirements.

The Phalaborwa Complex, as the geological feature is known, is a volcanic pipe, a feeder to a large volcano that erupted 2 049 million years ago. The pipe intruded through granites and greenstones of the Kaapvaal Craton, which forms the surrounding country. The area around Phalaborwa is studded with numerous small pointed koppies. These are satellite intrusions to the main pipe. Imagine a tree, side branches included, all neatly cut off at half the tree's height – that is how the intrusive complex would look without the surrounding granite. The main trunk is the Phalaborwa Complex itself and the side branches would represent the surrounding intrusions.

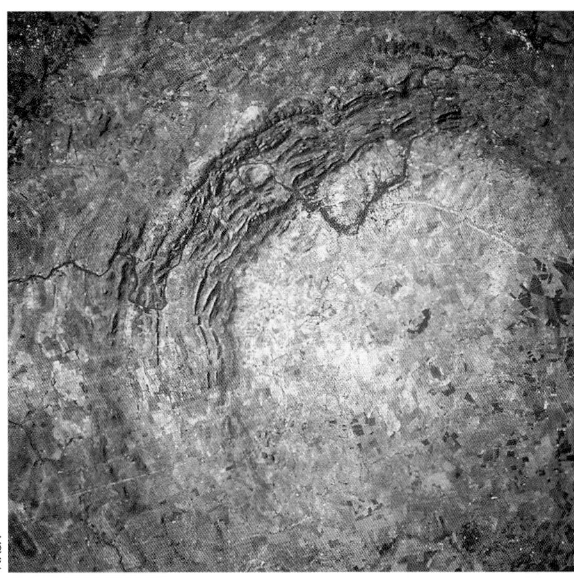
NASA

Figure 4.26 ***About 2 023 million years ago, a large asteroid struck the Kaapvaal Craton, forming a crater 300 km in diameter, the largest impact structure known on Earth. Erosion has since removed the crater, exposing the rocks that lay beneath the original floor. The semi-circular mountains of the Vredefort Mountain Land, seen here from the NASA space shuttle, surround the uplifted core of the crater. The southern half of the crater core is covered by younger rocks.***

The Phalaborwa Complex belongs to the alkaline family of igneous rocks, rocks that are rich in alkali metals and deficient in silica. The complex hosts a truly bizarre variety of rocks, including one first recognised at Phalaborwa and known as foskorite. It also contains carbonatite, a very unusual igneous rock made of calcium carbonate, chemically the same as limestone. The different rock types are arranged in concentric cylinders in the main pipe and in some of the satellites. Only the central core, which consists of carbonatite, contains copper, and is surrounded by phosphate-rich foskorite. Thus, two major and totally different ores are mined side by side at Phalaborwa.

THE VREDEFORT CATASTROPHE

The exceptional and largely continuous record of rock formation on the Kaapvaal continent from 3 600 to 2 060 million years ago indicates that geological processes operating in this early period of Earth history were basically similar to what we see today, with the main driving force being the movement of crustal plates above slowly convecting mantle. However, a little more than two billion years ago, a geological event took place in the centre of the continent that bears no relation to these plate tectonic processes; a large explosion occurred that, quite literally, shook the world. The little Free State town of Vredefort near Parys, some 120 km southwest of Johannesburg, is located at ground zero.

Geologists have known about the unusual character of the rocks around Vredefort for more than a century, but the real cause was only confirmed in the 1990s. The most obvious features, visible from an aircraft and especially from space, are near-perfect semicircular ranges of hills in an arc about 80 km in diameter. This is called the Vredefort Mountain Land (**figure 4.26**). More than a century ago, geological mapping showed that this Mountain Land is due to the presence of resistant quartzite layers of the

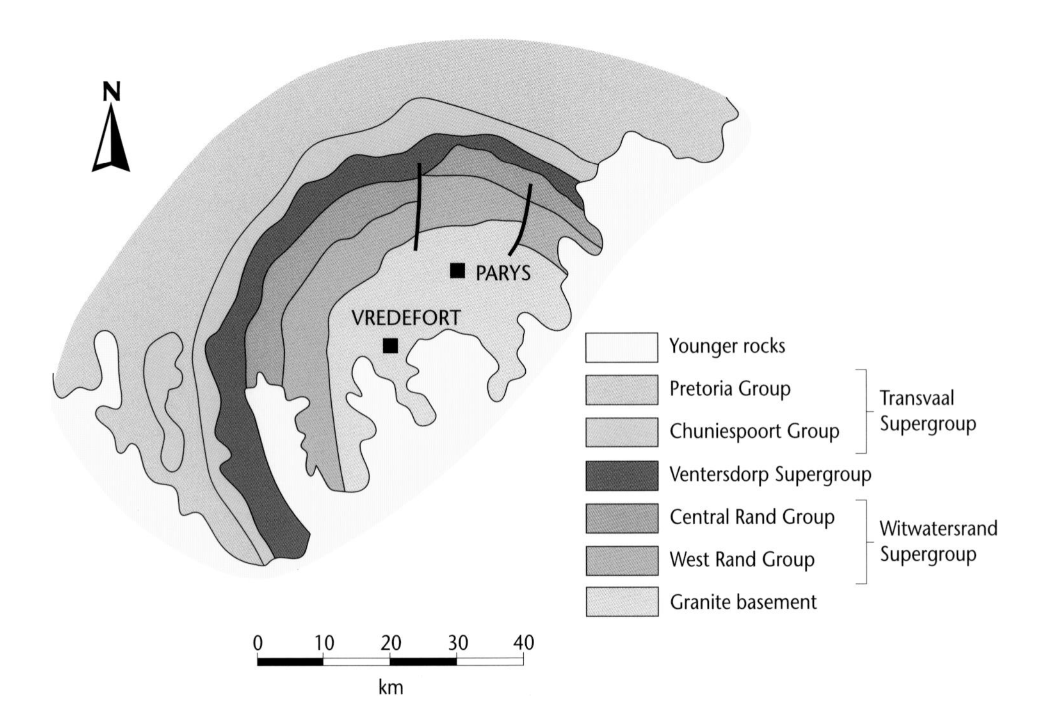

Figure 4.27 *A geological map of the core of the Vredefort structure. The southern half is covered by younger rocks. Originally, horizontal layers were pushed up and eroded, forming concentric rings.*

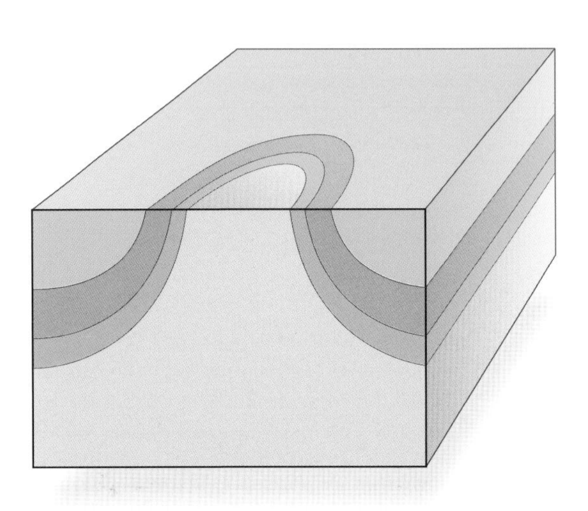

Figure 4.28 *This diagram illustrates how the rise of underlying rock through overlying rock layers produces the circular distribution of the layers, a characteristic feature of a dome.*

Witwatersrand Supergroup that produce ridges, with intervening softer mudstones that produce valleys.

What is unusual is that these sedimentary layers, which are normally gently inclined elsewhere in the region, are turned on end in this case. Although these layers are hidden to the south beneath younger rocks of the Karoo Supergroup (**figure 4.27**), geological and geophysical mapping has shown that the rocks of the Witwatersrand, Ventersdorp and Transvaal Supergroups encircle a core of Archaean granites and greenstones that are somewhere between 3 400 and 3 200 million years old. Such a geological structure, in which the oldest rocks are found in the centre, is called a **dome** (**figure 4.28**).

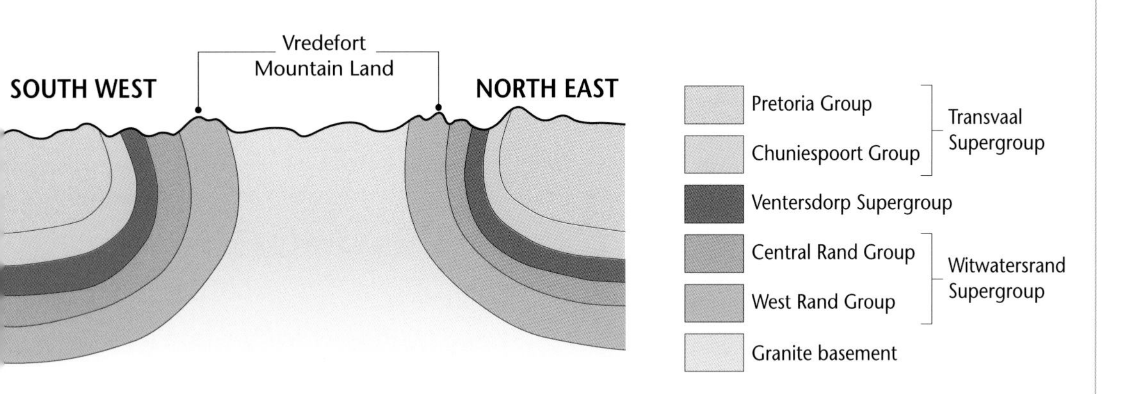

Figure 4.29 ***This geological cross-section through the Vredefort Dome illustrates the overturned layers of Witwatersrand, Ventersdorp and Transvaal Supergroup strata.***

Creation of the Dome

The creation of the Vredefort Dome involved the upturning of some 20 km of rock layers. The basement granite punched through the overlying, horizontally stacked rocks of the Dominion, Witwatersrand, Ventersdorp and Transvaal Supergroups, overturning the layers in the process (**figure 4.29**). The sheer magnitude of this uplift points to the unusual origin of the dome. Spectacular fracturing and jointing in the layered rocks, as well as in the core granite itself, indicate that the granite was solid at the time and that the process must have been extremely rapid. Among the features generated are highly unusual conical fracture surfaces, known as shatter cones (**figure 4.30**), which have been noted at many, but very localised, sites around the world.

Figure 4.30 ***Many rocks in the Vredefort area have been fractured along concentric, conical surfaces, as illustrated here. These are known as shatter cones and resulted from an intense shock wave that passed through the rock mass.***

Many fractures are filled by rock fragments set in a dark, fine-grained or even glassy matrix (termed **breccias**). These breccias are of two types. **Pseudotachylite** breccias occur in veins from less than 1 mm to as much as 100 m thick, and are the most spectacular example of this rock type known (**figure 4.31**). Pseudotachylite is the name given to the dark matrix of the breccia, which resembles a volcanic glass known as tachylite. It is prefixed 'pseudo' because Vredefort is not a volcanic environment. Pseudotachylite is so abundant in the Vredefort Dome that it is difficult to find a rock that does not contain at least a small vein of the black material, as can be seen in the polished slabs of Vredefort granite that line many of the pillars at Johannesburg International Airport.

Roger Gibson

Figure 4.31 *The rocks at Vredefort contain veins of black, glass-like material known as pseudotachylite, which is often charged with partly rounded or angular fragments of the surrounding rock. Pseudotachylite forms as intense shock waves cause local melting of the rock.*

The second breccia type, the Vredefort **granophyre**, occurs as large dykes several kilometres long and many metres wide. In both breccia types, the matrix was derived by instantaneous melting of rock, followed by rapid cooling. Pseudotachylite has the same composition as the surrounding rock and contains fragments of the host rock, indicating that it was produced by local melting. In contrast, the granophyre dykes have compositions resembling no known rock type and their fragments come from far afield, indicating large-scale movement of the melt. Chemical evidence indicates that they contain a small amount of extraterrestrial material. Dating of the pseudotachylite and granophyre has established that the Vredefort event occurred 2 023 million years ago.

Roger Gibson

Figure 4.32 *This microscopic view of a rock from the Vredefort area shows the shock damage to the crystal structures of minerals in the rock (the grid-like pattern in the central grain, which is 0.2 mm long).*

In addition to large-scale fracturing and brecciation of rocks, there is also evidence of severe damage to the minerals in the rocks on a microscopic scale (**figure 4.32**). This damage arises from the collapse of the crystal lattice structure (the atomic scaffolding of crystals) in response to tremendous pressure in an extremely powerful shock wave that must have passed through the rocks. In places, the pressures were so great that the common mineral quartz has been compressed into unusual ultra-dense forms known as coesite and stishovite. Stishovite only becomes stable at great pressure, a pressure equivalent to the weight of a column of rock at least 300 km high, and therefore does not normally occur in the Earth's crust.

How did it happen?

Vredefort is clearly a site where major damage has occurred to the rocks. The gross structure indicates that the basement rocks rose up rapidly from below during this event, punching through the overlying rocks. This movement was associated with an immense explosion that compressed and smashed the rocks in the vicinity, causing localised melting (pseudotachylite) and shatter cones in the process.

The theory of plate tectonics offers no explanation for the rapid rise of basement and the release of such a gigantic amount of energy at a single place, as clearly happened at Vredefort. But plate tectonics is not the only large-scale geological process operating in the Solar System: one only has to look at the surface of the Moon through a pair of binoculars to see the effects of another important geological process – asteroid and cometary bombardment. The impact of a large asteroid could certainly generate the intensity of shock wave that is indicated by the damage in the rocks around Vredefort, including extensive, shock-induced melting. It would also account for the extraterrestrial component in the granophyre. But why the upward movement of the basement if the impact came from above?

It is tempting to compare the circular shape of the Vredefort Dome, with its largely flat core region and surrounding ring of rugged hills, with photographs of well-preserved impact craters such as Tswaing north of Pretoria or those on the Moon (**figure 4.33**).

***Figure 4.33A** A typical lunar crater showing the slumped rim, flat crater floor, and the uplifted central cone known as the central uplift.*

Terence McCarthy

***Figure 4.33B** The well-preserved terrestrial crater at Tswaing near Pretoria.*

IMPACT CRATERS

The impact of material from space is the most widespread geological process in our Solar System. The Moon shows the scars of hundreds of thousands of such impacts in the form of craters. Although many of these craters date from the early stages of formation of our Solar System, impact is still occurring today, but mostly by microscopic particles.

The Earth, too, is subjected to this constant rain of extraterrestrial material; each day several tons of extraterrestrial debris enters Earth's atmosphere. Most of these bullets from space (they travel at 10–70 km per second, compared with a rifle bullet that typically travels at 100–300 m per second) are asteroids – rocky fragments that are scattered between the orbits of Mars and Jupiter. However, some are comets consisting of loosely consolidated aggregates of dust and ice that periodically enter our Solar System from outer space. The smaller fragments melt and disintegrate as a result of friction with the atmosphere, producing shooting stars, but larger fragments of asteroid may survive to hit the Earth's surface.

If they are not too large, these fragments – called **meteorites** – may survive largely intact after hitting the ground. However, those meteorites and comets with diameters of more than a few tens of metres are not sufficiently slowed down by the atmosphere before hitting the Earth's surface, so they impact with tremendous energy release. This creates a characteristic circular crater filled with broken and fused rock. Investigation has shown that a 50-m-diameter meteorite or comet would release as much energy as a 10-megaton nuclear bomb (one megaton has the equivalent explosive power of one million tons of TNT) and form a 1-km-wide crater, while a 10-km-diameter body is likely to produce a crater more than 150 km in diameter.

Large impact craters are more than simple holes punched in the ground by a falling rock. At the point of impact the momentum of the object is transferred to the ground and the meteorite and target rocks are melted or even vaporised almost instantly. The impact creates a powerful shock wave that passes outwards from the point of impact. Travelling at supersonic speeds, the shock wave passes through the rock in the vicinity of the impact site within a few seconds. As it passes, the compressed rock beneath the crater rebounds upwards, expelling a huge amount of material and carving a large hole. The steep walls of the hole rapidly slump inwards, approximately doubling the original crater diameter and greatly reducing its depth. Collapsed material from the walls mixes with melted and fractured rock in the crater and with ejected fragments that fall back from above. The result is a fairly flat crater floor. Material blasted from the crater also falls outside, forming an ejecta blanket around the crater.

In craters larger than a few kilometres in diameter, the centre becomes raised above the surrounding floor during the rebound. This is because interaction between the shock wave and the Earth's crust in large craters creates a zone of enhanced rebound directly below the point of impact. This generates a dome-shaped central uplift during rebound that is between one-quarter and half the diameter of the crater. An analogy is provided by a drop of water falling onto a water surface. First, a hole with a crown-like ring of droplets forms, analogous to the initial impact. This is quickly followed by closure of the hole and a spout of water centred on the point of impact, analogous to the formation of the central uplift.

Such a comparison suggests that the Vredefort structure is a crater 70–80 km wide, but this is incorrect. The Vredefort Dome, in fact, represents only the central uplifted core of a much larger impact crater (*see* 'Impact craters', above).

Such central uplifts are common in large craters on the Moon and typically have diameters from one quarter to one half of the crater itself. Applying this relationship to the Vredefort Dome results in an original diameter for the Vredefort

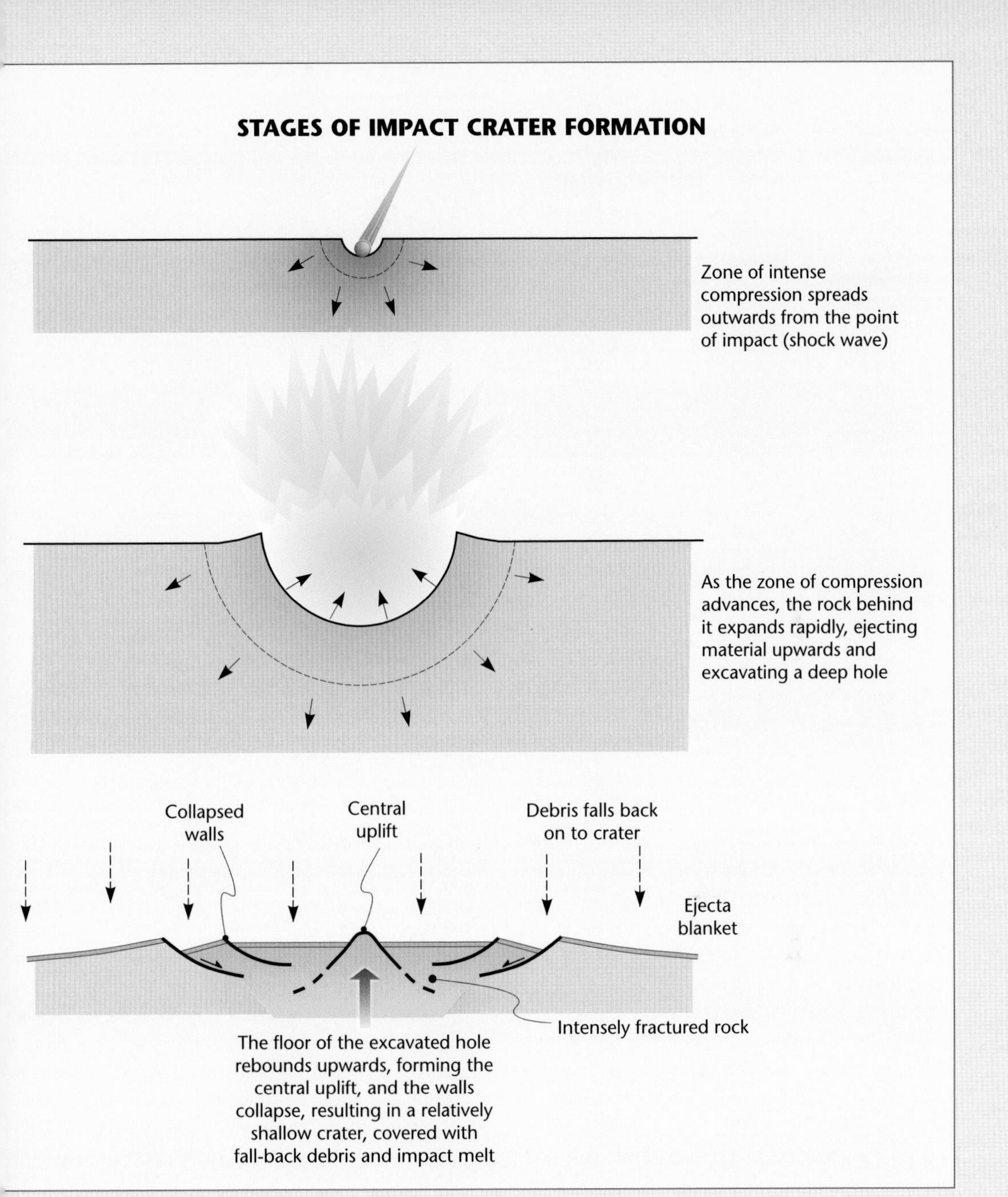

impact crater of between 150 and 300 km. Current geological estimates suggest the crater was probably closer to 300 km in diameter, making this the largest known impact crater on Earth. It is also the oldest known.

Johannesburg today occupies a position near the rim of the original crater (**figure 4.34**), but the crater itself is no longer there. In the 2 000 million years since its formation, erosion has removed the crater and its impact breccias, leaving only the

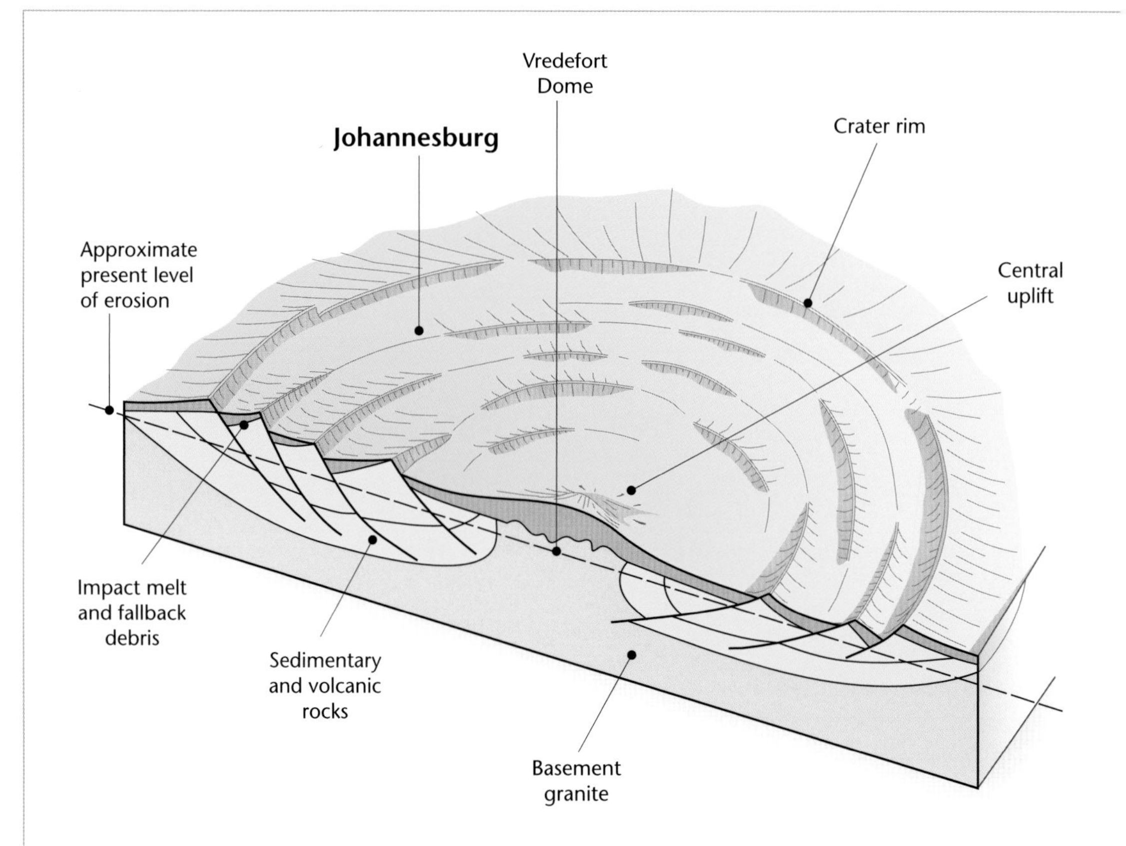

Figure 4.34 ***The semi-circular Vredefort Dome is the deeply eroded root zone of the central uplift of a very large crater, perhaps as much as 300 km in diameter. This reconstruction illustrates what the crater may have looked like before erosion.***

deeper layers that lay beneath the actual crater floor (**figure 4.34**). Similar fates have befallen many of the world's impact craters, including many that formed after the Vredefort event; however, the sheer size of the Vredefort impact has guaranteed that some evidence still remains today (*see* 'Vredefort fact file', right).

The formation of the central uplift at Vredefort created a circular, trough-like feature around it as the surrounding rocks were drawn downward and inward towards the centre by the rise of the central uplift. Erosion of the upper portions of the crater is complete, but this circular trough remains because of its depth, and is marked by the crescent-shaped region of Transvaal Supergroup rocks around the Vredefort core (**figure 4.19**). This infolding had an extremely important consequence: it protected the underlying Witwatersrand Supergroup rocks from subsequent erosion. Had the Vredefort impact not occurred, it is very likely that much of the Witwatersrand Basin, and particularly the gold-bearing upper Central Rand Group, would have been destroyed by erosion and South Africa would not have had a major gold mining industry. The preserving effect that the structure has had on the basin is the reason why the Vredefort Dome occupies a central position in the Witwatersrand Basin (**figure 4.4**).

The impact on life

Large impacts may result in mass extinction of species. Several such mass extinctions have been recorded, although only the famous dinosaur extinction has, to date, been convincingly linked to an impact. The Vredefort event, far larger than that which ended the dinosaur reign, could have had similar effects. However, at that stage life forms were very primitive and the food web was probably very simple. Species that relied on photosynthesis, such as those that built the stromatolites, could potentially have been severely affected because dust from the impact would have blotted out the Sun for many months.

To what extent they could remain dormant for this period is unknown, but it is likely that most would have been able to survive the long-term effects of the impact. After all, severe impacts were not new to the cyanobacteria: they had survived the numerous cosmic impacts that occurred during earlier times, such as those recorded in the spherule layers at Barberton (*see* Chapter 3). Anaerobic species that relied on exhalations of sulphur and other chemicals at volcanic vents for their energy source (*see* Chapter 6) would have been largely immune to the impact. Hence, the Vredefort event, spectacular as it must have been, may well have had little effect on the course of the evolution of life.

WESTERN MARGIN OF THE CONTINENT

Some time around 2 000 million years ago, the western edge of the fused Kaapvaal-Zimbabwe Craton lay along a northeasterly line extending from the Northern Cape through central Botswana and western Zimbabwe (**figure 4.9**). The western edge of the Kaapvaal Craton had been the site of deep-water deposition of iron formation, part of the Transvaal Supergroup. Beyond this to the west lay oceanic crust.

Commencing around 1 900 million years ago, shallow-water marine sediments – represented today by limestones, dolomites, quartzites and mudstones – began to accumulate along this margin, indicating a change in geological environment from relatively deep-water conditions to a shallow continental shelf. Terrestrial deposits in the form of red-coloured sandstone that formed in riverine settings also accumulated at this time as the sea became shallower and the shoreline built seaward. These deposits collectively form the **Olifantshoek**

VREDEFORT FACT FILE

Age: 2 023 million years old (oldest preserved impact structure on Earth)
Original crater dimensions: ±250–300 km wide, ±5 km deep (the Vredefort Dome is only the central uplift of the structure)
Excavated volume: ±70 000 km^3
Energy release: 100 million megatons (one megaton is the equivalent of the explosive power of one million tons of TNT)
Total duration of event: ±4 min
Impactor diameter: ±10–15 km
Seismic event caused by impact: Magnitude 14 (100 000 times more powerful than the 1976 Tangshan earthquake; *see* Chapter 11)
Extent of ejecta blanket: 350-km radius (individual blocks up to 4-km diameter)
Environmental effects: Burning off of some atmosphere and water; secondary earthquakes, landslides and volcanism globally, tsunamis up to hundreds of metres high; rain of molten rock; global cooling (months to years) caused by upper atmosphere dust blanket; possible global acid rain.

Note: Although evidence suggests the Vredefort impact structure is the largest on Earth, there are impact craters on the Moon that are up to six times larger. Traces of such impact structures may yet be found on Earth, or they may have been destroyed by subsequent geological activity. Recent research has suggested, however, that the very existence of our Moon is related to a glancing blow between Earth and a 3 000-km-diameter rogue planetoid that, had it hit head-on, may have destroyed our planet in its infancy!

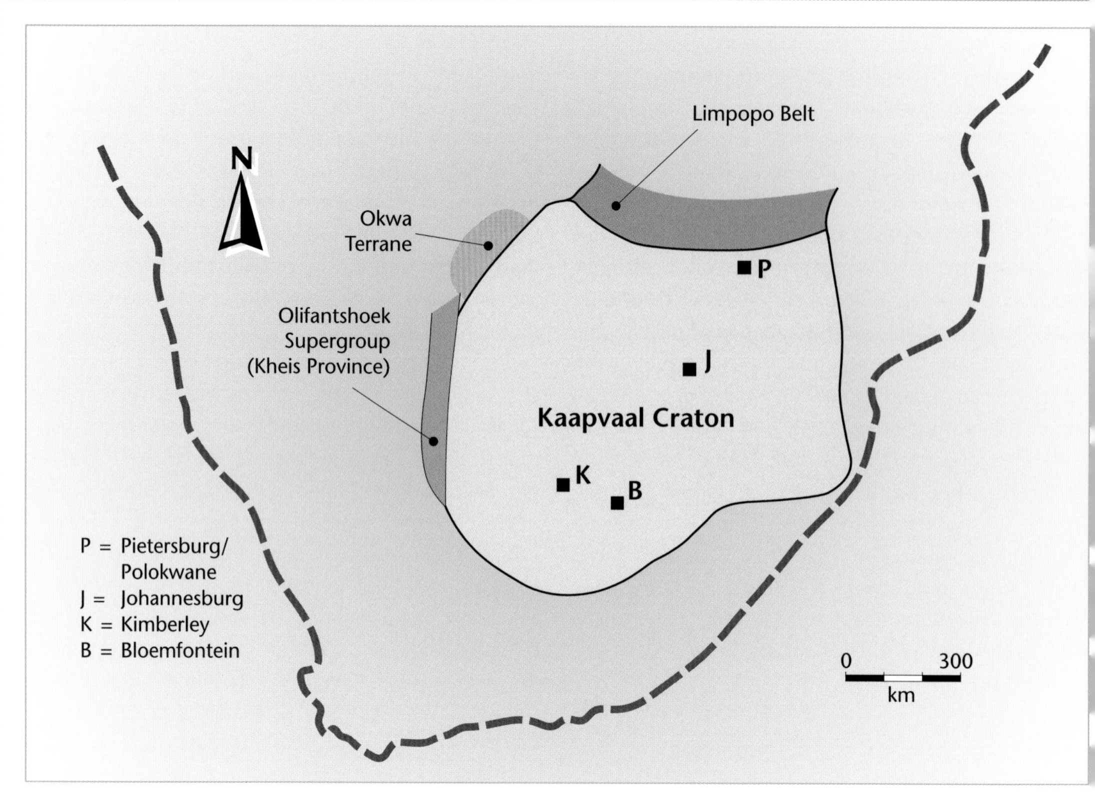

Figure 4.35 ***Deep-water sediments had collected along the western margin of the Kaapvaal Craton during formation of the Transvaal Supergroup. As a result of a change in geological environment some 2 000 million years ago, shallow-water sediments began to accumulate along this margin, forming the Olifantshoek Supergroup of the Northern Cape and the Okwa Terrane of Botswana. This shallowing of the sea may have been related to the approach of a continental mass from the west towards the western margin of the Kaapvaal-Zimbabwe Craton. (The coastline of South Africa and the locations of cities are shown for reference.)***

Supergroup (**figure 4.35**). Similar deposits were forming along the western margin of the Zimbabwe continent at this time (Magondi and Piriwiri Groups) and in central Botswana (Okwa Terrane).

Around 1 800 million years ago, this margin appears to have been involved in a major collision known as the **Ubendian event**, in which the Olifantshoek sedimentary rocks became intensely folded and squeezed eastwards, partly onto the Kaapvaal Craton (**figure 4.36**). The older Transvaal Supergroup rocks were also folded during this compression. Similar compression occurred along the Zimbabwe Craton margin. Further to the west, several island arcs and slivers of oceanic crust were amalgamated onto the continent, and granite magmas intruded into the crust.

This major compressional event appears to have been related to the closing of an ocean and the collision of the Congo Craton with the Kaapvaal-Zimbabwe Craton (**figure 4.37**). The resulting region along the western edge of the Kaapvaal Craton, characterised by sedimentary rocks that have been metamorphosed by heat and pressure, is known as the **Kheis Province**. The Ubendian event created a wide platform of continental crust along the western margin of the Kaapvaal-Zimbabwe Craton, the **Ubendian Belt** (**figure 4.37**) – a metamorphic belt similar to the Limpopo Belt between the Zimbabwe and Kaapvaal Cratons. Details of the history of this event remain obscure and controversial, however, because of subsequent developments that overprinted and largely obliterated the earlier history.

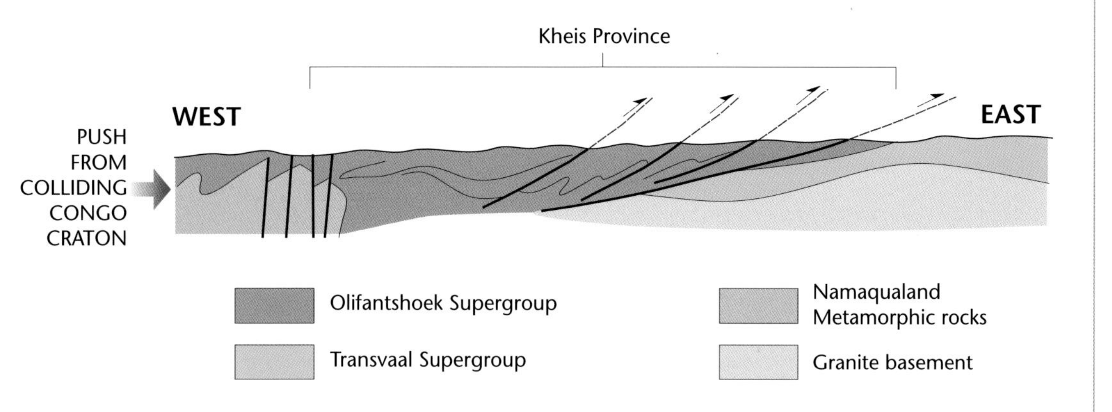

Figure 4.36 *About 1 800 million years ago, rocks of the Olifantshoek Supergroup became intensely folded and thrust eastwards onto the Kaapvaal Craton. At the same time, granite magmas intruded into the crust in this area, producing a belt of folded and metamorphosed rocks along the western margin of the Kaapvaal Craton. This is known as the Kheis Province.*

AN OXYGENATED ATMOSPHERE

In the interior of the Kaapvaal Craton, the run-up to the Ubendian period of continental growth seems to have been characterised predominantly by erosion, but how much material was stripped away is difficult to establish. The depth of erosion was probably quite variable over the continent. Nevertheless, the thick piles of sedimentary and igneous rocks deposited during the preceding 1 000 million years, although reduced in size, were not completely removed, and formed a geologically quite heterogeneous floor on which the last great sedimentary deposit of the Proterozoic Eon was deposited.

The age of this new deposit has not been well established because of the scarcity of suitable materials for dating, but appears to have formed in the interval between 1 900 and 1 700 million years ago. The sedimentary rocks deposited in this period form the spectacular scenery of the Waterberg and

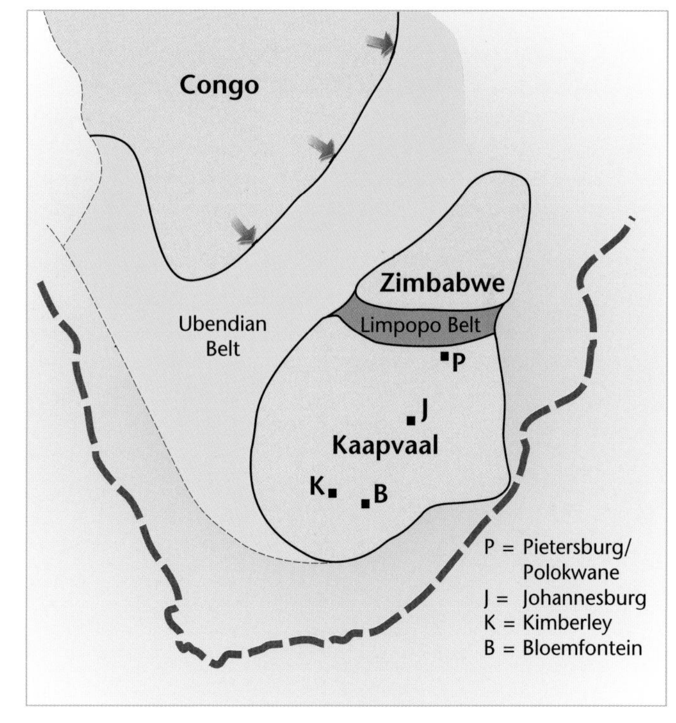

Figure 4.37 *The folding and granite intrusion that took place along the western margin of the Kaapvaal Craton 1 800 million years ago may have been caused by the collision of the Congo Craton with the Kaapvaal-Zimbabwe Craton, and is known as the Ubendian event.*

Figure 4.38 ***The resistant sandstones of the Waterberg and Soutpansberg Groups give rise to spectacular scenery in the Limpopo and North West Provinces. These rocks were deposited about 1 800 million years ago on a vast alluvial plain that covered much of the Kaapvaal Craton. This period of deposition may have been initiated by the collision of the Congo and Kaapvaal-Zimbabwe Cratons.***

Soutpansberg mountains (**figure 4.38**). The **Waterberg Group** of rocks today occurs in several separate regions: in the Limpopo Province extending northwards and northwestwards from the town of Bela-Bela/Warmbaths as far as eastern Botswana; and to the north of Middelburg in Mpumalanga (**figure 4.39**). These separate patches probably originally formed a single sheet of sedimentary rocks that has since become fragmented as a result of erosion. Maximum thickness of over 7 000 m is attained in the Bela-Bela/Warmbaths area.

The rocks consist almost entirely of sandstone, with minor conglomerate layers that locally attain several metres in thickness. Mudstones are rare. The most distinctive feature of these rocks is their red colouration: the rocks are either totally permeated by a deep red iron oxide, or certain planes within the rock, such as bedding planes and cross beds, are stained red (**figure 4.40**). The rocks are chemically resistant and very hard, so they produce spectacular cliffs and mountainous topography.

The rocks forming the Soutpansberg mountains, the **Soutpansberg Group**, also consist of these red-coloured sandstones, but here they are underlain by voluminous basaltic lavas, with interlayered volcanic ash beds of rhyolite. These lavas erupted about 1 900 million years ago.

The rocks of the Waterberg and Soutpansberg Groups are the Earth's oldest so-called **red beds**, indicating that the rocks were deposited under an atmosphere that contained free oxygen, the product of thousands of millions of years of photosynthesis by cyanobacteria. The oxygen-bearing atmosphere converted trace amounts of iron in the sediment to the red, ferric oxide form, which stains and locally even cements the sand grains. Before this, oxygen produced by the cyanobacteria had been consumed by iron and manganese dissolved in the oceans and

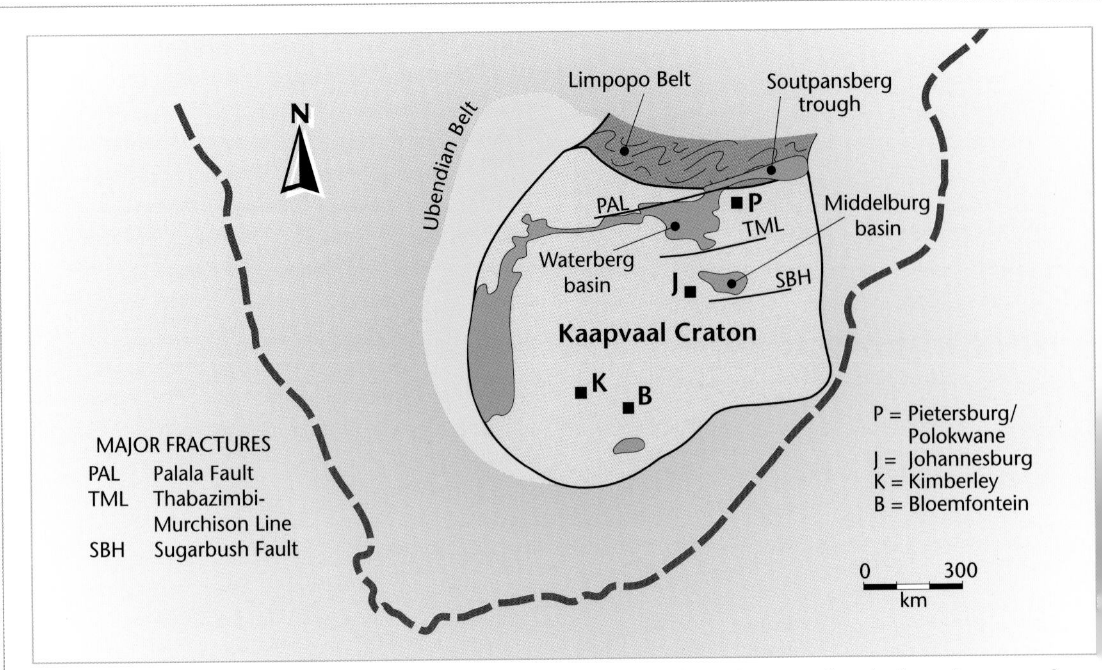

Figure 4.39 ***The originally extensive Waterberg and Soutpansberg Groups of rocks have been much reduced by later erosion and today occur only in restricted areas as shown in this map. (The coastline of South Africa and the locations of cities are shown for reference.)***

Morris Viljoen

Figure 4.40 ***The sandstones of the Waterberg and Soutpansberg Group are stained red by iron oxide, indicating that they were deposited under an atmosphere containing free oxygen. These rocks are the oldest so-called red-beds known on Earth, and record the first appearance of substantial quantities of oxygen in the Earth's atmosphere.***

precipitated as iron formations, and little free oxygen entered the atmosphere. But as this sink became exhausted, oxygen began to leak into the atmosphere, expressing its presence in the form of red beds.

Virtually the entire Waterberg and Soutpansberg Groups were deposited by rivers flowing on a vast plain, which may have extended over much of the Kaapvaal Craton. Many channels appear to have developed, which divided and rejoined in a braided network. Flow was predominantly to the south and southwest, suggesting that the sediment was derived from the north. Occasionally, strong winds shaped the sand into dunes up to 17 m high (there was still no land vegetation at that time) and there are indications that small lakes formed locally.

The cause of this period of deposition of sediment is not known, except in the case of the Soutpansberg Group, which shows the hallmarks of rifting in the form of mixed basalt and rhyolite volcanic activity, as well as large faults extending along the length of

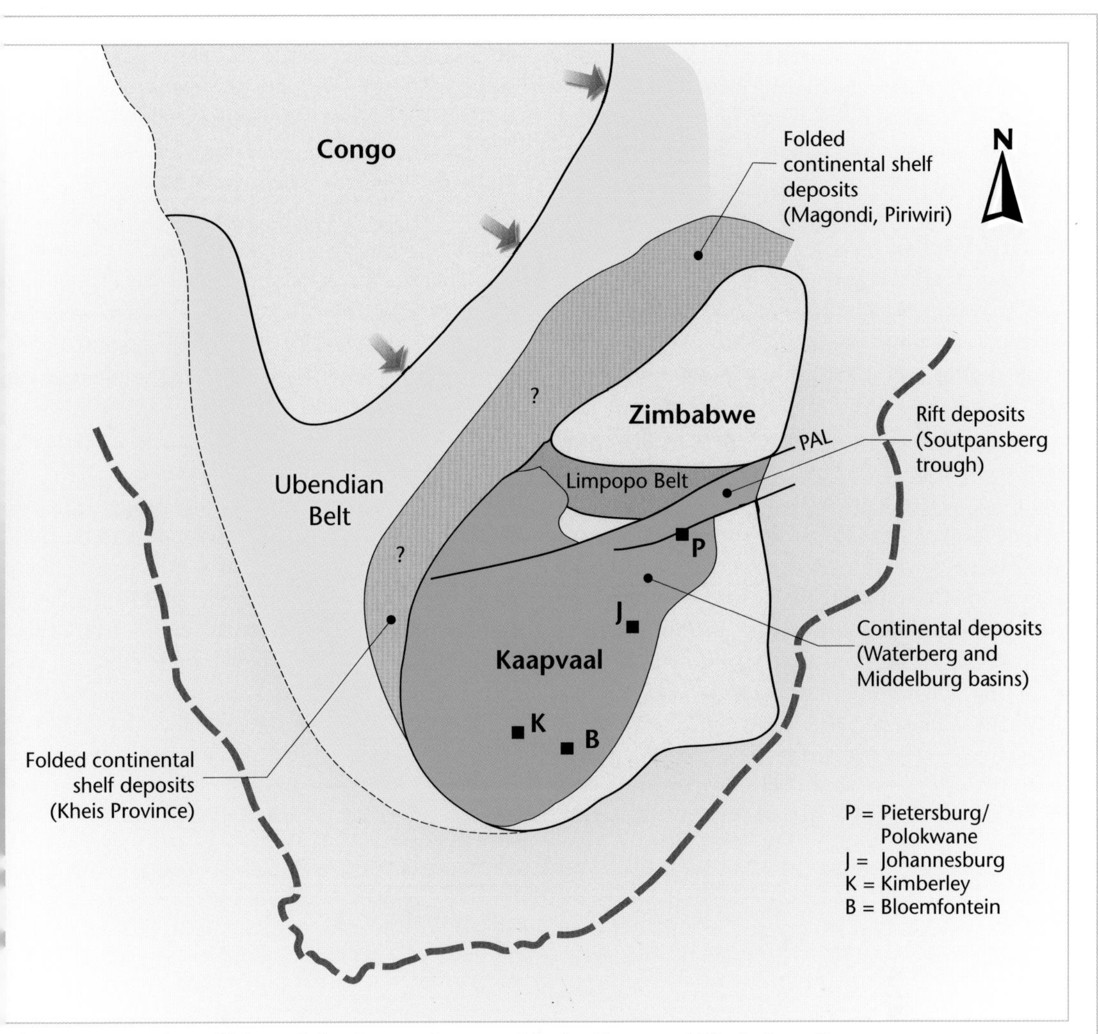

Figure 4.41 ***The collision of the Congo Craton with the Kaapvaal-Zimbabwe Craton may have induced rupturing of the latter continent and the development of rifts. The Waterberg and Soutpansberg Group sediments were deposited in the resulting depressions. Folding of the sedimentary rocks along the western continental margin may also have been caused by this collision.***

the mountain range and beyond to the west. It is possible that the entire riverine plain was a response to a period of incipient but widespread rifting of the continent, but only the Soutpansberg region developed the characteristics of a true rift. There are indications that other faults – notably the Thabazimbi-Murchison structure, the Palala fault and possibly the Sugarbush Fault – were activated during this rifting, creating asymmetric troughs that today accommodate the Middelburg and Bela-Bela/ Warmbaths remnants of the Waterberg Supergroup (**figure 4.39**).

Deposition of the Waterberg and Soutpansberg Groups seems to coincide with the Ubendian event in the west, and it is possible that collisional events in the west created rifts on the Kaapvaal-Zimbabwe Craton (**figure 4.41**), perhaps in a similar way to the formation of the Ventersdorp Supergroup.

The Waterberg and Soutpansberg Groups are the last of the major sedimentary accumulations to form on the Kaapvaal Craton in the Proterozoic Eon. In fact, apart from scattered igneous intrusions, no further major geological event is recorded on the craton until about 300 million years ago. Among these intrusions are some important rocks: diamond-bearing kimberlite pipes near Postmasburg (1 600 million years old) and at Cullinan (the Premier Mine, 1 200 million years old,) and the larger Pilanesberg Complex and Timbavati Gabbro.

PILANESBERG

The Pilanesberg, 50 km northwest of Rustenburg, is best known as a game reserve, but geologically it represents the root zone of a 1 200-million-year-old volcano. The rocks belong to the alkaline family and are rich in sodium and potassium. The Pilanesberg volcano is the largest of several alkaline igneous intrusions scattered across the North West and Mpumalanga Provinces.

The form of the intrusion can be likened to a car windscreen cracked by a stone: cracks radiate out from the point of impact, but close to the impact the cracks form concentric rings. In the case of Pilanesberg, impact came from inside the Earth, generated by magma trying to escape through brittle rock. The radiating cracks were filled to produce dykes that can be traced for up to 100 km to the southeast and northwest. Molten rock also filled the ring fractures, forming so-called ring dykes. The magma welled up to surface, creating a large volcano, while some magma crystallised more slowly inside the volcano, producing coarser-grained rocks.

The volcano intruded through rocks of the Bushveld Complex (**figure 4.21**). Erosion has stripped off most of the lavas, as well as the surrounding Bushveld Complex rocks, but some lava still remains. The alkaline rocks of the Pilanesberg Complex are chemically more resistant to weathering than the Bushveld rocks, so the Complex stands as a circular cluster of hills above the plain of the surrounding Bushveld.

THE ROCKS OF TIMBAVATI

In the southwestern portion of the Kruger National Park are several low ridges that appear to be composed of neatly piled boulders of dark grey to black rock. These rocks are known as the **Timbavati Gabbro** (**figure 4.42**). They are in fact dykes, formed when basaltic magma intruded through vertical fractures in the crust and crystallised slowly, forming fairly coarse-grained gabbro.

The magmas intruded about 1 100 million years ago. Dyke and sill intrusions of this age are very widespread. They occur in the northern part of the

Morris Viljoen

Figure 4.42 *Dykes of Timbavati Gabbro form boulder-strewn ridges that are usually seen in the Kruger National Park in Mpumalanga.*

Kruger National Park, in the Vredefort area, in eastern Botswana, and very extensively in the eastern highlands of Zimbabwe. Their widespread occurrence suggests that they represent a significant event in the history of the Kaapvaal-Zimbabwe Craton, one that must have produced extensive lava flows, perhaps covering much of the craton. Little of this lava-field remains, and small remnants associated with sandstones are confined to eastern Zimbabwe (the Umkondo Group). Rocks of this period have received little attention and their significance in the geological evolution of southern Africa is unknown.

KAAPVAAL CRATON – REMARKABLE GEOLOGICAL TERRANE

The Kaapvaal Craton is geologically unique. It preserves a record of geological events extending from 3 600 million years ago when the Earth's first continents were forming, through a number of rift events when the craton became inundated by the sea. This resulted in sediment accumulation: the Dominion Group 3 075 million years ago; the Witwatersrand and Ventersdorp Supergroups 2 970 to 2 700 million years ago; and the Transvaal Supergroup 2 600 to 2 100 million years ago. The last-mentioned period records the climax of the cyanobacteria, which gave the Earth its oxygenated atmosphere.

Moreover, the Kaapvaal Craton preserves two major and economically important igneous intrusions: the Bushveld and Phalaborwa Complexes 2 060 million years in age. It preserves the largest known meteorite impact structure on Earth, the Vredefort Dome, 2 023 million years old. Late rifting gave rise to the Waterberg and Soutpansberg Groups, 1 900 to 1 700 million years ago, which reveal an oxygenated atmosphere for the first time; and to the Pilanesberg Complex, a 1 200-million-year-old volcano. All of these rocks are incredibly well preserved; they appear today in virtually the same condition as when they were formed. They have been reduced in area to some extent by erosion, but the fact that large remnants remain suggests that the land surface lay close to sea level for close on 1 000 million years. The extensive preservation of these formations is a puzzling feature of the geological history of the region, given the fact that their cumulative thickness exceeds 20 km (*see* 'Why the continental crust is 35 km thick' on page 51).

Also remarkable are the quantity and variety of mineral deposits the rocks of the craton contain: the vast gold deposits of the Witwatersrand Supergroup, the largest on Earth; the platinum, chrome, vanadium and fluorspar deposits of the Bushveld Complex and the associated refractory mineral deposits, all of which are the largest known on Earth; the huge iron and manganese deposits of the Transvaal Supergroup; the diamond deposits of Premier mine, and the substantial apatite, copper and vermiculite deposits of the Phalaborwa Complex.

This variety and quantity of mineral deposits is without equal anywhere in the world. Little wonder that the Kaapvaal Craton has been described as the richest piece of real estate on Earth. But that is not all. In later times, the craton also became host to major diamond, titanium and coal deposits. These are discussed later in this book (*see* Chapter 9).

SUGGESTED FURTHER READING

Cairncross, B and Dixon, R. 1995. *Minerals of South Africa*. Geological Society of South Africa, Johannesburg.

Eales, HV. 2001. *A First Introduction to the Geology of the Bushveld Complex*. Council for Geoscience, Pretoria.

Eriksson, PG, Alterman, W, Catuneanu, O, Van der Merwe, R and Bumby, AJ. 2001. 'Major influences on the evolution of the 2.67–2.1 Ga Transvaal Basin, Kaapvaal Craton.' *Sedimentary Geology*, vol 14, pp 205–231.

Johnson, MR, Anhaeusser, CR and Thomas, RJ. 2005. *The Geology of South Africa*. Council for Geoscience, Pretoria.

Reimold, WU and Gibson, RL. 2005. *Meteorite Impact*. Chris van Rensburg Publications, Johannesburg.

Viljoen, MJ and Reimold, UW. 1999. *An Introduction to South Africa's Geological and Mining Heritage*. Mintek, Johannesburg.

Wilson, MGC and Anhaeusser, CR. 1998 (6th Edition). *The Mineral Resources of South Africa*. Council for Geoscience, Pretoria.

Sedimentary rocks of the Damara Belt in Namibia were deposited in a sea that closed during the assembly of the supercontinent Pangaea, causing folding of the sedimentary rock layers.

5
SUPER-CONTINENTS

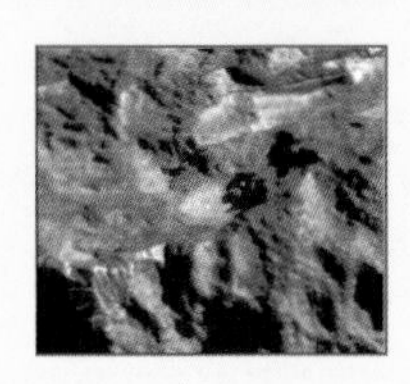

Guy Charlesworth

ROUTE MAP TO CHAPTER 5

Age (years before present)	Event
3 500 to 1 500 million	• Three continents formed during this period by amalgamation of small, old cratons. These continents were: Atlantica, consisting of parts of present-day West Africa and South America; Arctica, consisting of parts of present-day Canada, northern Siberia and Greenland; and Ur, consisting of parts of present-day South Africa, Madagascar, India and Australia. The Kaapvaal Craton formed the western extremity of Ur. Ur was a small continent 3 000 million years ago, but grew by amalgamating with other cratons: about 2 700 million years ago, the Zimbabwe Craton fused with Ur and the Kaapvaal Craton in particular, forming the Limpopo Belt of metamorphic rocks along their common boundary; 1 800 million years ago, the Congo Craton collided with Ur, fusing to the Kaapvaal-Zimbabwe Craton and forming the Ubendian Belt of metamorphic rocks along their common boundary.
1 600 to 1 400 million	• Rifting occurred along the Ubendian Belt between the Kaapvaal-Zimbabwe and Congo Cratons. Sediments were deposited in the rifts. Arctica became enlarged by the addition of parts of what are today the Baltic and Ukraine, western North America and eastern Antarctica to form a larger continent called Nena.
1 000 million	• Nena, Ur and Atlantica fused to form the supercontinent Rodinia. A mountain chain of global extent formed along the boundaries, underlain by metamorphic rocks that have since been exposed by erosion. This global metamorphic belt is known as the Kibaran or Grenville Belt. Metamorphic rocks of this age formed across South Africa to the south of the Kaapvaal Craton, known as the Namaqua-Natal Belt, and across Namibia and Botswana.
700 million	• Rodinia broke up into three major fragments, which dispersed around the globe. During the break-up, the portion that contained southern Africa also experienced rifting. The Kibaran metamorphic belt extending across Namibia and Botswana opened to form the Khomas Sea, while a seaway also opened up across the Namaqua-Natal Belt. Sediments accumulated in these seas.
600 to 300 million	• The fragments of Rodinia reassembled to form Pangaea. The Khomas Sea and the seaway across the Namaqua-Natal Belt closed during the assembly of Pangaea. The sedimentary rocks on the sea floors became folded and metamorphosed, and were intruded by granites. Today these form the Damara Belt of Namibia and the Malmesbury, Kango and Kirkwood Groups of South Africa. The granites are known as the Cape Granite Suite.

FORMATION OF SUPERCONTINENTS

The notion that the Earth's continents were joined together in the past is very old, but the first serious attempt to assemble them using geological principles was by Alfred Wegener in 1915 (*see* Chapter 2). He envisaged that all of the Earth's landmasses once formed a single continent, a supercontinent, which he named Pangaea. This notion has stood the test of subsequent scrutiny and is today widely accepted.

But how did Pangaea form? We saw in Chapter 2 that continents move across the globe conveyor belt-style, propelled by the ocean crust and its underlying lithosphere. This drags oceanic crust to destruction at subduction zones, producing new continental crust. New oceanic crust is created at ocean ridges, giving plates a helping push. Continents are moved along by these conveyors but cannot be subducted as they are buoyant. They plug subduction zones and fuse with other continental material conveyed there.

Creation of new continental crust at subduction zones, as well as the fusion process itself, has led to the gradual growth of continental crust and of individual continents over geological time. There is, of course, no theoretical limit to the size to which a continent can grow; there is no reason why most or even all of the continental crust on Earth should not occasionally be swept together to form a single landmass – a supercontinent. Just such an assembly created Pangaea about 300 million years ago.

Pangaea was short-lived. It broke apart, initially forming two large fragments, Gondwana and Laurasia, which underwent further fragmentation and dispersal to produce the present continents (**figure 5.1**). The fragments dispersed into an ocean, the ancestral Pacific (called Panthalassa), which continues to be consumed, while the Atlantic, in particular, grows in size. Projecting forward in time, it is not difficult to imagine what might happen in future. Australia is moving north and will merge with Southeast Asia, and the Pacific Ocean could close completely as the Americas collide with Asia-Australia, thus laying the foundations for yet another supercontinent consisting of the Americas, Australia and the Asia-Europe-Africa block.

The formation of Pangaea was a relatively recent event, geologically speaking. Sizeable continents have been around for at least 3 000 million years, so it is not unreasonable to suppose that supercontinents may have formed before Pangaea. Unfortunately, the further one goes back in time, the more fragmentary the evidence.

The existence of supercontinents prior to Pangaea is a hotly debated topic, as many scientists are somewhat sceptical of the evidence. Nevertheless, there are indications suggesting the possible existence of at least one older supercontinent that formed about 1 000 million years ago. It has been named **Rodinia**, from a Russian word meaning motherland. The events surrounding the formation and break-up, possibly of Rodinia and certainly of Pangaea, left important imprints on the geology of southern Africa. To place these in context, we need to review the life and death of the supercontinents.

Shaen Adey / IOA

Figure 5.1 ***The sedimentary strata along the southern Cape end abruptly at the coast, having been torn off during the opening of the Atlantic Ocean.***

A BRIEF HISTORY OF THE CONTINENTAL CRUST

In the preceding chapters, we tracked the formation and subsequent history of the Kaapvaal Craton from the earliest island arcs, their amalgamation into a stable continent, and the subsequent accumulation of sediments and other rocks on the basement provided by this early continent. The merger of the Kaapvaal, Zimbabwe and Congo Cratons greatly enlarged this early continent.

Similar processes were happening elsewhere on the globe as continents formed and enlarged. Some believe these continents ultimately merged to form Rodinia and subsequently Pangaea. John Rogers of the University of North Carolina at Chapel Hill and the late Raphael Unrug of Washington State University have provided succinct, although speculative (and slightly different) reviews of how this saga may have unfolded. There are also other interpretations of the evidence, but to avoid additionally complicating an already complex story, we will follow the Rogers-Unrug version.

Tracking the assembly, dispersal and reassembly of continents is no simple task. Even creating simplified diagrams is difficult because the continents of today were assembled from disparate pieces, fragments of other continents in the past – jigsaw puzzles within jigsaw puzzles. To make sense of this complexity, it is more convenient to begin at the end, paradoxical though this may seem. This makes it possible to relate the building blocks of continents to present day geographical areas, which provides a more comprehensible context.

In the maps compiled here to illustrate the history of the supercontinents, dashed outlines of modern continents (or parts thereof) have been included to assist readers to keep track of the pieces. These continental margins only came into existence with the fragmentation of Pangaea, but without them the story, complex as it is, would be impossible to follow. In reading this complex history, two things are important: follow the diagrams carefully; and remember that the Earth is a sphere, although we portray the continents on a flat map.

How can we recognise old supercontinents? The Earth's earliest continents, or cratons, contain the distinctive granite-greenstone combination of rocks, like the Barberton region described in Chapter 3. Later merging of these cratons produced metamorphic belts along their common boundaries, such as the Limpopo Belt (**figure 5.2**). Mapping of the cratons and their surrounding metamorphic belts, and measuring the ages of the belts, provides important information needed for the reassembly of past continents.

Grant Cawthorn

Figure 5.2 *Folded and compressed rocks, including these Sand River gneisses of the Limpopo Belt, were formed during the collision of the Kaapvaal and Zimbabwe cratons 2 700 million years ago.*

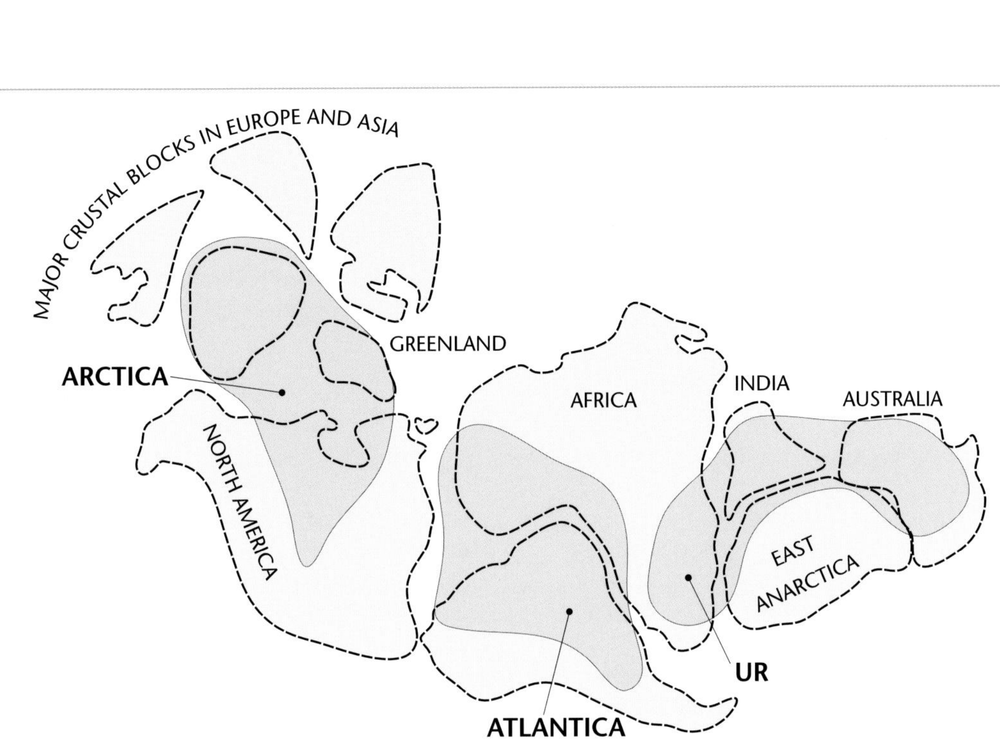

Figure 5.3 ***Some earth scientists believe that all of the Earth's continents are swept together from time to time, forming supercontinents. The most recent supercontinent to form was Pangaea, shown here, the assembly of which was completed about 300 million years ago. Contained within Pangaea were three large fragments of what some scientists believe was an earlier supercontinent. These were Ur, Atlantica and Arctica.***

Additional evidence is provided by palaeomagnetism preserved in rocks, as this provides an estimate of latitude. By determining latitude in rocks of different ages on a continent, the past movement of the continent can be determined. Knowledge of the past movements of different continents or parts of continents, as well as the ages of metamorphic belts, allows for the reconstruction of past continental amalgamations. The process is not without its problems, however, as the significance of rock ages is often ambivalent, and it is not possible to measure longitude from palaeomagnetism. For these reasons, sceptics maintain that an additional essential requirement in the reassembly process is an extremely fertile imagination. With these caveats in mind, let us see what Rogers and Unrug have to say about Rodinia.

The original continents

Rogers envisages three primary or initial continents: **Ur, Arctica** and **Atlantica**. The name Ur was chosen as it means the oldest or original; Arctica because the continent was composed of parts of northern Canada, northern Siberia and Greenland, most of which now lie in the Arctic Circle; and Atlantica because the fragments from which it was formed today lie astride the Atlantic Ocean. Each of these three continents grew over a protracted period of time from smaller fragments of continental crust, and may have remained as separate entities at least until the formation of Rodinia. Their positions in relation to Pangaea are shown in **figure 5.3**. Pangaea is used here rather than Rodinia, because it makes it easier to relate the three original continents to present geographical regions.

Ur is the oldest continent. Some 3 000 million years ago, Ur was just a narrow sliver that included the Kaapvaal Craton and the Pilbara region of Western Australia, as well as similar rocks that now form parts of Madagascar and India (**figure 5.4A**). Various additions to Ur occurred between 3 000 and 1 500 million years ago, such as the Zimbabwe and Congo continents, described in Chapter 4, leaving characteristic metamorphic belts such as

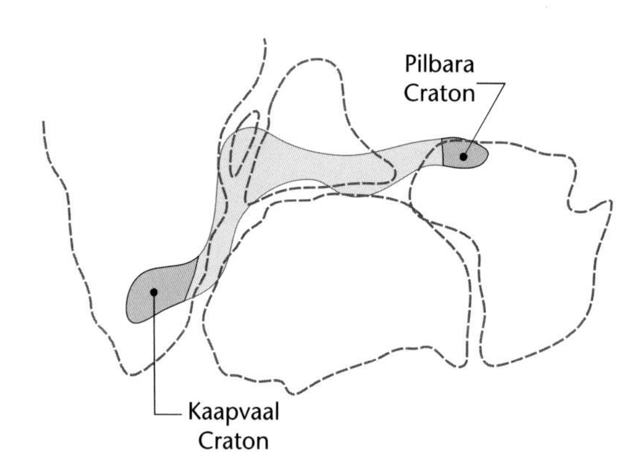

Figure 5.4A ***Ur is the Earth's oldest continent. About 3 000 million years ago, it consisted of a narrow sliver of granite-greenstone terrane, a craton, parts of which occur today in Madagascar, India and Australia. The Kaapvaal Craton formed its western extremity. (The dashed lines in the diagram show present-day continental outlines and are included only to provide a reference frame.)***

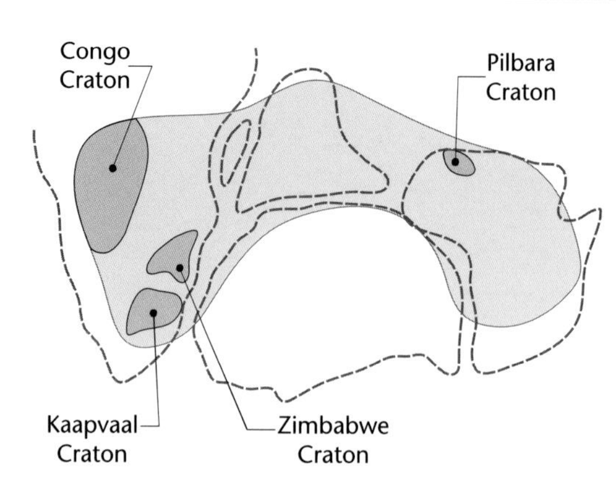

Figure 5.4B ***Over the period 3 000 to 1 500 million years ago, Ur grew by the amalgamation of other small cratons, such as the Zimbabwe Craton (which collided and fused with the Kaapvaal Craton 2 700 million years ago) and the Congo Craton (which amalgamated with the Kaapvaal-Zimbabwe Craton about 1 800 million years ago). The regions between the cratons are composed of new continental crust formed by the addition of island arcs, granites and sedimentary material as the cratons converged. The rocks in these regions have experienced metamorphism and form linear metamorphic belts. This diagram shows the extent of Ur 1 600 million years ago.***

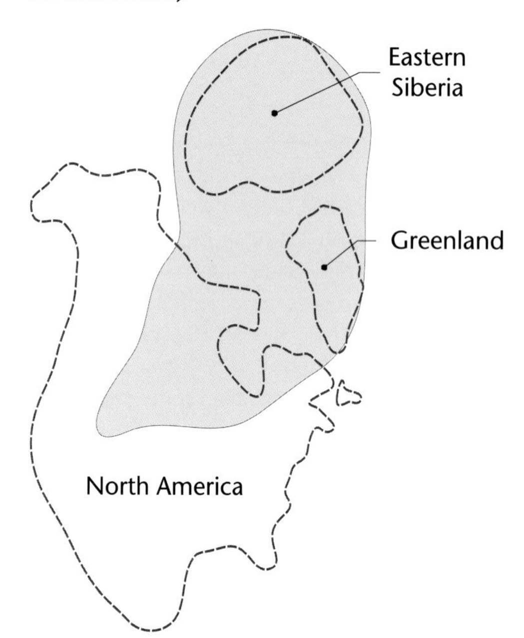

Figure 5.5 ***The continent of Arctica assembled between 2 500 and 1 500 million years ago, although some of its component parts are very ancient, dating back over 4 000 million years. It consisted of what are now portions of Siberia, northern Canada and Greenland.***

the Limpopo and Ubendian Belts along the sutures (**figure 5.4B**). It is metamorphic belts such as these that enable geologists to identify the boundaries between amalgamated fragments, while the ages of the rocks tell when the amalgamation took place.

Arctica consists of the ancient terranes of North America, Greenland and Siberia (**figure 5.5**). Some of these fragments are very old, dating back about 4 000 million years, but their assembly into Arctica seems to have taken place between 2 500 and 1 500 million years ago.

Likewise, Atlantica contains fragments of very ancient crust, but seems to have assembled around 2 000 million years ago (**figure 5.6**). Although Rogers favours three initial continents, Unrug, by contrast, did not regard Atlantica as a single continent, believing rather that it was represented by a number of small fragments that amalgamated somewhat later, during the formation of Rodinia.

The assembly of Rodinia

The next stage in the assembly was the enlargement of Arctica by the addition of smaller fragments of old continental crust that today form parts of the Baltic, Europe, Ukraine, western United States and eastern Antarctica. This occurred between 1 600 and 1 300 million years ago (**figure 5.7**), forming an enlarged Arctica, known as **Nena** (an acronym for northern Europe and North America). About 1 000 million years ago, Ur and Atlantica collided with Nena to form **Rodinia** (**figure 5.8**).

The sutures along which collision occurred are marked by distinctive metamorphic belts. These are global in extent – unlike the older belts, such as the Limpopo Belt, which are more localised. The belts of metamorphic rocks formed during the assembly of Rodinia are known in Africa as the Kibaran Belts, but elsewhere as the Grenville, after a region in eastern Canada. These suture zones include

Figure 5.6 ***The continent of Atlantica assembled about 2 000 million years ago from cratons that today form portions of South America and West Africa.***

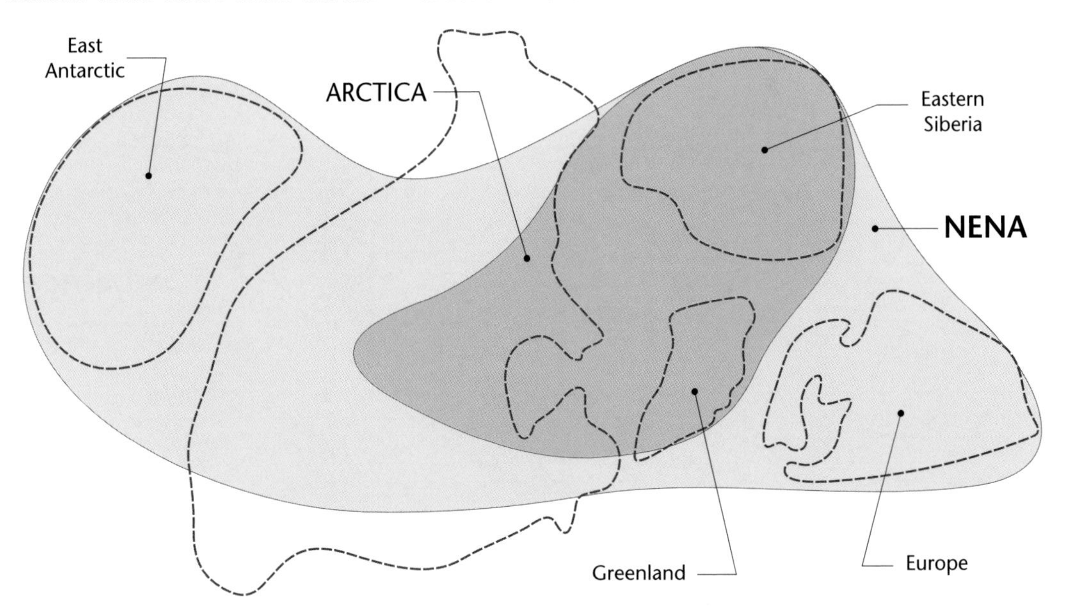

Figure 5.7 ***Between 1 600 and 1 300 million years ago, Arctica was enlarged by the addition of parts of what are today the Baltic and Ukraine, Europe, western North America and eastern Antarctica, to form Nena.***

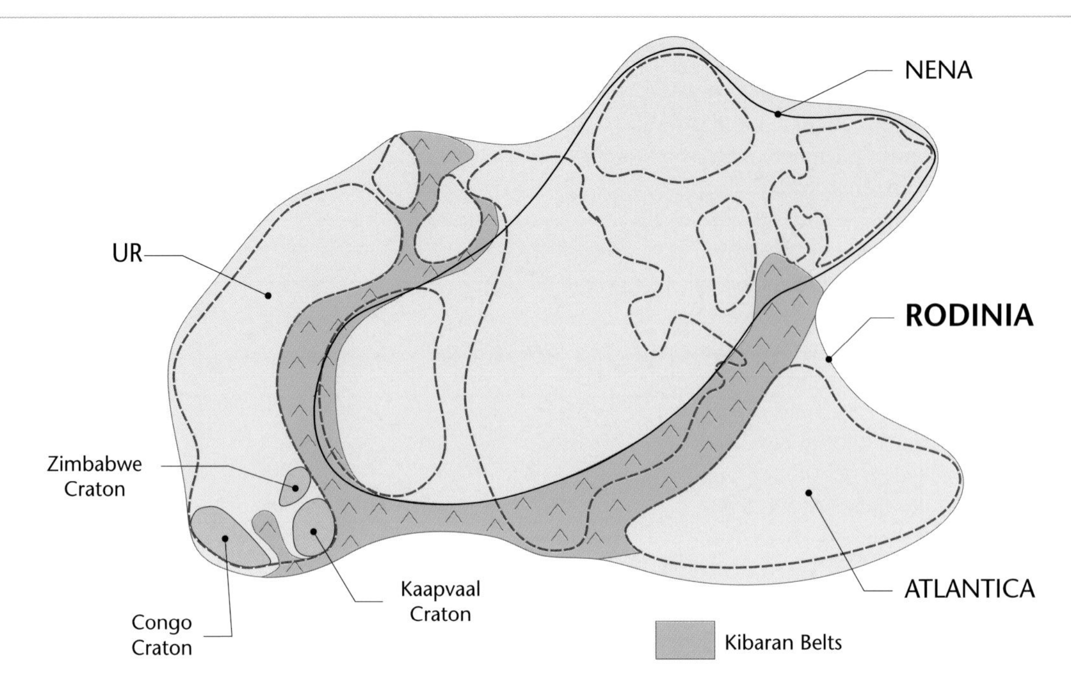

Figure 5.8 *About 1 100 million years ago, Nena, Ur and Atlantica were swept together to form the supercontinent Rodinia. The boundaries between these amalgamated continents would have formed impressive mountain ranges like the Himalaya and Andes, forming a global network of ranges. These mountains have since been eroded, exposing the metamorphosed rocks that formed their root zones. These are known as the Kibaran (or Grenville) Belts.*

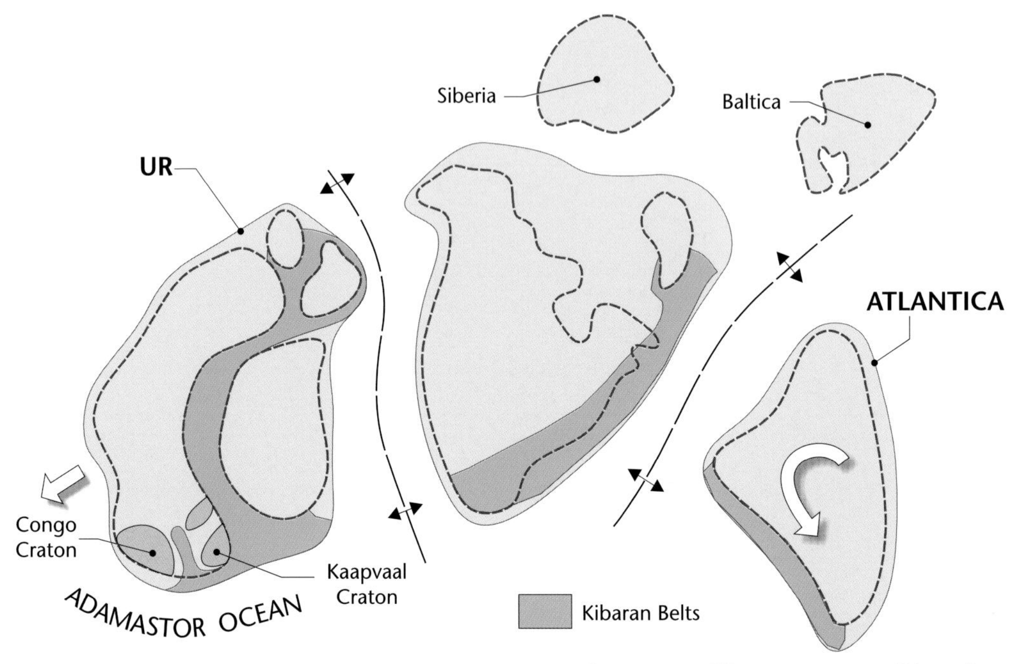

Figure 5.9 *Rodinia broke up into several major fragments about 700 million years ago. Rifting also occurred within the separating fragments.*

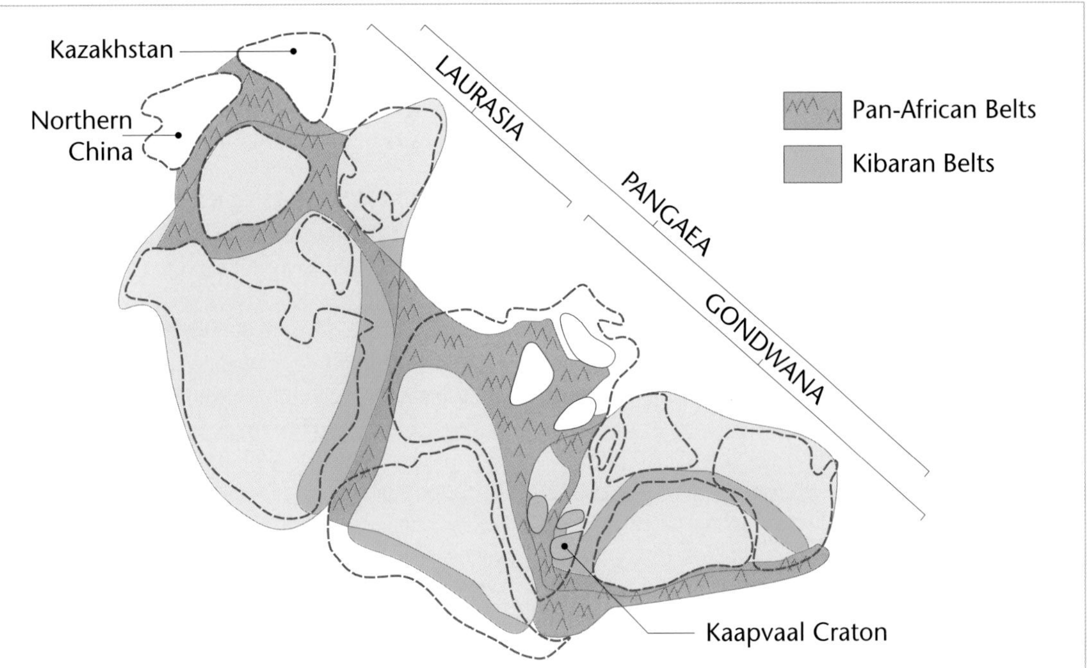

Figure 5.10 ***The fragments of Rodinia moved around the globe and began to reunite approximately 500 million years ago to form Pangaea. Continental fragments that today form parts of northern China and Kazakhstan, as well as parts of Arabia, were added in the process. Assembly was complete by about 300 million years ago. The boundaries between the fragments formed extensive mountain ranges, especially on the African continent. These ranges have since been removed by erosion, exposing their metamorphosed root zones, which are known as the Pan-African Belts.***

newly created crust in the form of granite intrusions and metamorphosed and folded sedimentary rocks deformed during the subduction process, as well as compressed slivers of oceanic crust and accreted island arcs. Rodinia thus grew not only by amalgamation, but simultaneously by the formation of new crust along the sutures. The Kibaran (or Grenville) Belt would have formed a huge mountain chain of global proportions, and far more impressive than modern-day global mountain chains such as the Cordillera of the western Americas, or the chain linking the Pyrenees and the Himalaya.

Rodinia appears to have survived as a supercontinent for a few hundred million years before it fragmented about 700 million years ago, breaking into several pieces (**figure 5.9**). These were: Ur, locked together with eastern Antarctica; Laurentia; Atlantica; Baltica (Europe); and Siberia. It is quite common, perhaps even normal, for continental break-up to occur along old sutures as these seem to be continued lines of weakness in the continental crust. Even rifts forming today, such as the East African Rift, follow old metamorphic belts marking the sutures between crustal fragments. The fragments of Rodinia dispersed around the globe. During this period of major fragmentation, localised rifts also developed within the larger fragments, but without leading to complete separation.

The assembly of Pangaea

The dispersed fragments began to reassemble about 600 million years ago. Some 500 million years ago Ur and eastern Antarctica, having moved around the globe, united with Atlantica, which had meanwhile rotated in an anticlockwise direction. This landmass was further enlarged by the addition of the Arabian continental mass to form what is known as **Gondwana**. Finally Gondwana, Laurentia, Baltica, Siberia and masses that today form parts of northern China and Kazakhstan combined about 300 million years ago to create Pangaea (**figure 5.10**). The metamorphic belts that formed along the sutures between the assembled fragments of Gondwana, about 500 million years in age, are known as the **Pan-African Belts** (**figure 5.10**), as they are particularly extensively developed on the African continent.

The story of Earth's last two supercontinents, as presently understood, is based on a huge database of detailed studies of geological terranes around the world, carried out by many geologists of different nationalities over many decades. The database is continually being updated as the ages of rocks, as well as their latitudes at different times (determined using palaeomagnetism), are more accurately determined. The detail, as with so much in the natural sciences, is subject to debate and differences of opinion, which arise largely from our lack of reliable information.

The history of the supercontinents will undoubtedly be refined in the future. The account presented here provides a broad framework of the way things may have happened, but there are many uncertainties. Metaphorically speaking, it is not cast in stone, although literally it is! The rock record simply needs more careful deciphering. But there are still many sceptics. While few question the former existence of Pangaea, Laurasia and Gondwana, many believe that Rodinia is simply fantasy.

The notion of ancient supercontinents has its strong proponents, who are continually revising and elaborating the hypothesis. This is perhaps best illustrated by the work of John Rogers himself, who recently presented evidence for the possible existence of a supercontinent even older than Rodinia, which he has named **Columbia**. This supercontinent may have formed about 1 800 million years ago, breaking up some 1 500 million years ago. Rogers and others have even suggested that supercontinent formation may be a cyclical process repeating every 500 to 750 million years.

In this cycle, all of the Earth's continental crust comes together to form one continent, counterposed by a single ocean. The supercontinent then fragments, and the pieces disperse into the single ocean, which gradually decreases in size, until ultimately it is completely consumed as the continental fragments reunite to form yet another supercontinent. Perhaps just such a process is happening today as the Indian and particularly the Atlantic Oceans, formed during the break-up of Pangea, grow at the expense of the Pacific Ocean. This could well lead to the formation of yet another supercontinent in the future.

THE IMPRINT OF RODINIA ON SOUTHERN AFRICA

The story of Rodinia provides a useful context for understanding geological events that affected southern Africa. Some 1 800 million years ago, the Kaapvaal-Zimbabwe continent was attached to similar rocks in the east, together constituting the continent of Ur (**figure 5.4B**). On the western side of the continent, a major accretion event had occurred, forming the Ubendian platform of continental crust, as described in Chapter 4. Unrug suggested that this accretion also involved the Congo continent, as described in this book (*see* Chapter 4). However, a collision of the Congo Craton with Kaapvaal-Zimbabwe at this time is at variance with the reconstructions proposed by Rogers, which brings these entities together some 1 000 million years later.

This illustrates the uncertainties in the reconstruction of early continent configurations. The current information is simply not good enough in some areas to be certain. Of course, that is not true everywhere, and where the data are good there is generally consensus about the reconstructions.

We will accept that the Congo Craton became attached to the Kaapvaal-Zimbabwe Craton 1 800 million years ago, forming a high mountain range atop a metamorphic belt. This range was denuded and reduced to normal continental elevation, exposing once deeply buried rocks at the Earth's surface. The area that is now Bushmanland then consisted of Ubendian metamorphic rocks including old island arcs and ocean floor sediments that had accreted in the run-up to the fusion of the Congo Craton and the Kaapvaal-Zimbabwe Craton.

The region experienced rifting between 1 600 and 1 400 million years ago. The new Ubendian continental crust thinned and subsided below sea level, and shallow-water sediments were deposited on this granitic basement, forming the **Bushmanland Group** of rocks (**figure 5.11**). It is likely that the rift extended northeastward along the Congo-Kaapvaal-Zimbabwe suture as well, because

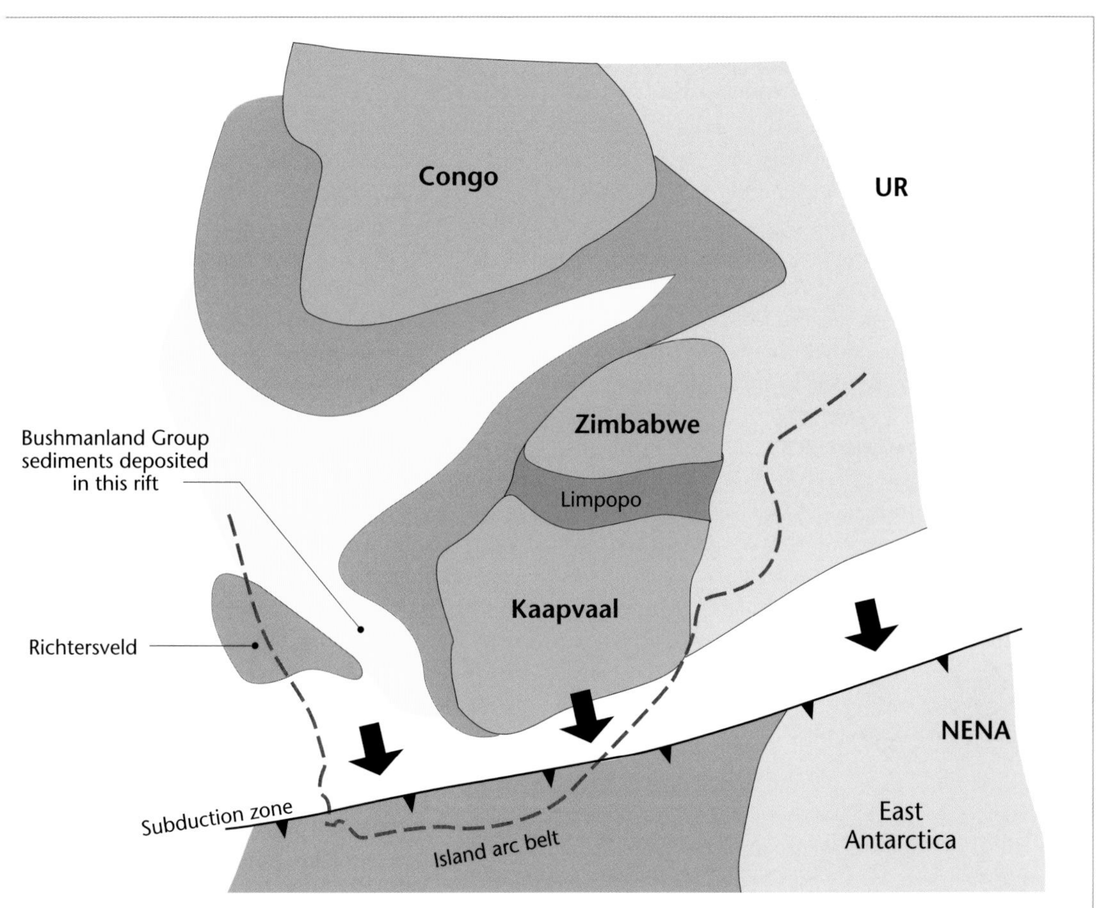

Figure 5.11 *Between 1 600 and 1 400 million years ago, what is today southern Africa, then part of the ancient continent Ur, began to experience rifting. The Congo Craton began to separate and a rift also developed across what is now Bushmanland, which widened to form shallow seas. Sediment accumulated on the continental shelves in these seas, and oceanic crust began to form across Bushmanland as the rift widened. During the assembly of Rodinia (1 100 million years ago), the sea separating the Congo and Kaapvaal-Zimbabwe Cratons closed, as did the sea across Bushmanland. The resulting mountains have since been removed by erosion, exposing a belt of metamorphic rocks across the southern portion of the Kaapvaal Craton (called the Namaqua-Natal Belt) and between the Congo and Kaapvaal-Zimbabwe Cratons. These belts form part of a global network of metamorphic belts known as the Kibaran (or Grenville) Belts (see* **figure 5.8***). (The outline of southern Africa is shown for reference.)*

similar rocks to the Bushmanland Group occur as far afield as northern Botswana – for example, in the Ghanzi area. In eastern Namaqualand, the continental crust ruptured completely and oceanic crust began to form (known as the **Areachap Terrane**). This oceanic crust appears to have been connected to an ocean basin, floored by oceanic crust, which extended across the entire southern margin of the Kaapvaal Craton as far as KwaZulu-Natal. The formation of the Ubendian Belt 1 800 million years ago and its subsequent rifting may be related to the assembly and fragmentation of the recently proposed supercontinent Columbia.

The rifting and separation of continental fragments in the Bushmanland and Namaqualand areas stopped about 1 300 million years ago as the southern margin of the Kaapvaal Craton approached a subduction zone. This marked the start of the assembly of the southern African portion of Rodinia. The ocean basin in eastern Namaqualand began to close and island arcs began to accrete onto the southern margin of the Kaapvaal Craton (**figure 5.11**). This closure culminated about 1 100 million years ago, during the **Kibaran** event. The rift in which the Bushmanland Group of sedimentary rocks had been deposited (**figure 5.11**), as well as the rift that had developed between the Congo and Kaapvaal-Zimbabwe Cratons, closed during this collision.

The sedimentary rocks in these rifts were crumpled and folded, forming part of the global Kibaran mountain chain (**figure 5.8**). The rocks involved in this event, including both the older Ubendian basement and the Bushmanland Group, were intensely folded and metamorphosed, and intruded by a multiplicity of granites. In KwaZulu-Natal, this event led to the accretion of island arcs onto the Kaapvaal Craton, while large slivers of ocean floor were thrust up onto the continent. Sedimentary rocks deposited on the former ocean floor were also folded during this compression.

Today these rocks form the complex geology between Margate and the Tugela River in KwaZulu-Natal and much of the area between Upington and the Atlantic coast in the west. They represent the roots of a vast mountain range that extended across the entire southern part of the country and to the northeast through Namibia and Botswana, forming part of the extensive Kibaran (Grenville) global mountain range (**figure 5.8**).

Now deeply eroded, that portion of the belt in South Africa is known as the Namaqua-Natal Belt. The area from Pofadder along the Orange River as far as the eastern Richtersveld consists of a relatively pristine remnant of the older Ubendian crust. Elsewhere these older rocks were reworked by the later Kibaran event, which obliterated all traces of their earlier history. The island arcs and associated granites that accreted onto the southern margin of the Kaapvaal Craton not only formed the Namaqua-Natal Belt, they also formed the Falkland Plateau and parts of the Antarctic Peninsula. Today, most of the Namaqua-Natal Belt is deeply buried beneath younger rocks; only the eastern and western portions are exposed on the surface.

Figure 5.12 ***The hills around Aggenys in the Northern Cape are formed by sedimentary rocks that were folded and metamorphosed during the assembly of Rodinia 1 100 million years ago.***

The Kibaran event has counterparts in many other parts of the world where rocks of similar age record an almost identical history of accretion and collision. These constitute a global system of metamorphic belts and mark the sutures along which the final assembly of Rodinia took place (**figure 5.8**).

The Namaqua-Natal Belt hosts several mineral deposits, most important of which are the copper deposits at Prieska and O'okiep (both now largely worked out), and the large copper-lead-zinc deposits at Aggenys. The Prieska and Aggenys deposits are thought to have formed at undersea hot springs during the rifting that preceded the Kibaran event, and are associated with metamorphosed volcanic and sedimentary rocks (**figure 5.12**). The O'okiep copper deposits, in contrast, occur in dyke-like igneous intrusions, and originated by the separation of metal-sulphur compounds from magmas that intruded the crust during the Kibaran event.

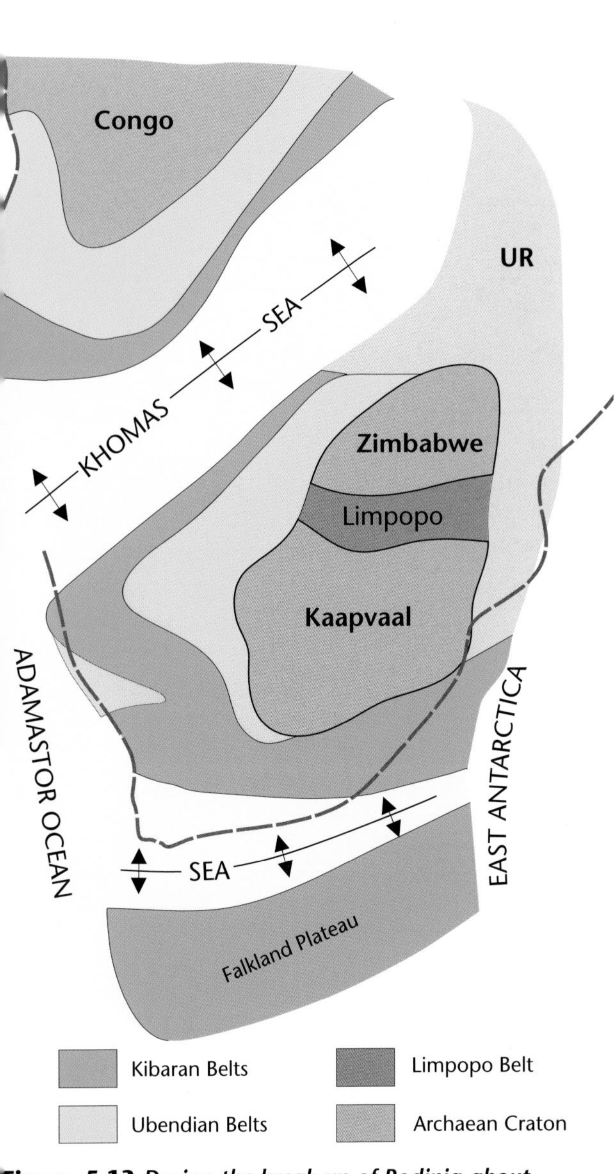

Figure 5.13 *During the break-up of Rodinia about 700 million yeas ago, rifts again developed between the Congo and Kaapvaal-Zimbabwe Cratons, and also across the Namaqua-Natal Belt, creating shallow seas, probably similar to the Red Sea today. Sediments accumulated in these shallow seas. As Pangaea began to assemble about 500 million years ago, these rifts closed and the sedimentary rocks were compressed, folded and metamorphosed. Granites intruded into these metamorphosed rocks, such as the Cape Granite. These metamorphic rocks form part of the Pan-African network of metamorphic belts.* **Figure 5.9** *gives the larger-scale context to this diagram.*

THE BREAK-UP OF RODINIA

Rodinia began to fragment around 700 million years ago when the continent of Ur, including east Antarctica, separated (**figure 5.9**). Ur developed internal rifts at about the same time in response to the general rifting of the supercontinent, and extension and even rupturing occurred along lines of weakness provided by the Kibaran Belts. An ocean basin reopened along the suture between the Congo and Kaapvaal-Zimbabwe Cratons, known as the Khomas Sea, which became floored by oceanic crust as it widened (**figure 5.13**). The Khomas Sea may have been similar to the modern Red Sea.

Thinning of the crust also occurred along the Namaqua-Natal Belt, allowing invasion of the sea over what is now the southern Cape. Gradually, sediments began to accumulate in these newly created depressions.

The sediments that accumulated on the thinned Namaqua-Natal crust were primarily marine in origin, although they include some riverine deposits such as conglomerates. The most common rock types are mudstones and sandstones, representing intermediate- to shallow-water accumulations. Also present are marine limestones, which today host the Cango caves.

The rocks formed in this period are not very well exposed, except in the extreme west of the country, as they are largely overlain by younger rocks. They have been assigned a variety of names because of their wide geographic separation. These include the **Malmesbury Group** in the west, the **Kango Group** in the Oudtshoorn area, and the **Kaaimans Group** in the Mossel Bay-Knysna area (**figure 5.14**).

Sediments also accumulated in the Khomas Sea in central Namibia. Initially, riverine deposits formed, as the rifting continent began to subside, together with volcanic rocks. Then, as the crust subsided

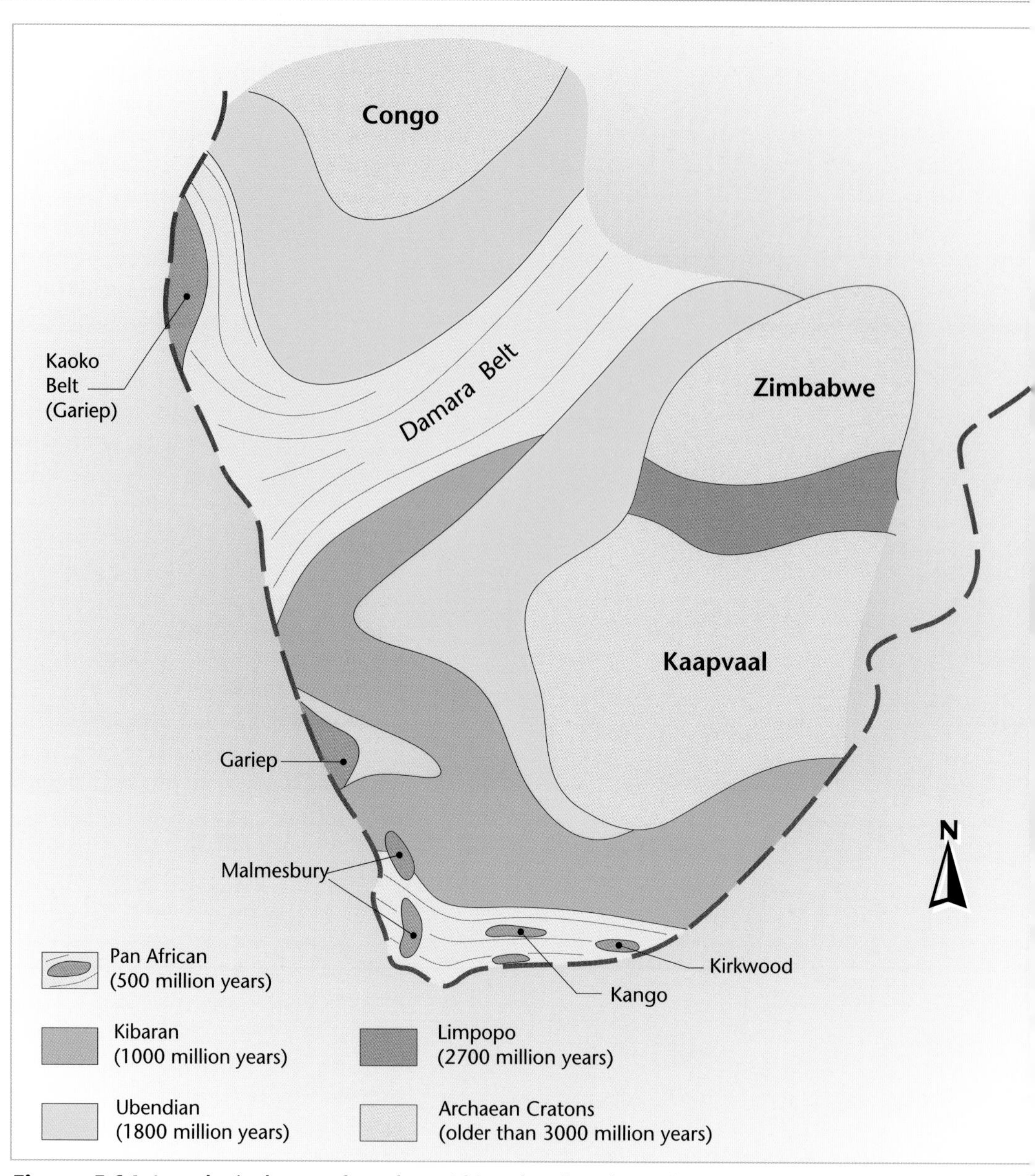

Figure 5.14 *A geological map of southern Africa showing the major metamorphic belts that surround the Kaapvaal Craton. These are the product of continental growth, and especially supercontinent formation. On its eastern side, the Kaapvaal Craton remained attached to the ancient rocks of the primordial continent Ur until the break-up of Gondwana about 160 million years ago.*

further, these deposits were followed by shallow-water marine sands, shelf muds and limestones (notably the **Otavi Group**). Some deeper-water sediments also accumulated along the continental slopes in the form of turbidites as the rift widened and oceanic crust began to form between the separating continental fragments. In addition, tillites formed – indicating glacial activity in spite of the fact that the continent at that time was close to the equator. This unusual occurrence led to the suggestion that Snowball Earth conditions prevailed at this time (*see* 'Snowball Earth' on page 119).

While these events were taking place in the interior of the continent of Ur, sedimentation was also occurring along its west coast into the so-called **Adamastor Ocean** (**figure 5.13**). Here, shallow marine sediments – including sandstones, shales and limestones – accumulated to form what is now known as the **Gariep Group**. Meanwhile, Ur was moving across the globe, steadily approaching the continent of Atlantica and closing the Adamastor Ocean.

THE MAKING OF GONDWANA

The remnants of Rodinia, including Ur and East Antarctica, finally made contact with Atlantica about 600 million years ago along a subduction zone directed towards the west beneath Atlantica. The collision seems to have started in the north, and the Adamastor Ocean closed progressively southwards, like a pair of scissors. Rocks of the Gariep Group were squeezed between the converging continents and became intensely folded. The crust thickened, the sediments became metamorphosed, and a mountain range formed along the suture. Granite magmas were produced as the crust thickened, and rose upwards, intruding the folded sediments.

Meanwhile, the interior of Ur also began to experience compression. A subduction zone formed along the northern margin of the Khomas Sea, and it began to close. Slivers of ocean floor were detached as the sedimentary deposits of the ocean were crushed between the approaching continents. Here, too, granite magmas intruded the buckled sediments. This buckled and metamorphosed assemblage of rocks is known as the **Damara Belt** (**figure 5.14**).

In a similar way, the sedimentary rocks to the south of the Kaapvaal Craton (Malmesbury, Kango and Kaaimans Groups) were folded and invaded by granites 600 to 500 million years ago. These granites can be seen intruding the metamorphosed Malmesbury Group sedimentary rocks at a famous locality at Sea Point in Cape Town, described by Charles Darwin in his account of the cruise of the research ship *HMS Beagle* in 1836. Known as the **Cape Granite Suite**, they form the rounded outcrops at many of the Cape's famous beaches such as Clifton, Camps Bay and Boulders Beach (**figure 5.15**), and inland such as Paarl Rock.

This event marked the formation of Gondwana and culminated about 550 million years ago as Atlantica and Ur became welded together. Known as the **Pan-African** event, this left a network of mountain ranges across the southern Cape and up what is now the West Coast, part of a global network of mountain ranges (**figure 5.10**). Erosion of these ranges started as soon as they began to form, and sediment was shed eastwards from the West Coast ranges and southwards from central Namibia into a shallow basin (*see* 'Sedimentation in foreland basins' on page 162). The resulting sedimentary rocks are known as the **Nama Group** and are still sporadically preserved in the Western and Northern Cape, and more extensively in Namibia.

KAAPVAAL – THE PASSIVE SPECTATOR

In summary, the period from 2 000 million to 500 million years ago saw major, globe-changing events that left an indelible mark on the geology of South Africa. Metamorphic belts formed along with

Figure 5.15 ***Rounded outcrops of 550-million-year-old Cape granites occur at many of the Western Cape's beaches, such as Boulders Beach near Simon's Town.***

SEDIMENTATION IN FORELAND BASINS

Subduction zones on the edges of continents are characterised by deep ocean trenches flanked by mountain ranges along the continental edge, such as we see today along the western margin of South America. The mountain ranges are an expression of thickened crust – thickened by compression, by the addition of sedimentary material scraped off the subducting slab, and by the intrusion of granite and diorite magma that rise from the mantle above the slab as it sinks and dehydrates (*see* Chapter 2). The Andes Mountain range typifies such thickened crust.

Thickening of the crust in this way has an important secondary effect in that it creates a localised load that depresses the Earth's crust, causing it to sag (the crust and upper mantle actually float on the underlying asthenosphere – *see* Chapter 2). The depression is particularly evident on the landward side of the mountain range, where it expresses itself as a sunken tract known as a **foreland basin** (**A**). In some cases, the floor of such a depression may sink below sea level, in which case it becomes a seaway. Rivers rising in the mountain range and sometimes on the landward side of the depression transport eroded sediment to the depression, where it is deposited either in water if such exists in the depression, or on extensive river floodplains if the depression is more or less filled with sediment. The accumulating sediment is generally coarser (sand and gravel) on the mountain range side of the foreland basin because of the greater topographic relief.

Flooded foreland basins are sometimes asymmetric, being deepest adjacent to the mountain front. If sufficiently deep, a steep slope may develop below sea level, and sediment cascading down the slope produces sedimentary deposits known as turbidites (*see also* 'Sedimentation in ocean trenches', page 82).

In areas of vigorous subduction, huge thrust faults arising from compression in the mountain belt advance landwards and may fold and displace the sediment that has accumulated in the foreland basin (**B**). Such foreland basins can be long-lived and often contain piles of sedimentary material spanning hundreds of millions of years and exceeding 10 km in thickness.

Where two continents collide – for example, India and Asia, which formed the Himalaya mountains – these depressions may form on both sides of the mountain range. A modern example of a foreland basin in this setting is the Indo-Gangetic plain of India, which is an area of active sediment accumulation along the southern foothills of the Himalaya. Although the floor of this depression is still subsiding, the huge amount of sediment shed from the Himalaya keeps the basin filled, resulting in a gradually thickening pile of sediment beneath the Indo-Gangetic floodplain.

mountain ranges, and the continental crust was substantially enlarged. The mountains were eroded and the crust ripped apart, forming rift basins that were invaded by the sea, only to close again, and forcefully, creating new mountain ranges.

At least two, possibly three, supercontinents formed and fragmented, all of which included the ancient Kaapvaal Craton. They have left this ancient craton surrounded by metamorphic belts: the 1 800-million-year-old Ubendian Belt in the west; the 1 100-million-year-old Kibaran Belt in the west and south; and the 500-million-year-old Pan-African Belt, also mainly confined to the south (**figure 5.14**). In addition to these three metamorphic belts, there is the older (2 700-million-year) Limpopo Belt that formed along the suture between the Kaapvaal and Zimbabwe Cratons.

In spite of all of this activity surrounding the Kaapvaal Craton, its rocks remained unaffected, passive spectators to these violent happenings. With the formation of Gondwana, that was about to change. However, before we continue, it is necessary to consider another important participant in our story: life.

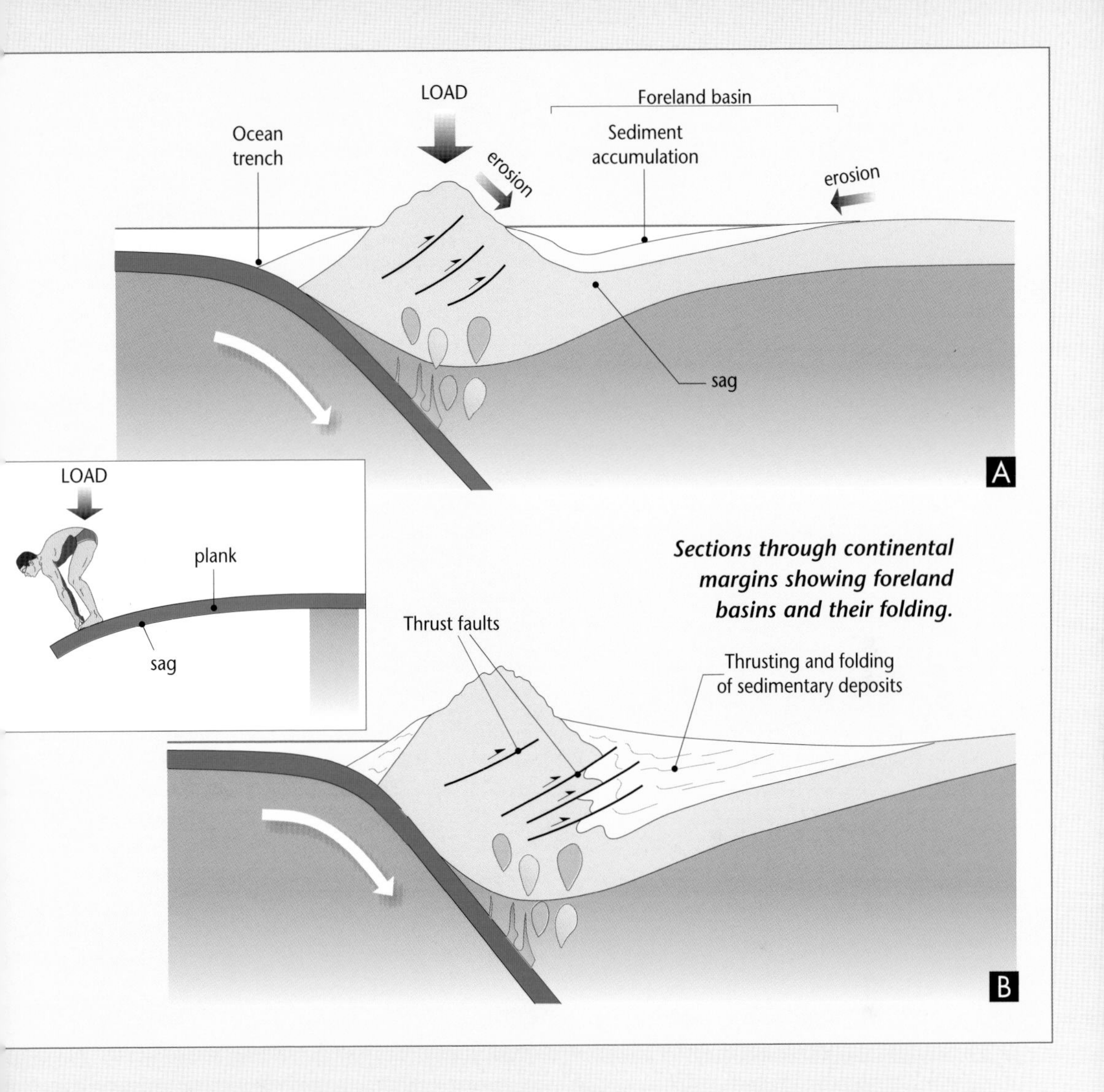

Sections through continental margins showing foreland basins and their folding.

SUGGESTED FURTHER READING

Johnson, MR, Anhaeusser, CR and Thomas, RJ. 2005. *The Geology of South Africa*. Council for Geoscience, Pretoria.

Redfern, R. 2000. *Origins*. Cassel and Company. London.

Rogers, JJW. 1996. 'A History of Continents in the Past Three Billion Years.' *Journal of Geology*, vol. 104, pp 91–107.

Thomas, RJ, Agenbacht, ALD, Cornell, DH and Moore, JM. 1994. 'The Kibaran of Southern Africa: Tectonic Evolution and Metallogeny.' *Ore Geology Reviews*, vol. 9, pp 131–160.

Viljoen, MJ and Reimold, UW. 1999. *An Introduction to South Africa's Geological and Mining Heritage*. Mintek, Johannesburg.

Walker, G. 2003. *Snowball Earth*. Bloomsbury, London.

THE STORY OF
EARTH & LIFE

6
EARLY LIFE

Heat-loving bacteria were among Earth's earliest living forms. They are still found around hot springs such as at Yellowstone National Park, USA, where they form slimy ribs and coatings on the sinter domes.

Terence McCarthy

ROUTE MAP TO CHAPTER 6

Age (years before present)	Event
About 4 000 million	• The first self-replicating organisms appeared at an undersea hot spring. They derived their energy from chemical imbalances between the hot spring water and sea water. They appear to have undergone rapid diversification, exploiting a variety of different chemical energy sources available in the hot spring environment. At that time, the atmosphere consisted mainly of carbon dioxide, with some nitrogen.
3 900 million	• By this time, photosynthesising cyanobacteria had probably evolved. Their appearance meant life was no longer confined to hot springs, but could exist anywhere where light, carbon dioxide and water were available. These bacteria consumed carbon dioxide and produced oxygen as a waste product. This oxygen was consumed by iron and manganese dissolved in the ocean. In addition, the cyanobacteria promoted the precipitation of calcium carbonate from sea water, further reducing the carbon dioxide content of the atmosphere. The composition of the atmosphere gradually changed from carbon dioxide- to nitrogen-dominated.
2 000 million	• Free oxygen began to accumulate in the atmosphere, creating a crisis for the anaerobic life dominating the planet. This gave rise to a symbiotic relationship between a respiring bacterium and an anaerobic one, producing the first organisms with mitochondria. These became the ancestors of animals and fungi. A symbiotic relationship developed between the mitochondrial-bearing species and cyanobacteria, giving rise to chloroplasts inside the cells. These symbionts were the ancestors of plants.
720 million	• A major global glaciation occurred, recorded in the tillites of the Ghaub Formation, Otavi Group.
600 million	• Early sponge-like creatures evolved, and the first fossils of the sponge-like Otavia are preserved in the Auros Formation, Otavi Group.
590 million	• A second major global glaciation occurred. This and the earlier glaciations may have provided the impetus for the radiation of species.
545 million	• Fossils of the soft-bodied multi-celled organisms of the Ediacaran or Naman fauna were preserved as body impressions. These seem to be an extinct side-line group very different from animals of later times.
540 million	• The complexity and abundance of multi-celled organisms increased rapidly (the so-called Cambrian Explosion). At this time all the major groups (phyla) of invertebrates that exist today came into being.

ABOUT LIFE

Decay is inherent in all component things. We take this fundamental principle for granted, as it so pervades our daily lives. Machines wear out and break down. Compressed air escapes from a punctured tyre and it takes energy to return it to its compressed state; gas simply will not do it on its own. Even the expanding Universe is responding to this principle. Some cosmologists predict the end to our Universe, the so-called heat death, as completely dissipated in the vacuum of space, like gas released from a balloon: isolated, dead particles, without any meaningful interaction. This principle is enshrined in a fundamental Law of Nature, the Second Law of Thermodynamics, which holds that systems spontaneously increase their state of disorder or entropy. So important is this principle that, in the opinion of Stephen Hawking, it even gives direction to the flow of time.

Living things are an anomaly in the context of the Second Law, in that they exist as self-organising systems that, through growth, produce order. Each of us grew from a minute, single cell. The instructions to create us were encoded in strands of atoms, DNA, in the nucleus of that single cell. These instructions were carried out faithfully over decades, creating everything that is us, capable of abstract thought and with the ability to manipulate the world around us. This is order of the highest degree.

There are other apparent forms of order in the Universe, such as the growth of crystals and the formation of planetary systems and galaxies. These are but responses to fundamental forces of Nature – electromagnetic and gravitational forces – and happen spontaneously. Their order is different from that seen in living things. Unlike the formation of ice crystals, which happens billions of times a second in a storm cloud, life does not arise spontaneously. In fact, there is good reason to believe it happened only once in the entire history of Earth – that all living things are descendants from a common ancestor.

In terms of energy, a growing organism is the equivalent of a balloon inflating itself. Organisms achieve this remarkable feat by borrowing energy from (or increasing the entropy of) their environment. Of course, it is only a loan and has to be repaid, a debt settled by death and subsequent decay. Life is perhaps the most remarkable thing in the Universe, after the Universe itself.

ON THE ORIGIN OF LIFE

How did life arise? The short answer is that we do not know. But, as with all fundamental questions, there is certainly no shortage of informed speculation. And the way these speculations have developed is by detailed study of the variety of life forms, both extant and fossil, in the myriad habitats they occupy. Even this approach is fraught with difficulty; it is perhaps akin to trying to work back to the abacus by studying the workings of modern computers.

Life seems to have come into existence about 4 000 million years ago. At that time the Earth was probably largely covered by oceans. The atmosphere consisted mainly of carbon dioxide and water vapour, probably with some carbon monoxide and nitrogen. There were probably traces of oxygen, produced by the breakdown of water molecules in the upper atmosphere under the influence of solar radiation. Plate tectonics had started, and subduction zones existed, as did undersea hot springs (hydrothermal vents). The plates were probably smaller than today and plate motion more vigorous because of a hotter mantle.

The Moon was present and the Earth experienced tides. Meteorite bombardment was still a common phenomenon. Some continents existed, and chemical decomposition of rocks by the atmosphere (weathering) was underway, adding substances like iron, silica and calcium to the oceans. The sea probably had a composition similar to today except that it would have been very anoxic. The interaction of sea water with hot rocks in hydrothermal systems probably had a strong influence on its composition, adding iron, manganese, sulphur and other substances.

Although the Sun's radiance was less intense than today, by about 30%, the Earth was probably warmed by a greenhouse effect, since carbon dioxide dominated the atmosphere. Surface temperature may have been quite high, perhaps in the range 50°C to 80°C, and boiling of the oceans – perhaps even complete vaporisation – would have occasionally occurred as a result of impacts by large meteorites.

Atmospheric pressure was higher than today, probably in the range 30 to 50 bars (compared to about one bar today). This is equivalent to the pressure encountered on Earth today at water depths of between 300 and 500 m – great enough to crush the average World War II submarine.

There may have been an assortment of organic compounds (including amino acids) in the early ocean, delivered to Earth by certain types of carbon-rich meteorites known as carbonaceous chondrites, and by comets. The status of nitrogen in the early atmosphere is uncertain, and this element may have been combined with oxygen as nitrate or with hydrogen as ammonia, both dissolved in the sea, with some free nitrogen in the atmosphere.

Energy to power early life

Organisms rely on natural energy imbalances to obtain energy. Sulphur, for example, will react with oxygen to form a sulphur-oxygen compound, giving off energy. If the oxygen is in the form of a gas, the energy is given off as heat as the sulphur burns. There are bacteria that have developed ways of controlling this chemical reaction, making it happen within their cells and extracting the energy in a form useful to themselves. Our bodies do essentially the same thing. We obtain energy by reacting carbohydrates with oxygen, and extract the energy in a useful form. This can also be done in air – carbohydrates burn. In fact, we burn them to measure the maximum amount of energy they can usefully give us – their so-called kilojoule or calorific value, which we can look up in books on nutrition. Once the reaction is complete and the energy has been extracted, the reaction products are discarded. Waste products are an inevitable consequence of life.

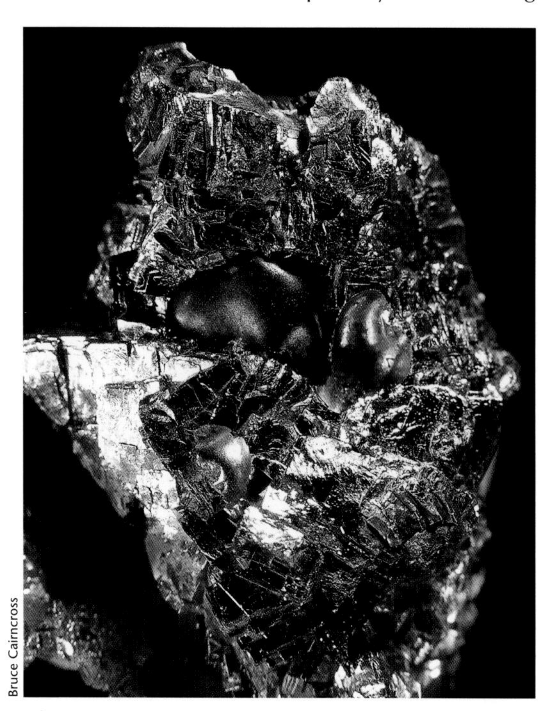

Bruce Cairncross

Figure 6.1 ***The mineral pyrite, an iron sulphide, may have provided the template for the first self-replicating molecules – the start of life. This pyrite crystal from a Witwatersrand conglomerate has a carbon nodule attached to its surface. (The image is twice natural size.)***

Chemical imbalances (disequilibria) are common – dry grass, for example. Put a match to it and it will go up in flames as it reacts with oxygen in the air to produce the stable combination of water and carbon dioxide. But in the absence of a match, dry grass will persist indefinitely in the presence of oxygen.

The precursors to life must also have derived their energy from natural chemical reactions, in all probability relying on chemical imbalances that existed between rocks and the oceans. The most likely place where life emerged was close to hot springs (hydrothermal vents – *see* Chapter 2) on the ocean floors. Water that was present in fractures in the rocks of the ocean floors became heated by volcanic activity and equilibrated with the rocks, in the process dissolving oxygen-deficient compounds. This water then emerged at these hot springs into an environment supplied with oxygen by a slow trickle from the breakdown of water molecules in the upper atmosphere. The chemical disequilibrium between the sea water and the hot spring water probably provided the energy to start life.

Life's common ancestry

Life as we know it probably began as a self-replicating organic molecule. Perhaps a distant echo of that molecule is carried today in all living things. All life shares the common use of a few molecules, just four in fact, which are linked together in a chain to form the blueprint of life – the genetic code. We use a combination of two digits to store and manipulate information in our computers: 0 and 1, *on* and *off*, the binary system. Life somehow chose to use a combination of four digits to store information. Most organisms today use a molecular chain of these four molecules, called DNA, to carry the code. All also make use of a simpler version to read the blueprint and to actually do the work of running life, especially the manufacture of proteins. This is RNA. It is surmised that the first organism probably used this simpler form both to carry the information for replication and to execute it.

All living things also consist of cells, little sacks formed by a membrane, which contain the chemicals of life, as well as the DNA molecule that holds the instructions about what to do with the chemicals. The DNA is capable of replicating itself and the entire sack. All living things have another peculiar commonality. Amino acids (the building blocks of proteins), nucleotides (the building blocks of DNA and RNA) and simple sugars come in two distinct forms – so-called left-handed and right-handed forms (like a pair of gloves). All living organisms use the left-handed form of amino acids and the right-handed form of nucleotides and sugars in their metabolism. The common use of the same four molecules in the genetic code, as well as the use of the same forms of amino acids, nucleotides and sugars, suggests that all living things descended from one single, self-replicating system.

Life probably arose only once. Each one of us represents the end of an unbroken line of ancestors extending back over 4 000 million years to that first self-replicating organism. Every animal, insect and plant likewise represents the end of an unbroken line extending back to that same organism. Her name is LUCA, the Last Universal Common Ancestor.

The first replicating RNA could not have had a blueprint, however, and must have arisen by some other process. Some scientists believe that replication may have occurred on the crystalline surface of a mineral, possibly a clay mineral or iron sulphide (**figure 6.1**), which provided the template. Others speculate that another type of mineral, called zeolite, provided the site for replication. Zeolites have cavities inside their crystal structures, in which molecules can come together and unite. They are widely used as catalysts in the petroleum industry for this reason. Some scientists maintain that the cell membrane is the more important thing, and must have come first, because the chemicals of life need to be contained before they can react with each other. Yet others believe that proteins rather than RNA came first, a view originating with a famous experiment by Stanley Miller and Harold Urey, in which amino acids were synthesised by electrical discharges in an assumed early Earth atmosphere (water, methane, ammonia and hydrogen).

In reality, while the subject is frequently discussed, no one really knows how life started, what the essential chemicals were, or where they came from. Even the most primitive living organism uses an incredibly sophisticated chemistry, so far removed from any conceivable, simple, self-replicating molecule that it is impossible to relate life to such a molecule. Perhaps in desperation, the famous British biologist Sir Francis Crick resurrected the notion of panspermia, that life on Earth was seeded from outer space, an idea first proposed by the Swedish chemist Svante Arrhenius. Of course, this does not solve the problem of life's origin; it merely defers it elsewhere.

THE EARLIEST LIFE

While the origin of life remains enigmatic, scientists have learnt much about its subsequent development on Earth through the study of living organisms, many of which appear to be extremely primitive. These may in fact be living fossils like the coelacanth or cycads, little different from their ancient ancestors that formed the first living cells.

The most primitive form of life is considered to be a group of bacteria known as the **archaea** or **archaebacteria**. These are single-celled organisms without a nucleus (**prokaryotes**). Archaebacteria occur in a variety of habitats. Many are thermophiles (heat loving) growing around hot springs in volcanic regions and thriving in very acidic conditions. Some inhabit the deep oceans, living around hydrothermal vents along the mid-ocean ridge system. Here they obtain their energy from the chemical imbalance between the hot water and the ocean. They use chemical reactions – such as the oxidation of hydrogen sulphide emerging from the vents to sulphate, the oxidation of iron and manganese, or the reduction of sulphate to sulphide – to power their metabolism.

These bacteria have acquired an impressive name – chemolithoautotrophic hyperthermophiles. This translates as follows: *chemo* = chemical foods; *litho* = sourced in rock; *autotrophic* = to

Terence McCarthy

Figure 6.2 ***Aerobic photosynthesising bacteria (cyanobacteria) use carbon dioxide and water to manufacture carbohydrates. Their appearance meant that life could exist away from hot springs. They nevertheless remain heat tolerant, and their green, slimy colonies can survive water temperatures up to 70°C, such as these at Yellowstone National Park.***

supply energy without other agencies; *philes* = lovers of; *hyper* = extreme; and *thermo* = heat.

A branch of these archaebacteria have adapted to living in extremely saline conditions, the so-called halobacteria. Another group, the methanogens, live in very oxygen-deficient, waterlogged environments, either hot or more ambient, and release methane gas, deriving their energy from fermentation. These bacteria are common today in swamp environments, where they produce the streams of swamp gas that constantly bubble from the black mud. They also thrive in the gut of cattle.

Most of the archaebacteria live under oxygen-deficient conditions, and many prefer hot environments, perfect adaptations to the hydrothermal vents of the early Earth. These simple organisms probably formed the rootstock of all life, including humans. We may still carry a memory of our ancient hydrothermal origins in our need for certain essential trace elements such as copper, zinc, cobalt, nickel, molybdenum and selenium, which form part of important molecules called enzymes, used in protein manufacture. These trace elements are extremely rare in our everyday environment, but are relatively common in the hot, sulphur- and metal-rich water that emerges from hydrothermal vents. It must have been in this environment that these enzymes first evolved.

Euan Nisbet of the University of London has used this line of reasoning to propose that life arose around hydrothermal vents in island arc environments because metals such as zinc, copper and molybdenum are more abundant in these systems.

There is a second group of bacteria, known as the **eubacteria**, which evolved from the archaebacteria early on in life's history. The difference between the eubacteria and the archaebacteria is subtle, involving differences in the structure of their DNA. The primitive forms of eubacteria are also heat lovers and probably evolved in the hydrothermal arena. However, they have widened both their characteristics and their domain, and they played a major part in life's subsequent history. It is this branch of the family that developed photosynthesis and the processes of denitrification and nitrogen fixation that changed the world forever.

PHOTOSYNTHESIS

Photosynthesis is arguably the most important chemical process on Earth because it powers almost the entire global ecosystem. The essentials of the process are simple – carbon dioxide and water combine to make carbohydrate and oxygen gas – but the execution is a masterpiece of chemistry. Carbon dioxide and water will not react with each other spontaneously and are stable together. Energy has to be added to get them to react. This energy comes from light and is captured by the remarkable and extremely complex molecule chlorophyll.

How this molecule and its associated chemical pathways arose is another mystery in the story of life. Euan Nisbet has suggested that it may have originally developed as a heat sensor to protect bacterial cells in the hydrothermal environment. Another possibility is that it evolved from compounds used in the fermentation process. The most primitive form of photosynthesis involves carbon dioxide and hydrogen sulphide, used by the purple sulphur bacteria for carbohydrate manufacture. The process does not release oxygen and is termed anaerobic photosynthesis. Energy to drive it comes in the form of infrared radiation (heat), so it is suited to the hot, hydrothermal environment where it most probably evolved. Later, in the **cyanobacteria** (**figure 6.2**), the process came to involve carbon dioxide and water, using more energetic radiation in the form of red light. This is termed aerobic photosynthesis and produces oxygen as a waste product.

The appearance of aerobic photosynthesising bacteria was extremely important because it meant that life was no longer restricted to hydrothermal vents. Instead of relying on anoxic sites or sites where there was a natural chemical disequilibrium, life could now survive just about anywhere. The only food required was a steady supply of carbon dioxide and water, which were both abundant, and light, which was freely available from the Sun. As a consequence, the cyanobacteria multiplied and colonised the global oceans. All life generates waste. In the case of the cyanobacteria, the waste product was oxygen, a gas that was poisonous to most of the other bacteria cohabiting with the cyanobacteria.

Photosynthesising bacteria evolved very early in the story of life; there is chemical evidence to suggest that they were present on Earth 3 800 million years ago. The oldest rocks to contain visible signs of their presence are those from Barberton, where fossilised bacterial cells have been discovered in cherts (**figure 6.3**). In addition, there is evidence in the rocks at Barberton of bacterial colonial living in the form of stromatolites (**figure 6.4**). Fossilised bacteria have also been found in rocks of the same age in the Pilbara region of Western Australia.

THE OXYGEN CRISIS

During the Archaean Eon cyanobacteria became the dominant life form. They thrived in the carbon dioxide-rich environment and pumped out oxygen. When bacterial cells die, their tissue could potentially react with oxygen, re-forming carbon dioxide, so there should be a balance without net accumulation of oxygen. However, not all of the dead bacterial cells were available for reaction with oxygen; some were buried in ocean sediment and were removed from possible reaction. As a

Maarten de Wit

Figure 6.3 ***A scanning electron microscope image of a 3 400-million-year-old chert from the Barberton Mountain Land showing fossilised, spherical, single-celled bacteria. Each bacterium is about one micron in diameter.***

consequence, there was a net gain in oxygen in the environment. But there were other oxygen consumers as well, particularly iron and manganese dissolved in the oceans. These steadily consumed the oxygen as it was produced, precipitating iron and manganese as oxides in the sediment on the ocean floors (*see* Chapter 4).

Carbon dioxide was steadily being removed from the Earth's atmosphere by photosynthesis, but there was another very efficient mechanism removing carbon dioxide, and this was the precipitation of calcium carbonate in the oceans. Carbon dioxide dissolves in sea water, which also contains calcium; together these form the sparingly soluble mineral calcite. The stock of calcium was continually replenished by weathering of rocks exposed to the atmosphere and by hydrothermal vents. As the area of the continents grew during the Archaean, calcium was added in increasing amounts to the sea, and made a substantial contribution to the removal of carbon dioxide from the atmosphere.

This removal, although essentially an inorganic process, was – and still is – mainly mediated by life, and rocks formed by precipitation of calcium carbonate in the absence of life are rare. Photosynthesising organisms removed dissolved carbon dioxide from the water in which they lived, causing precipitation of calcium carbonate, which accumulated on their slimy colonies, eventually producing stromatolites (*see* Chapter 4). New carbon dioxide was continually being added to the atmosphere by volcanoes. In addition, some of the carbon dioxide locked in sediment was recycled back into the atmosphere at subduction zones. Nevertheless, the overall effect of all this was a steady decline in atmospheric carbon dioxide throughout the Archaean.

At the same time, the nitrogen concentration in the atmosphere may have been rising as nitrogen in the oceans was consumed by the cyanobacteria for protein manufacture and released into the atmosphere as nitrogen gas by denitrifying bacteria. At some point, life must have become nitrogen limited, as most of the global stock of nitrogen accumulated in the atmosphere. At this stage, the cycling of nitrogen by bacteria would have become critical to the maintenance and productivity of life on Earth.

Carl Anhaeusser

Figure 6.4 ***Stromatolites from the 3 500-million-year-old Barberton greenstone belt were formed as colonies of photosynthesising cyanobacteria consumed carbon dioxide from the ancient oceans, causing calcium carbonate to precipitate. This stuck to the slimy bacterial colony, preserving its shape.***

The Archaean Eon saw a radical change in the composition of the Earth's atmosphere from one that was dominated by carbon dioxide to one dominated by nitrogen. The Sun was steadily becoming hotter, but the greenhouse effect of carbon dioxide was declining. Methane, a powerful greenhouse gas produced by anaerobic bacteria, may have compensated in part for this decline.

Atmospheric pressure was probably also steadily declining towards its present value because of carbon dioxide removal from the atmosphere. The cyanobacteria reached their peak between 2 600 and 2 400 million years ago, during widespread deposition of banded iron formations and carbonate rocks such as those found in the Transvaal Supergroup (*see* Chapter 4). By about 2 000 million years ago, the reservoir of iron and manganese in ocean water had been depleted by precipitation on the sea floor, while the production of oxygen continued steadily. A crisis loomed for the anaerobic life that dominated the planet as the oxygen concentration in the atmosphere began to rise sharply.

THE EUKARYOTES

Oxygen, being a very reactive gas, was a potential source of energy. New types of bacteria evolved to exploit this resource and became oxygen respiring, the so-called purple bacteria. Although important, they were not to become the main players as the crisis unfolded. Instead, a completely new type of life emerged. It involved a symbiosis between eubacteria and archaebacteria to form the **eukaryotes**, which have cells with nuclei and organelles (discrete bodies within the cell which fulfil specific functions).

Exactly how this happened is not known, although there is little dispute about the outcome. It may have come about as heat- and carbon dioxide-loving archaebacteria, critically stressed by the rising oxygen and falling carbon dioxide levels, began to associate with purple bacteria because they locally depleted the oxygen concentration and produced carbon dioxide. Eventually, an archaebacterium captured a purple bacterium by completely enveloping it in its cell membrane. The assimilated purple bacterium kept the oxygen level low inside the archaebacterium and also supplied carbon dioxide. In return, it got fed. Over time, this symbiotic couple's descendants became completely integrated, perhaps cementing the union by an exchange of genes – something that bacteria readily do. The pair remained together, but separate. The purple bacteria became the mitochondria that provided the energy needs of the symbiotic pair.

The originator of this theory of the symbiotic origin of eukaryotes, Lynn Margulis, believes that a similar symbiotic association may have given the eubacteria tails to enable them to swim, possibly even before the symbiosis that gave rise to the mitochondria. This first stage of symbiotic relationship may be the origin of the cell nucleus. These bacterial symbionts are the ancestors of animals and fungi.

A symbiotic association between early mitochondrial-bearing cells and cyanobacteria gave rise to chloroplasts, the bodies in plant cells where photosynthesis takes place. These were primitive algae and became the ancestors of plants. Earth's oxygen crisis thus had an important outcome, for by giving rise to mitochondrial- and chloroplast-bearing cells, efficient producers and users of oxygen emerged and the way was clear for the evolution of multicellular, higher life forms (**figure 6.5**). The anaerobic bacteria that had dominated the planet for more than 1 000 million years were relegated to isolated places where oxygen could not reach.

THE OXYGEN SHIELD

The appearance of oxygen in the atmosphere had another important spin-off. When oxygen molecules, which contain two atoms of oxygen bonded together, are exposed to ultraviolet radiation from the Sun, the bond breaks and two highly reactive oxygen atoms are produced. These react with other molecules of oxygen, producing ozone, which is a molecule containing three oxygen atoms. When an ozone molecule is struck by ultraviolet light, it reverts to an oxygen molecule, releasing a free oxygen atom. This promptly reacts with another oxygen molecule to again produce ozone.

Oxygen in the upper atmosphere is thus continuously being cycled in an oxygen-ozone loop. The importance of this cycle is that it absorbs ultraviolet radiation, which is harmful to cells. The appearance of free oxygen in the Earth's atmosphere thus resulted in a radiation shield, enabling living organisms to live initially in the intertidal zone between the high- and low-tide levels, and ultimately on land.

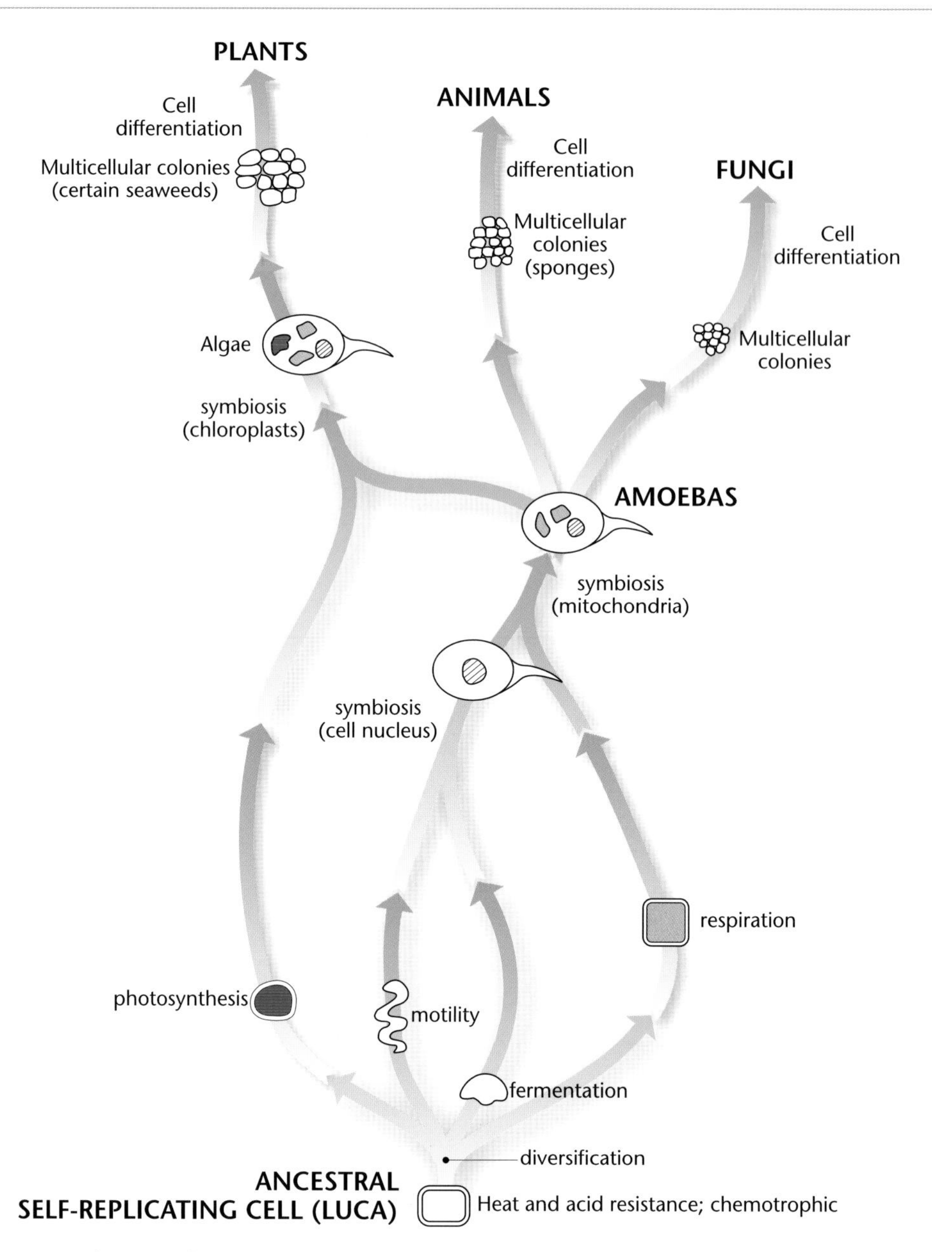

Figure 6.5 *A schematic diagram illustrating the diversification of early life from the ancestral self-replicating organism, LUCA (Last Universal Common Ancestor). The first living organism probably arose at a hot spring on the ocean floor where it obtained energy from the chemical imbalances between hot spring water and ocean water. Diversification followed, leading to varieties of simple bacteria that possessed different characteristics: the ability to swim; and the ability to obtain energy by fermentation, photosynthesis and respiration. Symbiotic relationships developed between these simple forms leading to cells with nuclei, and with mitochondria (the power plants of cells) and chloroplasts (where photosynthesis takes place). These new varieties formed the ancestors of the higher life forms, namely animals, plants and fungi. Many of the primitive life forms still survive today – living fossils unchanged over thousands of millions of years.*

THE EARTH'S ATMOSPHERE

In the early Archaean, Earth was warmed by the greenhouse effect of its carbon dioxide-rich atmosphere. By about 1 800 million years ago, most of the carbon dioxide was gone, buried as calcium carbonate and organic carbon. The oxygen content of the Earth's atmosphere was high, but less than today, while nitrogen probably formed the dominant constituent, as it still does.

Some believe that there may have been a second pulse of oxygen increase about 1 000 million years ago, which brought oxygen to its present level. Cycling of atmospheric gases was in full swing: nitrogen through nitrogen fixing and denitrifying bacteria, oxygen through respiration, decay and photosynthesis by assorted bacteria. Carbon was also being cycled, and some kind of balance was probably established between the addition of carbon dioxide from volcanoes, decay and respiration, and its removal by photosynthesis, precipitation of calcium carbonate and burial of carbon in sediment. It is not known, however, what the concentration of carbon dioxide in the atmosphere was.

In spite of rising energy output from the Sun, Earth's temperature had remained within a relatively narrow range, and still does to this day, 1 800 million years later. This stability suggests that Earth has some kind of a thermostat that helps it regulate its temperature. Some believe that life does the regulation. This is not to say that life does so in a planned way, which it obviously cannot do. Rather, life processes respond to trends in temperature in ways that may produce a counteraction, reversing the trend.

Several processes are involved in the putative control operation. Algae in the oceans release dimethyl sulphide, which reacts with oxygen to produce tiny droplets of sulphuric acid. These act as seeds for water droplet formation, resulting in clouds, as sulphuric acid has a strong affinity for water. In the event of the Earth warming, algal growth is stimulated and more dimethyl sulphide is produced, resulting in greater cloud cover. The clouds reflect more of the Sun's energy and the temperature falls. If the Earth starts to cool dramatically, photosynthesis slows and carbon dioxide removal declines, as does cloud cover. Carbon dioxide released from volcanoes by respiration and decay accumulates in the atmosphere, increasing the greenhouse effect, restoring temperature.

A non-living thermostat has also been suggested. This basically involves changes in the global rate of weathering as temperature rises and falls. The belief is that increased weathering of rocks, induced by a rise in temperature, releases more calcium, which removes carbon dioxide from the atmosphere, and vice versa, thereby regulating the temperature. Such a non-life-based thermostat would, however, be very slow to respond, unlike the life-based version, so the latter would produce a much more sensitive control of temperature.

Life undoubtedly has shaped the composition of the atmosphere. What would it be like without life? Two celestial neighbours provide possible answers to this question. The atmospheres of Venus and Mars have compositions that are fairly similar, consisting of about 96% carbon dioxide and about 3% nitrogen. Argon and various other gases make up the remainder. The atmospheric pressures are quite different, that on the surface of Venus being about 90 bars, while Mars has a very thin atmosphere, exerting a pressure of only about 0.01 bar. The temperatures are quite different too. The surface of Venus is uniformly hot, about 460°C, while Mars, lacking a thick blanketing atmosphere, is freezing cold, averaging about -50°C. Earth's atmosphere is quite different, consisting of 78% nitrogen and 21% oxygen, the remainder being mainly argon.

The isotopic composition of hydrogen in the atmospheres of the three planets indicates that all have lost significant amounts of water since they formed – Venus the most and Earth the least. It is estimated that Earth could have lost an amount of water equivalent to about 1 km ocean depth. Water was lost as a result of interaction with solar radiation in the upper atmospheres of the planets, which broke the water molecules, releasing hydrogen, which diffused into space. Towards the end of the Hadean Era, it is likely that these three planets had similar atmospheres, dominated by carbon dioxide and water vapour, with small amounts of nitrogen. Their atmospheres probably also contained traces of oxygen produced in the upper atmosphere. The oxygen would have been steadily consumed as it was produced by reaction with rocks and also with iron and other compounds dissolved in the oceans. Iron forms a red oxide when it reacts with oxygen (i.e.

rust) and this is possibly the reason for the surface colour of Mars, the Red Planet (**figure 6.6**).

While these planets probably initially had similar atmospheres, their subsequent evolution was quite different. Venus seems to have become overwhelmed by a runaway greenhouse effect; it lost all of its water and its surface is now a furnace. Mars lost its atmosphere possibly as a result of a combination of factors: it is small and its weak gravitational field would have held the atmosphere less securely; it has no magnetic field, so its atmosphere is more susceptible to bombardment by the Solar Wind, which is a constant stream of high-energy-charged particles emitted by the Sun; and because it is small, geological activity may have ended long ago, terminating addition of gas to the atmosphere. Earth evolved into a remarkable planet with an atmosphere very different from that with which it started. The effects of life have been profound. Without life, the Earth would probably have become similar to Venus.

MULTICELLULAR LIFE

Until relatively recently, fossil evidence for life older than about 540 million years – the start of the Cambrian Period – was unknown. What the early geologists observed in the rocks deposited at this time was the sudden appearance of a wide variety of animals with hard body parts that could become fossilised, including snail-like creatures (molluscs), brachiopods (two-shelled animals) and arthropods (crabs, insects, spiders, etc.), especially trilobites. This sudden appearance of diverse animals became known as the Cambrian Explosion.

In the early 1800s, most geologists interpreted this to signify the start of life – the Act of Creation. Charles Darwin realised that this phenomenon posed a problem for his Theory of Evolution. He noted that among the trilobites of the early Cambrian were many different species. His theory required that they should all have evolved in a slow, step-wise manner from a common ancestor. He therefore surmised that:

> *during these vast, yet quite unknown periods of time* [before the Cambrian], *the world swarmed with living creatures. To the question why we do not find records of* [life in] *these vast primordial periods, I can give no satisfactory answer.*

We now know Darwin was correct: the primordial periods did swarm with life. But how did the ancestors of the diverse Cambrian fauna arise?

NASA

Figure 6.6 ***The surface deposits on Mars are red due to the presence of oxidised iron (rust) – perhaps a result of loss of water from the planet.***

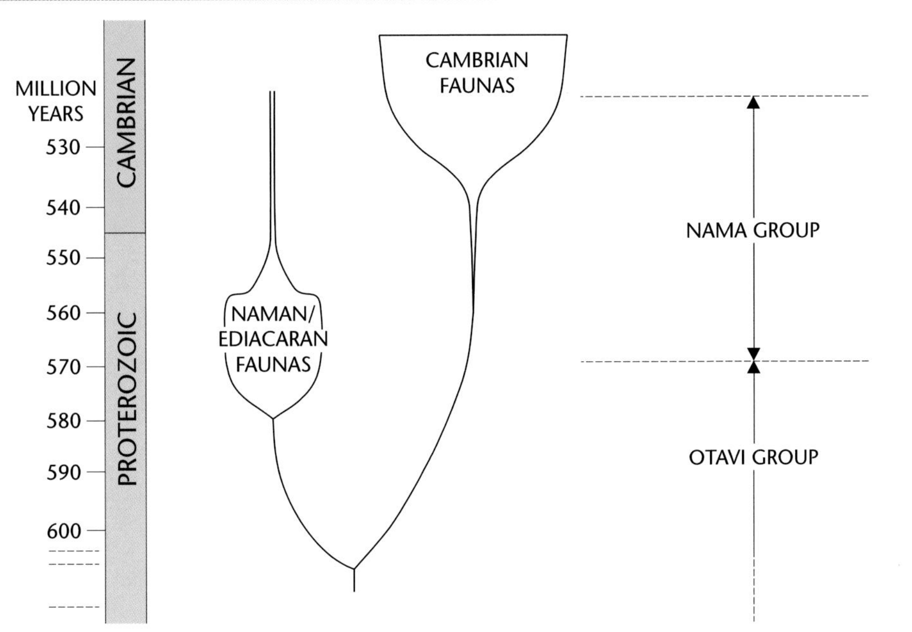

Figure 6.7 ***Between 600 and 500 million years ago, two major radiations of multicellular animal life occurred, the Ediacaran (Naman) and the Cambrian. All modern life evolved from these early ancestors.***

The eukaryotes today form four major groups of organisms: animals, plants, fungi and the protists (amoebas, algae). The first three of these are multicellular in form, with tissue differentiation or specialised tissues. The origins of these complex forms of life are steeped in mystery. The most likely is by colonial growth, when unicellular organisms divided, but the progeny failed to separate. Some seaweeds, consisting of algal colonies, perhaps provide a glimpse of the ancestral plants, and sponges of the ancestral animals. Over many generations, certain parts of these colonies began to assume particular functions, serving the interests of the colony as a whole. Perhaps in this way more complex organisms arose.

THE NATURE OF ANIMALS

The Precambrian fossil record is dominated by stromatolites, but they underwent a decline in diversity during the Late Precambrian and this has been attributed largely to the rise of primitive multicellular grazers. The nature of the origin and evolution of animals remains a puzzle, but recent new fossil discoveries, as well as recent morphological and molecular data, have greatly increased our understanding of the distant origin of animals. The earliest recognisable fossils of animals date to about 610 million years, although their ancestral lineages were certainly much older, but until now their fossils elude us.

When we trace the evidence for past animals, or **Metazoa**, back in time, the closer we approach their origins, the more difficult it is to separate their representatives from those of other once-living organisms such as plants. So what are the basic characteristics of an animal? A useful definition is that an animal is a multicellular heterotroph – something that feeds on other living organisms or their remains. Being multicellular, it is separated from single-celled, heterotrophic protozoans such as amoebas. But fungi also conform to this definition, so it must be added that animals develop from blastular embryos (embryonic stage involving a hollow ball of cells) – something that can be established easily in the case of living animals, but with great difficulty, if at all, when dealing with fossils.

A CRITICAL TIME FOR ANIMAL EVOLUTION

During the 20th century, evidence accumulated to indicate that the period between 600 and 500 million years ago was crucial for the evolution of animals. As shown in **figure 6.7**, two important evolutionary radiations took place in this interval: the so-called Ediacaran or Naman fauna, and the Cambrian Explosion of animal life. These were preceded by two glacial episodes that appear to have been very intense, apparently even affecting the tropical latitudes and resulting, according to some authorities, in Snowball Earth conditions (*see* 'Snowball Earth' on page 119). This was also when the supercontinent of Gondwana was being assembled from its various parts, three of which are of particular relevance to this account of early animal evolution in the southern African context.

Southwestern Africa, in particular Namibia, straddles two ancient continents, the Kaapvaal-Zimbabwe Craton in the south and the Congo Craton in the north. But in the early days of Gondwana, the old continent of Atlantica lay very close in the west, separated from the African mainland by the Adamastor Ocean (*see* **figure 5.13**). The Gondwanan assembly of these three continents is recorded in the Damara Belt of Namibia (*see* Chapter 5). Of particular relevance to the history of early animals are the Otavi and Nama Basins situated on the Congo and Kaapvaal-Zimbabwe Cratons respectively (**figure 6.8**).

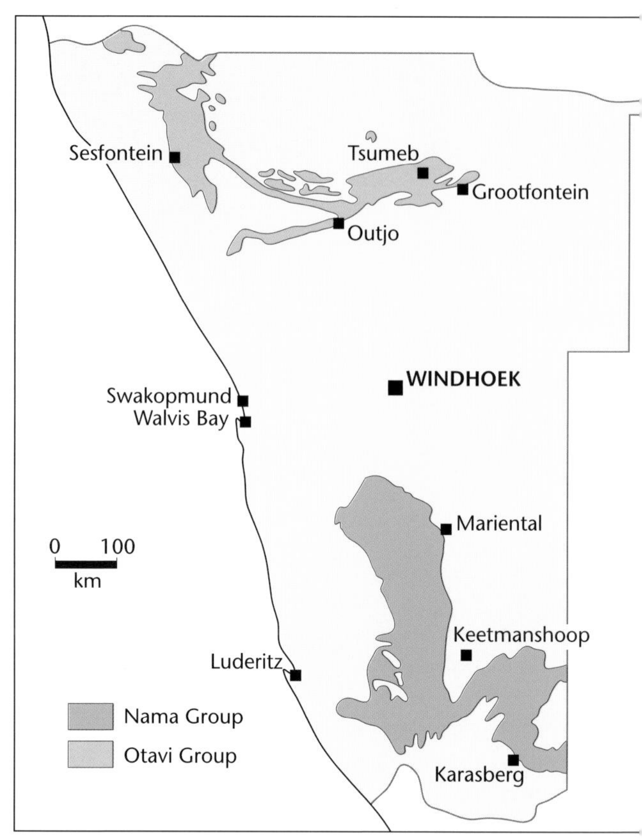

***Figure 6.8** Geological map showing the distribution of rocks of the 750- to 550-million-year-old Otavi Group and the 549- to 539-million-year-old Nama Group. These rocks contain fossil evidence for the first appearance and diversification of multicellular animals.*

ANCIENT ANIMALS OF THE OTAVI GROUP

Sedimentary rocks of the Otavi Group span a time interval of approximately 750 to 550 million years and were deposited in shallow waters along the northern coast of the Khomas Sea (*see* **figure 5.13**). Glacial tillites have long been known to be part of the Otavi Group, but recent fieldwork has shown that there is evidence for at least two – possibly three – distinct glacial episodes in this rock sequence. Furthermore, since the Khomas Sea (Otavi Basin) lay close to the equator during those remote times (which has been determined by palaeomagnetic studies), refrigeration had presumably been intense.

Immediately above each of the Otavi glacial deposits are very distinctive cap carbonates. To explain these, as well as the evidence for intense refrigeration, Paul Hoffman and his colleagues at Harvard University applied the Snowball Earth concept to these Namibian glacial deposits. On the basis of carbon isotope abundances in the limestones concerned, they postulated that during the glacial periods biological activity in the frozen oceans collapsed for millions of years, while runoff from the continents all but ceased.

The frigid conditions ended abruptly when volcanic outgassing raised the atmospheric carbon dioxide level, resulting in severe greenhouse conditions and rapid warming, during which the cap carbonates were laid down in response to the high carbon dioxide content of the atmosphere. It now appears that there were two major glacial episodes worldwide around this time: the earlier one (Chuos Formation of Namibia) about 720 million years ago; and the younger episode (Ghaub Formation of Namibia) about 590 million years ago. There may well have been other less intense glacial episodes as well, but their resulting deposits lack the distinctive cap carbonates.

It is perhaps significant that the first traces of recognisable multicellular animal life have been found in rocks deposited shortly after the Ghaub glaciation. It is tempting to speculate that it was the harsh frigid conditions of this and the Chuos glacial event that stimulated animal evolution, perhaps through extensive fragmentation of the geographic ranges of ancestral organisms involved. Recent molecular evidence suggests that ancestral animal lineages may go back even beyond the time of the Otavi sediments, so it would be most interesting if a glimpse could be gained of some of these small, soft-bodied ancestors as fossils in dolomites of the Otavi Group.

Stromatolites are abundant in Otavi dolomites, a product of photosynthesising cyanobacteria. Although a search has been made for evidence of multicellular life in carbonates of the Otavi Mountain Land and further east in the Kaokoveld, nothing conclusive has been found. Part of the reason for this could certainly be that most of these rocks have been folded and metamorphosed, with the result that delicate structures have not survived.

In recent years, striking fossils of metazoan embryos have been found in China in phosphate-rich limestones of a rather younger date and it is known that phosphatisation can provide the ideal medium for the preservation of delicate organisms. A locality where limestones contain an abundance of unusual objects that have been replaced by calcium phosphate has recently been found in Namibia and may have provided the perfect conditions for the preservation of early metazoan life. The locality is situated on the flat, calcrete-covered Etosha plain. Here the limestones are still horizontal and have not suffered metamorphism. Careful geological investigations have shown that these limestones belong to the Auros Formation, situated below the Ghaub glacial unit, making them about 600 million years old.

The most common phosphatised objects found so far in these limestones strongly resemble small sponges. Their discoverer, Bob Brain, has recently given them the informal name of Otavia. When the limestone is dissolved in acetic acid, which is the best way to expose these minute fossils, each Otavia survives as a double-walled hollow bag of calcium phosphate between 0.3 and 3 mm long. It has several large openings on raised mounds and numerous smaller ones (**figure 6.9**). The large orifices, reminiscent of a sponge's oscula, lead directly into the bag's interior, while the smaller ones open into a peripheral labyrinth between outer and inner walls in the same way as the ostia of sponges do (oscula and ostia are openings that allow passage of water).

Another point of view is that although Otavia has a sponge-like appearance it could be a highly deceptive pseudo-fossil in which layers of calcium phosphate have formed around clusters of carbonate grains held together by an algal film and termed microphytolites. The jury is still out and the investigation continues.

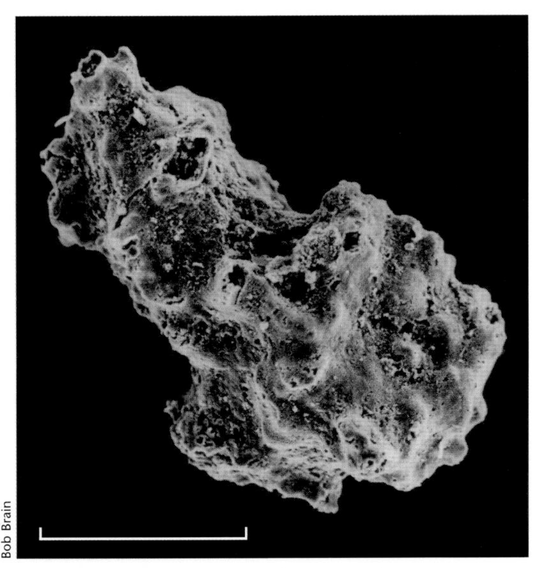

Bob Brain

Figure 6.9 ***A scanning electron microscope image of Otavia, believed to be an early sponge and possibly the oldest known multicellular organism on Earth (650 million years old). This was discovered by Bob Brain in the rocks of the Otavi Group in northern Namibia. (The scale bar in the bottom left-hand corner of this image represents a length of 0.1 mm.)***

FOSSILS OF THE NAMAN OR EDIACARAN RADIATION

The rocks of the younger Nama Group (*see* Chapter 5), consisting of alternating sandstones and limestones, cover a very large area in southern Namibia (**figure 6.8**). They were deposited in a foreland basin between 549 and 539 million years ago. The basin extends from the Orange River in the south almost as far as Windhoek in the north, and from the western escarpment to beyond the South African border in the east. Between 1908 and 1914 geologists discovered impressions of soft-bodied organisms in Nama quartzites of southern Namibia. These were later described by the German palaeontologist G Gürich as five fossil organisms new to science, including *Rangea schneiderhöhni* and *Pteridinium simplex* (**figure 6.10**). They have since been the focus of numerous studies.

In the interim, during 1946, Australian geologist RC Sprigg discovered impressions of soft-bodied organisms in the Late Proterozoic Pound Quartzite in the Ediacara Hills north of Adelaide, South Australia. He described some of them as among the oldest direct records of animal life in the world, observing that they all appear to lack hard parts and to represent animals of very varied affinities.

Figure 6.10 **Pteridinium** ***(A) and*** **Rangea** ***(B) were two ancient life forms that lived in southern Africa about 550 million years ago. (The scale bars are in centimetres.)***

Since then, a wide variety of taxa, based on body-fossil impressions, has been described from the Flinders Ranges of South Australia, and the term **Ediacaran fauna** has been proposed for the life forms that immediately pre-date the Cambrian Explosion.

Assemblages of impressions of soft-bodied organisms similar to those from Namibia and South Australia have since been found in various parts of the world, though it is somewhat ironic that these should be referred to as the Ediacaran fauna, since the first examples were found and described from Namibia, long before they were dicovered in Australia.

There is probably no other group of fossil organisms that has generated such a diversity of opinion as to their affinities as has this fauna. The German palaeontologist Adolf Seilacher wrote in 1989 that these organisms showed:

> *a unique, quilted type of biological construction that had no counterpart in the modern, or even the Phanerozoic biosphere.*

Three years later he assigned this fauna to a new Kingdom, the Vendobionta, which he defined thus:

Immobile foliate organisms of diverse geometries that were only a few millimetres thick, but reached several decimetres in size. A shared characteristic is the serial or fractal quilting of the flexible body wall, which stabilised shape, maximised external surface and compartmentalised the living content. Since no organs can be recognised, this content is thought to have been a plasmodial fluid, rather than multicellular tissue.

Seilacher was inclined to regard these creatures as representing a failed experiment in the evolution of animal life. However, there is wide divergence of views as to the affinities of these organisms. For example, *Dickinsonia* (the largest specimens were about 1 m across) has been variously classified by different researchers as: Phylum Coelenterata – jellyfish, sea anemones, etc; Phylum Platyhelminthes – flat worms; Phylum Annelida – segmented worms; Phylum Petalonomae – specially created for these organisms; Vendozoa – informal name created for these organisms; Phylum Vendobionta – specially created for these organisms; Kingdom Vendobionta – specially created for these organisms; and Kingdoms Plantae/Fungi – lichens.

In 1997 an impression of another remarkable quilted organism was found in Nama sandstones dated at about 543 million years old, and thus the youngest example of this fauna known. It was discovered on the farm Swartpunt in southwestern Namibia and was given the name *Swartpuntia germsi* (the latter in honour of Gerard Germs, a pioneer in the study of the Nama Group). Each oval petaloid is remarkably similar to the Australian form *Dickinsonia* in appearance, but several of these are attached to a central stalk, by which the organism was anchored to the sea floor.

Fossils of this remarkable organism seem to lend weight to the concept of the Vendobionta as an extinct group, very different from any of the animals known from later times. In fact, it has been suggested that these were unique creatures living in the tranquil waters of the so-called Garden of Ediacara, deriving their nutrients from symbiotic green algae or absorbing them directly from the water. Whatever their mode of life, they certainly disappeared rapidly from the scene, possibly when the first effective animal predators appeared.

THE LINEAGE OF ANCESTRAL ANIMALS

Between 600 and 500 million years ago two radiations occurred (**figure 6.7**) – that of the Ediacaran fauna and also of the true animals, whose explosive development occurred in mid-Cambrian times about 530 million years ago.

In 1972, Gerard Germs demonstrated that the Nama sediments of Namibia contain fossils that formed part of an early animal lineage, long before the Cambrian Explosion. In that year he described *Cloudina*, from Nama limestones. This animal consisted of small calcareous tubes (**figure 6.11**) with a distinctive cone-in-cone structure. Although fossils of *Cloudina* are widely scattered through the limestones, there is one locality in the northern Nama Basin where *Cloudina* fossils are found in a

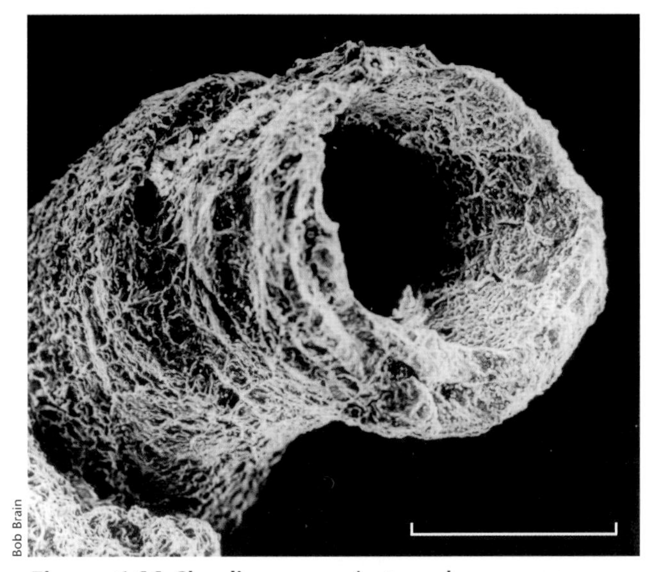

Figure 6.11 Cloudina, *an ancient coral or sea anemone, which lived in the oceans in southern Africa about 550 million years ago. This is one of the earliest organisms showing development of a shell, in this case tube-like. The tube is thought to have housed the body of an invertebrate, whose tentacles protruded from the opening. (The scale bar represents 0.2 mm.)*

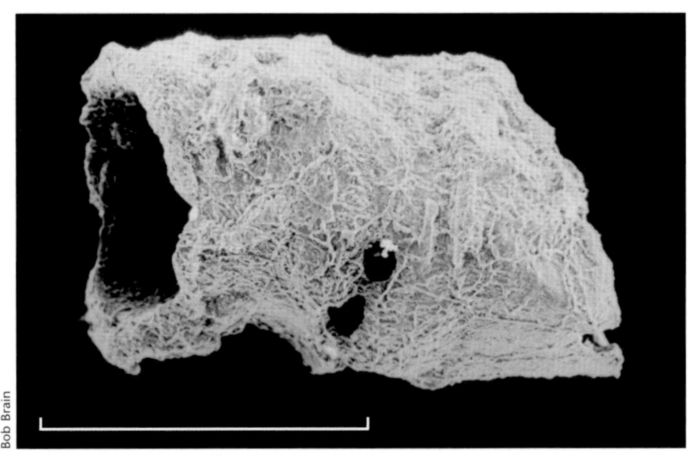

Figure 6.12 ***These small holes in the shell walls of* Cloudina *were possibly made by an early predator that bored through the shell to gain access to the soft animal body within. (The scale bar represents 0.5 mm.) The sudden appearance of hard body parts, which start occurring as fossils in rocks from the Cambrian Period, is believed to be the result of the rise in predation that started about 540 million years ago.***

spectacular thicket, or bioherm. Each individual tube is thought to have housed an animal of probable coelenterate affinities (possibly a sea anemone or coral), whose tentacles protruded from the open end. The other end was presumably attached to the algae-rich biomat on the ocean floor.

A specially significant feature of *Cloudina* is that it made use of biomineralisation (shell-like structure) in the construction of its tube long before this became a commonplace activity during the run-up to the Cambrian Explosion. As recently as the year 2000, fossils of a second biomineralising organism were found in Nama limestones and named *Namacalathus hermanastes*. The calcified part of each of these organisms took the form of a goblet-shaped cup, about 1 cm across, open at the top and with a number of holes around the sides, but with a slender stalk below that served for attachment to the surface of an algal reef. Zoological affinities remain uncertain but, as with *Cloudina*, the tentacle-bearing Coelenterata could be a likely option.

The calcification of an organic body-wall, as is present in both *Cloudina* and *Namacalathus*, could well have been a survival strategy in response to the attentions of early animal predators that were evolving at that time. *Cloudina* fossils are also known from other parts of the world and at one locality in China they have been phosphatised, which allows their removal from the limestone matrix with acetic acid. During a study of these *Cloudina* tubes it was noticed that many of the tube-walls had been perforated, presumably by some species of predator that had bored into them to feed on the soft tissue within.

Until recently, studies on Namibian *Cloudina* fossils were restricted to examining thin or polished sections because specimens did not survive acetic acid treatment. But recently silicified *Cloudina* fossils have been found by Bob Brain, and a number of these also showed evidence of predatorial boring, as seen in the Chinese example (**figure 6.12**). This has stimulated research to discover the ancient predator. Brain has identified several candidates in thin sections of rocks from this locality, but has yet to find a well-preserved individual.

Research in other parts of the world has demonstrated that the Precambrian-Cambrian boundary at about 540 million years ago is characterised by a rapid build-up of small, shelly fossils. In Namibia, sediments of the Fish River Subgroup cover this critical time period, but fossils have not yet been found in these rocks. For reasons that are still hotly debated, the Cambrian Period was also the time when representatives of virtually all the living phyla (groups) of animals made their appearance in the fossil record.

After the Cambrian period 540 million years ago, fossils show a rapid increase in complexity and abundance. Simple skeletal tubes appear first, and these are followed by early brachiopods (lamp shells), molluscs and arthropods, especially trilobites. Some extraordinary Early and Middle Cambrian deposits have been found containing fossils that record remarkable detail of the organisms, even preserving soft parts. One of these localities is the Burgess Shale of the Canadian Rockies.

More recently, equally well-preserved and somewhat earlier fossils have been found in the Chengjiang sediments of China, showing that the mineralisation of skeletal parts – the origins of which we can see in the Nama limestones – was by then a commonplace procedure. This Cambrian locality also includes fossils that may be very early fish.

In just 65 million years – possibly even less – the Metazoan radiation progressed from sponges to all the major groups of multicellular organisms found today. The rest of the Cambrian radiation involved the diversification of all these groups. Comparative developmental studies have shown that different multicellular animals have remarkably similar regulatory elements. For example, similar genes in both mice and flies control body patterning, eye and heart formation, segmentation and appendage differentiation. This suggests that all Metazoan animals evolved from some common ancestor, and the fossil evidence points to their initial radiation some time toward the end of the Precambrian Period.

The diversification of multicellular life began well before the start of the Cambrian Period. Later, during the Cambrian, many animals simultaneously began to cover themselves with hard surfaces amenable to fossilisation, resulting in the Cambrian Explosion. This is undoubtedly a reflection of the rising prevalence of predation. Euan Nisbet, in his book *Living Planet*, reminds us that in the long history of human conflict it is not the weapons that endure to record past wars, but the fortifications, like the city walls of Old Jerusalem, the castles of Europe, or the Great Wall of China. So, too, was it with early animal life.

THE IMPRINT OF LIFE

For nearly 3 500 million years, the Earth's biota appear to have consisted only of simple, single-celled organisms living in the oceans. Simple they may have been, but their effect on Earth was profound. Life shaped the Earth's atmosphere and may even have played a significant role in maintaining the surface temperature within a narrow range – a range that was suitable for the continued existence of life. During this long period the scene was set and conditions became suitable for the survival of more advanced life forms.

Evolution of multicellular organisms with tissue differentiation began only about 700 million years ago. Why their appearance took so long is unclear. Perhaps the answer lies in the composition of the atmosphere, in particular the oxygen abundance. Oxygen in the atmosphere may have had to reach a critical level before multicellular life could survive, a situation only attained about 1 000 million years ago.

Other factors could also have played an important role. Multiple glaciations between 750 and 590 million years ago may have isolated populations, leading to species diversification. The appearance of sexually reproducing organisms may have been important, as sexual reproduction increases genetic diversity in populations and accelerates evolutionary change. Once multicellular organisms did appear, they rapidly evolved into the great profusion of forms we find in the fossil record and living today.

SUGGESTED FURTHER READING

Casti, JL. 1989. *Paradigms Lost.* Avon Books, New York.

Fortey, R. 1997. *Life: an Unauthorized Biography.* Harper Collins, London.

Johnson, MR, Anhaeusser, CR and Thomas, RJ. 2005. *The Geology of South Africa.* Council for Geoscience, Pretoria.

Kasting JF. 2004. 'When Methane made Climate.' *Scientific American*, vol. 291, pp 52–59.

Knoll, AH. 2003. *Life on a Young Planet.* Princeton University Press, Princeton.

Lane, N. 2002. *Oxygen: the Molecule that Made the World.* Oxford University Press, Oxford.

Lovelock, J. 1988. *The Ages of Gaia.* Oxford University Press, Oxford.

Margulis, L. 1998. *The Symbiotic Planet.* Weidenfeld & Nicolson, London.

Nisbet, EG and Sleep, NH. 2001. 'The Habit and Nature of Early Life', *Nature*, vol. 409, pp 1 083–1 091.

Nisbet, EG. 1991. *Living Planet.* Chapman Hall, London.

Schopf, JW. 1999. *Cradle of Life.* Princeton University Press, Princeton.

THE STORY OF
EARTH & LIFE

7

THE ROCKS OF GONDWANA

IOA

The legacy of Gondwana in southern Africa is epitomised by the rocks of the Karoo.

ROUTE MAP TO CHAPTER 7

Age (years before present)	Event
500 million	• The supercontinent Gondwana had consolidated and the mountain chains along its sutures were being eroded, exposing the Pan-African metamorphic rocks. Gondwana lay in the southern hemisphere, but was drifting slowly northwards, passing over the South Pole in the process. What is now the southern Cape lay about 1 500 km from the margin of southern Gondwana.
450 million	• Rifting began across what is today the southern Cape, causing thinning of the crust and invasion by the sea, giving rise to the Agulhas Sea across the southern Cape. Sediment accumulating in the sea produced the Cape Supergroup.
310 million	• A subduction zone developed along the southern margin of Gondwana, and the interior of the supercontinent began to experience compression. This initiated folding of the Cape Supergroup rocks, and what was previously the Agulhas Sea became the Cape Mountains. A large depression formed along the northern margin of this mountain range in which sedimentary rocks of the Karoo Supergroup began to accumulate. At that time, the southern African portion of Gondwana lay over the South Pole and was covered by ice. Extensive glaciation occurred. As Gondwana moved northwards from beneath the icecap, the ice melted and retreated. Thick accumulations of glacially derived sediment were deposited (Dwyka Group). As the ice disappeared, large rivers formed and discharged sediment into the Karoo Sea from the Cape Mountains in the south and the Cargonian Highlands in the north, depositing the Ecca Group of rocks. Extensive swamps developed on the deltas at river-mouths in the north, giving rise to South Africa's coal deposits. The Karoo Sea gradually filled with sediment, derived mainly from the Cape Mountains in the south, and the deltas gave way to extensive flood plains bordering meandering rivers. These deposits form the Beaufort Group.
251 million	• A major mass extinction occurred, making 96% of species extinct. The cause is unknown. Meandering river deposits gave way to braided river deposits, possibly due to massive erosion following the die-off of land plants. Conditions returned to normal as the vegetation recovered. Only a few animal species survived. Gondwana continued its northward drift, and conditions became more arid, producing numerous salt pans. Some larger rivers meandered across the region, draining the higher rainfall areas to the south. Conditions became still more arid, and a vast sand sea formed (Stormberg Group of rocks).
182 million	• A mantle plume rose beneath southern Africa, the crust ruptured and vast quantities of lava poured out on surface, covering most of the region, in places to a depth exceeding 1 600 m. Remnants of these lavas form the mountainous regions of Lesotho. This event marks the start of the break-up of Gondwana.

THE IMPORTANCE OF GONDWANA

The surface of the Earth has in all likelihood witnessed the formation and dispersal of several supercontinents during its 4.6-billion-year history (*see* Chapter 5). However, it is the formation and fragmentation of Pangaea and its offspring Laurasia and Gondwana – which were the most recent of the supercontinents – that is most important for the present configuration of the world's continents, as well as for life as we know it today. Pangaea came into existence at about the time that land-dwelling life forms emerged, and it was on this supercontinent that the radiation of multicellular life into all the higher groups of terrestrial animals occurred. This common ancestry of the world's animals is evident on all of today's continents.

The rocks of the continents that once comprised Gondwana (that is, South America, including the now submerged Falkland Plateau, Africa, Australia, Antarctica, Madagascar and India – **figure 7.1**) contain over 60% of the world's economically important minerals. In addition, sedimentary and volcanic rocks deposited on the supercontinent today cover some two-thirds of South Africa and largely shape its landscape. These rocks are also valuable in preserving a unique record of the evolution of reptiles and mammals.

THE AGULHAS SEA

By 500 million years ago, Gondwana had consolidated (*see* Chapter 5), and the mountain ranges that had developed along the sutures between the assembled fragments (the Pan-African belts; **figure 5.10**) were being eroded, exposing rocks that had formed deep below the surface, such as the Cape Granites. What is today South Africa lay close to the margin of Gondwana, snugly tucked between the southern tip of South America and east Antarctica

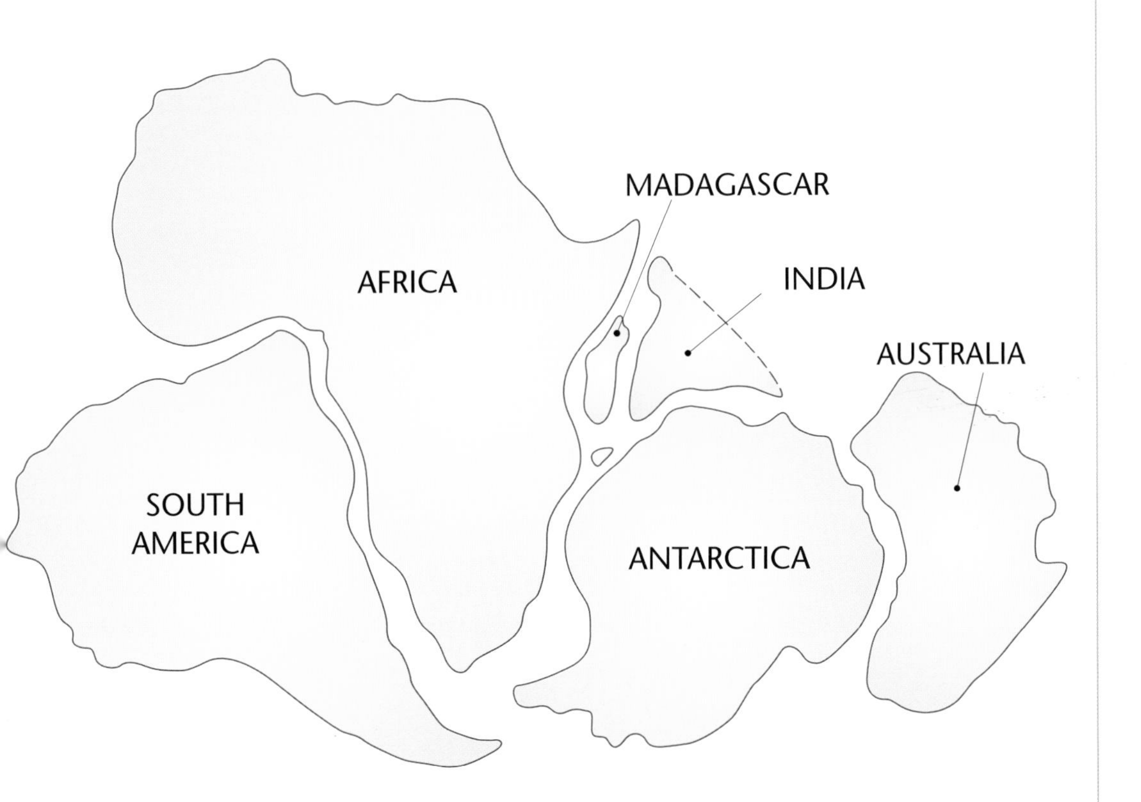

Figure 7.1 *The supercontinent Gondwana, which assembled between 700 and 500 million years ago, was made up of all of the continents of the southern hemisphere, together with India.*

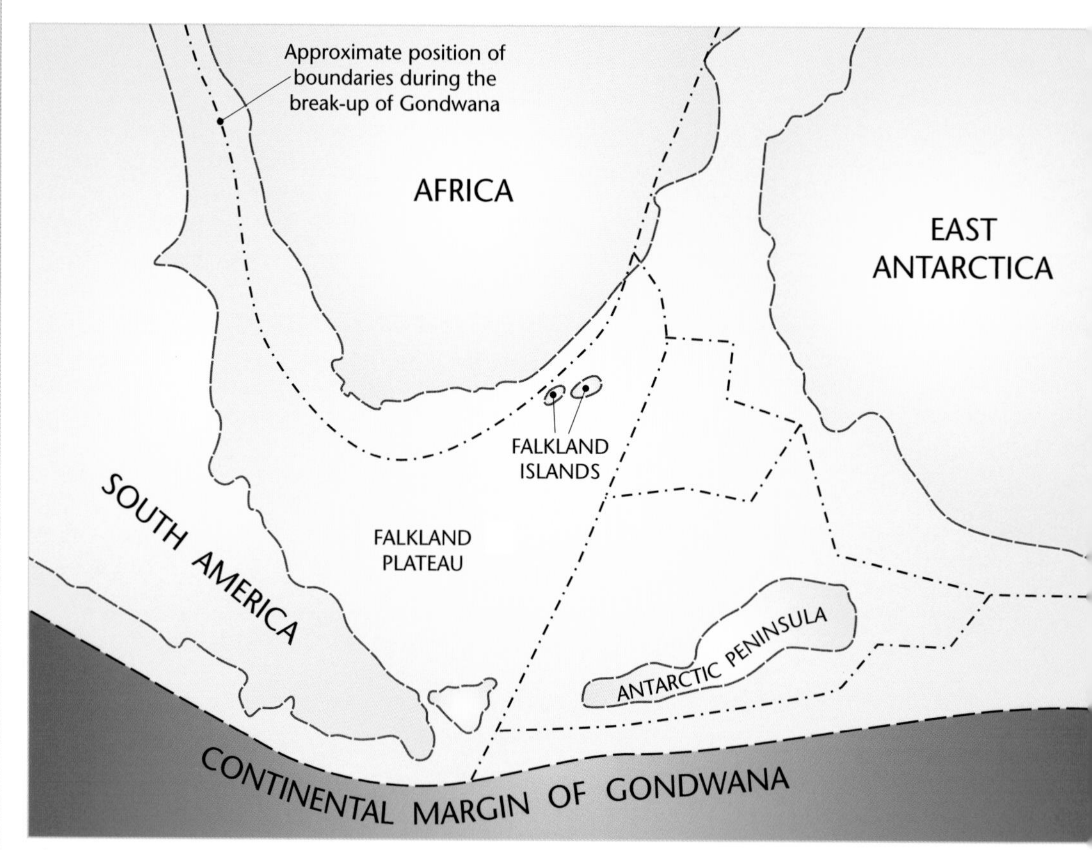

Figure 7.2 ***Some 500 million years ago, what is today southern Africa lay nestled between South America, the Falkland Plateau (now submerged) and East Antarctica, about 1 500 km from the margin of Gondwana.***

(**figure 7.2**). At that time, Gondwana was located in the southern hemisphere, and what is today northwest Africa (Morocco) was situated over the South Pole. The continent was moving northward, sliding slowly over the pole; by 450 million years ago what is now central Africa lay over the pole.

Rifting across the southern Cape

During northward drift, the supercontinent began to experience internal tension and the Pan-African belts of the southern Cape (**figure 5.14**) in particular responded by stretching and thinning, initiating rift valleys. The first rift to form developed in a north-south orientation, northeast of the present position of the Cape Peninsula. Rivers diverted into the rift and began to deposit sand and gravel to form the **Klipheuwel Group** (**figure 7.3**). Occasional incursions of the ocean resulted in the accumulation of shallow marine sands. The Klipheuwel Group has not yielded any fossils, but the rocks contain arthropod tracks, indicating that invertebrates were around at the time.

Continued stretching of the now deeply eroded Pan-African belts created a major rift. This extended eastwards across the southern portion of South Africa to beyond the present position of Port Elizabeth, with a northeasterly branch extending into present-day KwaZulu-Natal (**figure 7.3**). Rapid flooding of the rift occurred, leading to the formation of a seaway across the southern Cape, known as the **Agulhas Sea**. This sea was floored by rocks of the Klipheuwel Group in the west and by eroded Pan-African rocks further east (Cape Granites and associated metamorphosed sedimentary rocks, including the Malmesbury, Kango and Kaaimans Groups, *see* Chapter 5).

Figure 7.3 *About 500 million years ago, a rift began to develop along what is today the southern Cape coast. This was initiated in the west, but later extended across the entire region. The sea flooded this rift, forming the Agulhas Sea as the crust thinned and subsided. Initially sedimentation was confined to the western and eastern ends of the rift, forming the Klipheuwel and Natal Groups of rocks. As rifting continued, the depression enlarged and more widespread sediment accumulation took place, forming the Cape Supergroup.*

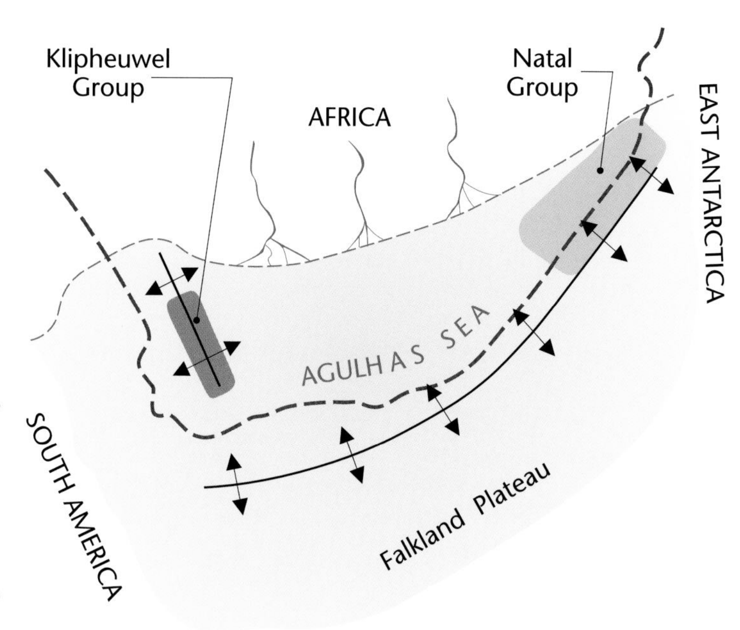

The sedimentary rocks deposited in the Agulhas Sea are known as the **Cape Supergroup**. The north-easterly branch received mostly coarse sediments, which today form the strata of the **Natal Group** (**figure 7.3**) and may have formed at about the same time as the Klipheuwel Group further west. Rocks of the Natal Group are especially well exposed in road cuttings between Durban and Pietermaritzburg, particularly at the Marian Hill Toll Plaza, as well as to the west of Melmoth and in the Oribi Gorge area (**figure 7.4**).

The rift continued to deepen, and as much as 8 km of sedimentary rocks accumulated in the Cape Supergroup. These rocks can be broadly subdivided into three units of differing age, environment of formation, and fossil content. They were deposited over a period ranging from about 500 to 330 million years ago (the Ordovician to the Carboniferous; *see* 'The geological time scale' on page 71).

Figure 7.4 *Sandstones deposited during the opening of the Agulhas Sea today form the cliffs flanking Oribi Gorge in KwaZulu-Natal.*

In the vicinity of Cape Town, rocks of the Klipheuwel Group are absent. Here the Cape Supergroup lies directly on an eroded surface of 550-million-year-old Cape Granites. The basal part of the

Erhardt Thiel / IOA

Figure 7.5 *The floor of much of the Agulhas Sea consisted of granites and metamorphic rocks formed during the assembly of Gondwana. The contact between these floor rocks (Cape Granite in this case) and the marine deposits of the lower Cape Supergroup, is well exposed along Chapman's Peak Drive on the Cape Peninsula.*

Figure 7.6 ***The sedimentary rocks that form Table Mountain in Cape Town are part of the Cape Supergroup, and were deposited in the Agulhas Sea between 500 and 460 million years ago. In the foreground are the metamorphic rocks of the Malmesbury Group.***

Cape Supergroup sequence, known as the **Table Mountain Group**, and its contact with the underlying granite, is rather well exposed along Chapman's Peak Drive on the Cape Peninsula (**figure 7.5**). The lowermost portion of the group makes up the imposing 1 098-m-high mountain for which Cape Town is so famous (**figure 7.6**). Although these rocks now form an inspiring flat-topped mountain, they originated in a shallow sea, into which numerous sandy, braided rivers disgorged their sediment load. The marine nature of this sequence is well documented by the presence of shallow-water animal tracks – notably those of trilobites – as well as by the character of the sedimentary rocks.

Sometime in the period between about 440 and 420 million years ago (the Ordovician, extending into the Silurian Periods), Gondwana experienced an extensive glacial period, during which the ice cap expanded greatly and tillites were deposited (Pakhuis Formation). At that time, what is now central Africa lay over the South Pole. Melting of the glaciers as the ice retreated resulted in the deposition of fine muds (Cedarberg Formation) in shallow embayments and glacial lakes. The fine-grained sediment was supplied by winnowing of the pre-existing glacial deposits of the Pakhuis Formation. These fine-grained sediments preserve a diverse assemblage of marine invertebrate fossils, including bivalves (mussels), gastropods (snails), cephalopods (octopus and related species) and a small eurypterid (aquatic arachnids), as well as a giant conodont, a distant ancestor of the vertebrates, and a newly discovered primitive jawless fish. Modern representitives of this group include lampreys and hagfish.

Deepening of the Agulhas Sea

About 400 million years ago (early in the Devonian Period), further extension accompanied by rapid subsidence deepened the Cape trough, bringing about the deposition of deeper-water, fine-grained sediments of the **Bokkeveld Group**. In contrast to the predominantly sandy sediments of the Table Mountain Group, the overlying Bokkeveld Group consists largely of mudstones. These rocks weather more quickly than the sandstones of the remainder of the Cape Supergroup and consequently form valleys rather than mountains. Weathered mudstones of the Bokkeveld Group today form the fertile soils on which the vines of the winelands of Stellenbosch, Paarl, Worcester and Robertson are cultivated.

It is within the fine-grained rocks of the Bokkeveld Group that the bulk of the well-known invertebrate fossils of the Cape Supergroup have been found. These fossils include a variety of marine organisms, with a particular diversity of brachiopods, as well as trilobites, molluscs (bivalves and gastropods), echinoderms (including starfish, crinoids and the extinct blastoids and cystoids), ostracods (small arthropods) and foraminifera (**figure 7.7**). In addition to this rich invertebrate fauna, fish with jaws (placoderms) were also present.

The rocks of the Bokkeveld Group grade upward into the more sand-rich deposits of the overlying **Witteberg Group**, so named for the impressive mountain range in the area to the south of Laingsburg. These rocks were laid down about 370 to 330 million years ago in diverse depositional settings ranging from rivers, fresh and brackish water lakes, to deltas and the shallow marine realm. Due to their more sandy nature, these rock strata do not normally preserve fossils as well as those of the underlying Bokkeveld, but the assemblage that is preserved includes primitive fish, extinct species of sharks, brachiopods (lamp shells), bivalves, a giant eurypterid sea scorpion (more than 1 m long), plants and numerous animal tracks.

MOUNTAINS RISE FROM THE SEA

The top of the Witteberg Group marks the end of sedimentation in the Agulhas Sea – an end brought on by a change from an environment of stretching, thinning and subsidence of the crust, to one of compression and crustal shortening and thickening. The cause of this change appears to have been the development of a subduction zone along the margin of Gondwana some 1 500 km to the south, which resulted in a continental margin much like

Figure 7.7 *Trilobites (**A**) were very common marine animals around 400 million years ago, whose fossils are found in rocks of the Cape Supergroup, as were starfish (**B**).*

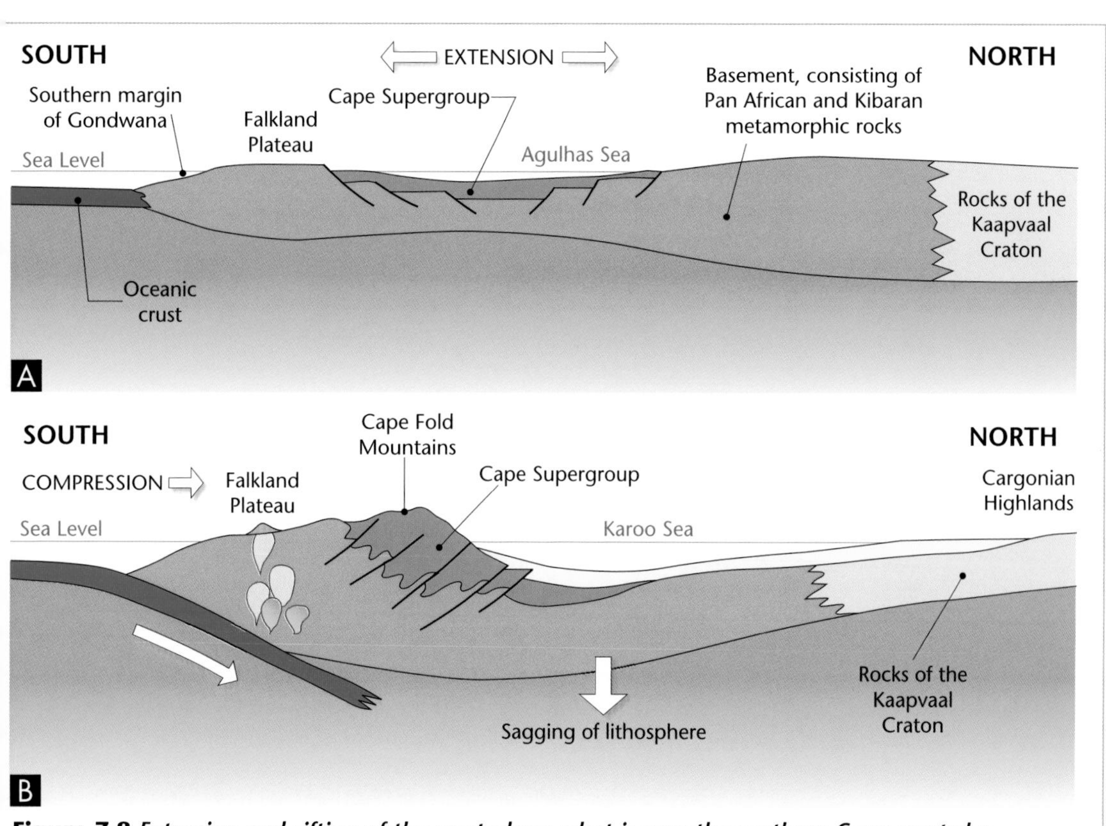

Figure 7.8 *Extension and rifting of the crust along what is now the southern Cape created a depression, the Agulhas Sea, in which rocks of the Cape Supergroup were deposited. This is shown in the cross-section through the margin of Gondwana (**A**). About 330 million years ago, a subduction zone developed along the southern margin of Gondwana, causing compression in the interior. Sedimentary rocks of the Cape Supergroup began to fold, forming a mountain range (**B**). Rhyolite magma rising from the subduction zone produced volcanic activity. The weight of this growing range caused the lithosphere to sag and a depression formed along its northern side, in which an inland sea formed – the Karoo Sea.*

the western United States today. The sedimentary rocks on the floor of the Agulhas Sea began to buckle as the crust was compressed (**figures 7.8, 7.9**). The crust thickened and a mountain range began to rise where formerly a sea had existed. This range was composed of sedimentary rocks previously deposited on the floor of the Agulhas Sea, and formed the mountains of the embryonic **Cape Fold Belt**.

Today, the rocks of the Cape Supergroup show a linear, but in detail, complex distribution (**figure 7.10**), a consequence of the intricate folding of the layers. The crust began to sag under the load of this mountain range and a depression or basin developed on its northern flank (**figure 7.8**; *see also* 'Sedimentation in foreland basins' on page 162).

Figure 7.9 *Folded layers of quartzite of the Table Mountain Group can be seen at Meiringspoort.*

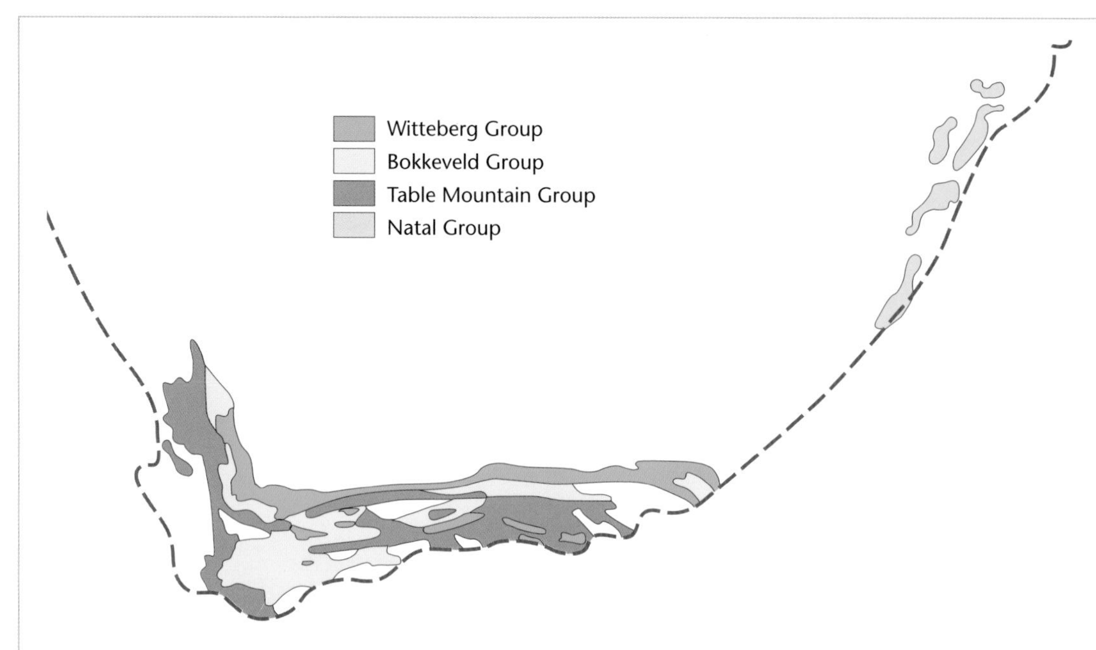

Figure 7.10 *This geological map shows the present-day distribution of the rocks of the Cape Supergroup, which largely lie in an east-west belt across the southern portion of the country. In detail, they form a complex pattern due to intense folding of the originally horizontal layers.*

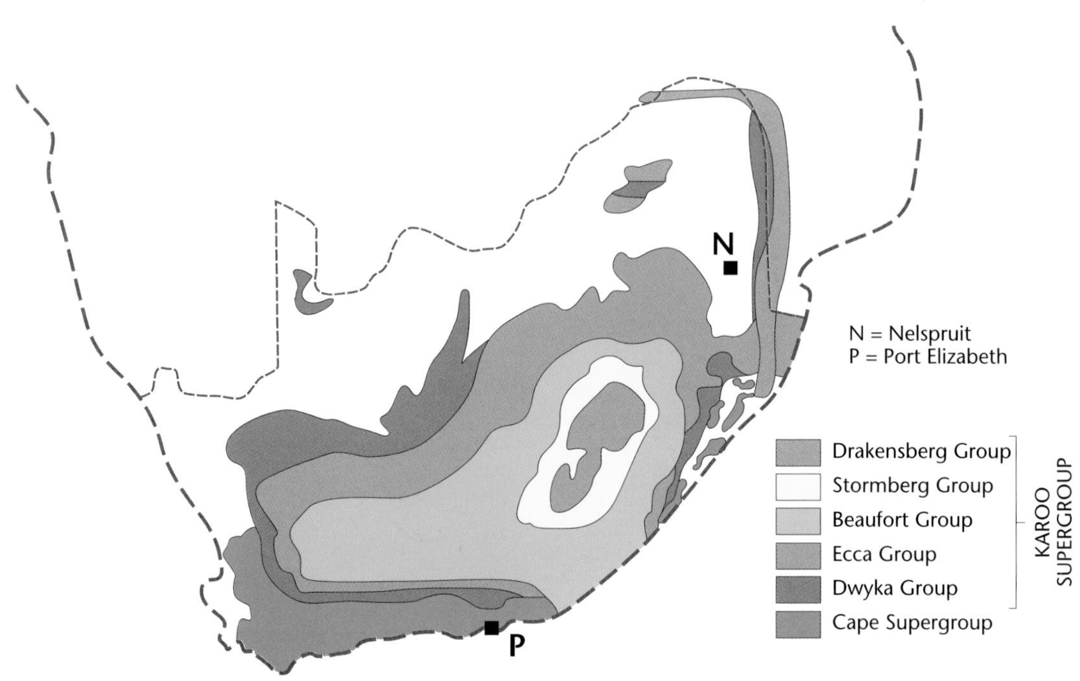

Figure 7.11 *This geological map shows the present-day distribution of the rocks of the Karoo Supergroup in South Africa, which consist mainly of sedimentary rocks. These rocks cover some two-thirds of the country and were deposited in the inland Karoo Sea, which formed behind the rising Cape Mountain range (see* **figure 7.8***).*

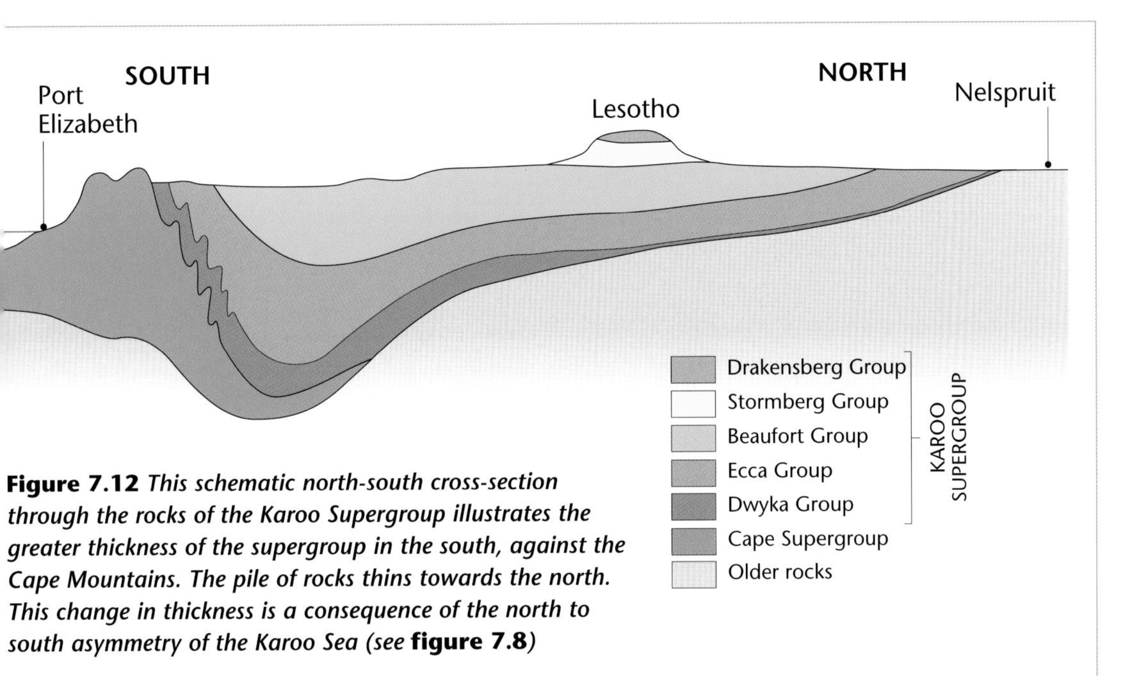

Figure 7.12 *This schematic north-south cross-section through the rocks of the Karoo Supergroup illustrates the greater thickness of the supergroup in the south, against the Cape Mountains. The pile of rocks thins towards the north. This change in thickness is a consequence of the north to south asymmetry of the Karoo Sea (see* **figure 7.8***)*

A completely new cycle of sediment deposition was about to begin, forming the **Karoo Supergroup**.

The rocks of the Karoo Supergroup today cover approximately two-thirds of the land surface of South Africa (**figure 7.11**) and form a thick pile of dominantly sedimentary strata that were deposited over the period 310 to 182 million years ago (Late Carboniferous to the Middle Jurassic Periods). Because of the nature of the mountain-building episode in the Cape Fold Belt, as well as the nature of the underlying lithosphere, the Karoo sequence of rocks is highly asymmetrical. It is thickest in the south (maximum cumulative thickness of 12 km) and thins dramatically to the north (**figure 7.12**).

EMERGING FROM THE ICE

At the time the Cape Fold Mountains began to form, South Africa was located over the South Pole because of the steady northward drift of Gondwana. The ice sheet covering southern Gondwana was probably several kilometres thick. Glacial deposits (the **Dwyka Group**) were therefore the first sediments to be deposited in the developing Karoo depression. It was the recognition of these glacial sedimentary rocks in South Africa, India and South America that provided important early evidence in support of the Theory of Continental Drift (*see* Chapter 2).

The basin was deepest in the south along the Cape Mountain front (**figure 7.8**). In the northern part of the basin, where the water was shallow, the ice was grounded (**figure 7.13**) along an elevated region that today extends from the Northern Cape via the Witwatersrand to

Figure 7.13 *The terminus of this Icelandic glacier is marked by glacial deposits. Melt water streams emerging from the glacier have eroded channels through the glacial deposits.*

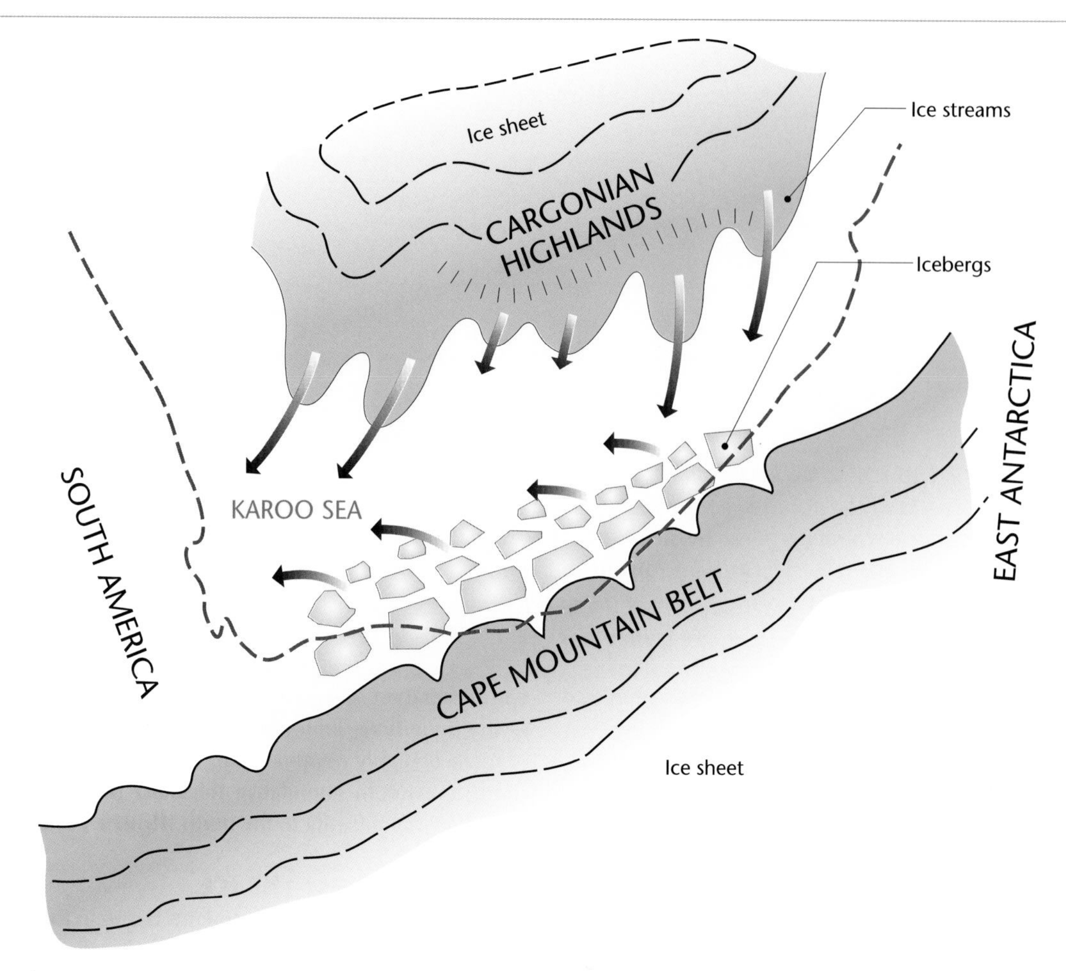

Figure 7.14 *At the time Gondwana assembled it lay in the southern hemisphere, but was gradually drifting northwards. About 300 million years ago, what is now southern Africa lay over the South Pole. As Gondwana moved northwards, the southern African region began to emerge from beneath the ice cap. The Cape Mountains had started forming by this time, creating the Karoo Sea to the north. Ice persisted for some time on the high ground to the south (Cape Mountains) and north (Cargonian Highlands) of the Karoo Sea.*

Mpumalanga, known as the Cargonian Highlands (**figure 7.14**). Here there is still evidence of the effects of these glaciers moving across the land surface. Striations caused by the glacial ice scouring the ancient bedrock over which it moved are known from the Kimberley area – e.g. the famous glacial pavements at Nooitgedacht near Barkly West (**figure 7.15**) and in KwaZulu-Natal, including in the grounds of the University of KwaZulu-Natal's Durban-Westville campus.

As the glaciers seasonally melted and retreated they left behind vast quantities of mud and large fragments of rock (**figures 7.13, 7.16**), which formed the characteristic, poorly sorted Dwyka tillite (**figure 7.17**). Unusual evidence of this glaciation can be seen on the banks of the Vaal River at Vereeniging. Here diamictite has filled previously existing solution cavities or caves in the dolomitic rocks of the Transvaal Supergroup (**figure 7.18**).

In the deeper southern part of the basin, against the slowly rising Cape Mountains, the large, deep inland **Karoo Sea** formed (**figure 7.14**). Glaciers emerging from the mountains floated out on this sea (**figure 7.19**). Large blocks of ice broke off the faces

Figure 7.15 *The striations on rock surfaces caused by the movement of the Dwyka ice sheet some 300 million years ago are well exposed at Nooitgedacht near Barkly West in the Northern Cape.*

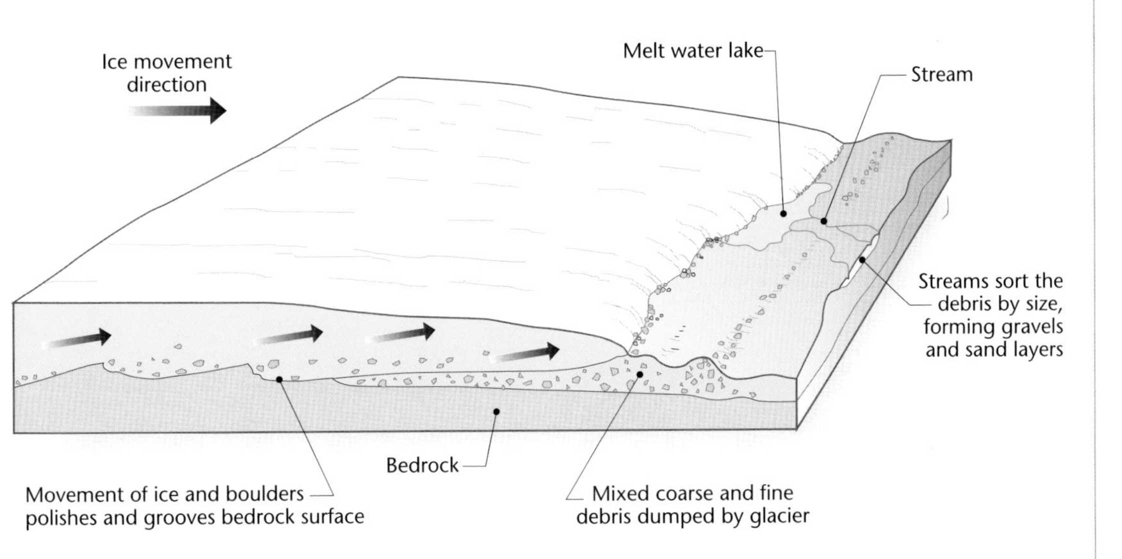

Figure 7.16 *On the higher ground lying to the north of the Karoo Sea (the Cargonian Highlands), the ice sheets were grounded, and their movement sculpted and striated the underlying rock surface. Debris in the form of boulders, sand and mud was deposited at their termini, as illustrated in this diagram. These glacial deposits form part of the Dwyka Group of rocks.*

Figure 7.17 ***A typical example of glacially deposited Dwyka tillite.***

Figure 7.18 ***Dwyka tillite fills a solution pocket in Transvaal Supergroup dolomite, exposed on the banks of the Vaal River near Vereeniging.***

of the glaciers and floated out as icebergs. As they melted, they released their unsorted sedimentary load, which sank to the floor of the inland sea, producing occasional dropstones (rocks dropped from melting icebergs) (**figure 7.20**). Several episodes of advance and retreat of the ice sheets have been identified in the southern Cape, which stacked different kinds of glacial deposit on each other.

The glacial deposits of the Dwyka Group are either very thin or non-existent on the Cargonian Highlands, which form the northern edge of the

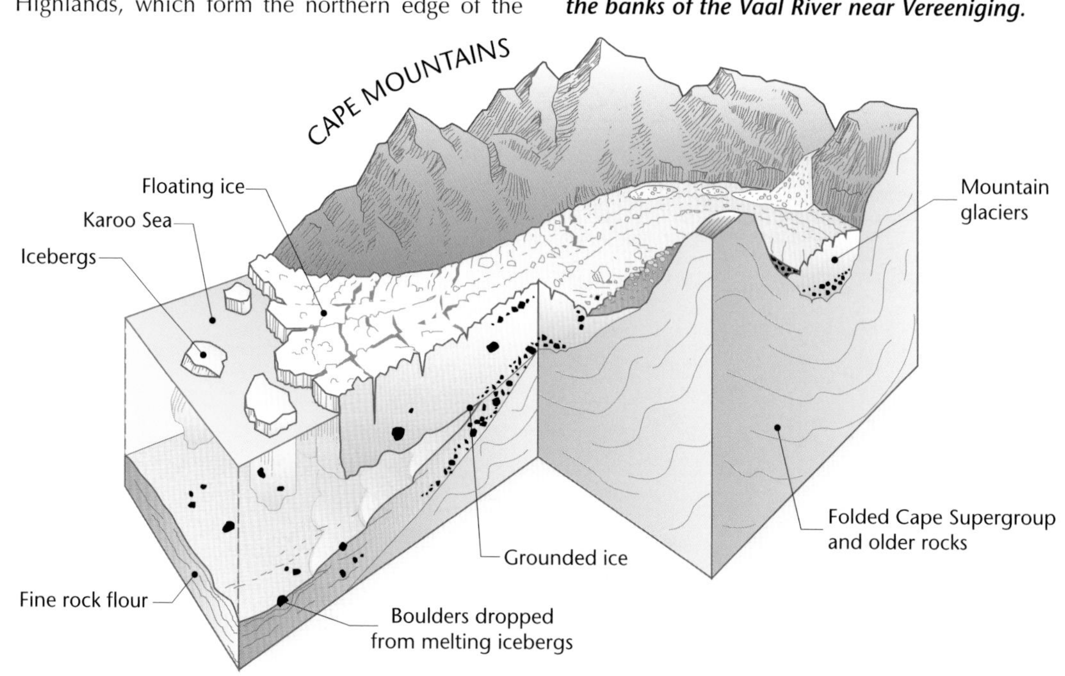

Figure 7.19 ***Along the southern Cape Mountains the Karoo Sea was deep, and glaciers floated out, where they calved, producing icebergs. As these icebergs melted, they dropped their load of sediment to the sea floor as a rain of rock flour and boulders, forming a glacial deposit up to 1 000 m thick.***

Figure 7.20 *Sedimentary deposits formed by melting icebergs consist of fine rock flour with occasional boulders, or dropstones.*

former Karoo inland sea. Here, grounded, flowing ice sculpted the land surface (**figure 7.16**), carrying away the debris. Along the Witwatersrand, for example, glacial deposits are very sporadically developed and, where present, are very thin. The deposits thicken towards the south, however, and in the southern Cape form a layer more than 1 000 m thick. They are particularly well exposed south of Laingsburg.

TEMPERATE FORESTS OF THE KAROO

Gondwana continued its northward drift, moving out of the polar region. The glaciers finally all melted, leaving a vast inland water body extending across South Africa and neighbouring regions of Gondwana (**figure 7.21**). This water body appears to have been connected to the open sea, but the tidal range seems to have been small. Perhaps it was similar to the Black

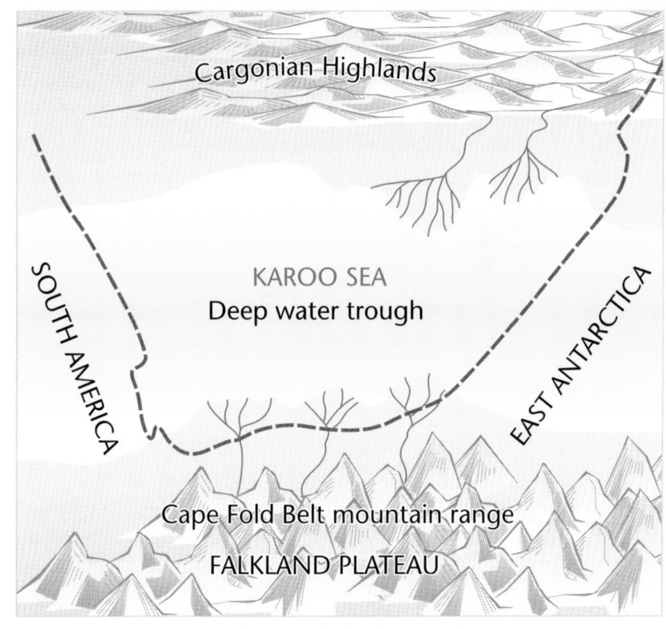

Figure 7.21 *As Gondwana drifted from the Antarctic region, all the ice melted and the Karoo Sea expanded. Rivers flowing into the sea from the Cargonian Highlands in the north and the Cape Mountains in the south deposited sediment, forming deltas along the coastline. These deposits are known as the Ecca Group.*

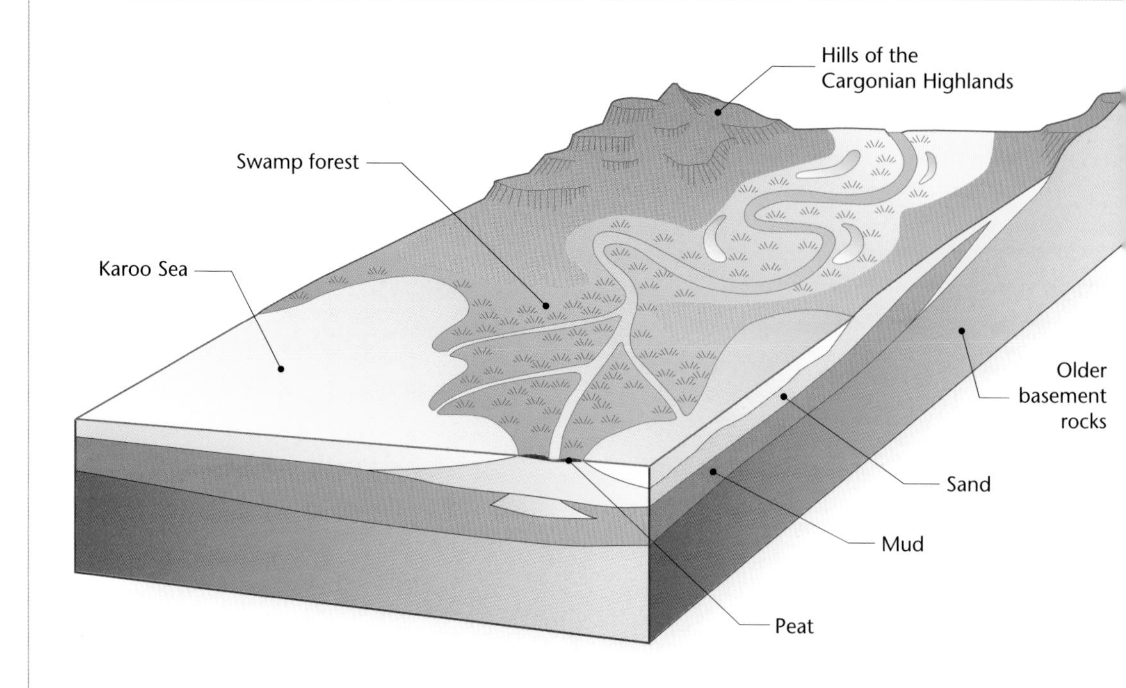

Figure 7.22 ***Along the northern coastline the Karoo Sea was shallow, and large, swampy deltas formed. Accumulation of plant material in these swamps gave rise to the coalfields of KwaZulu-Natal and the Highveld regions.***

Sea, with only a narrow connection to the ocean. The Cargonian Highlands formed the high ground to the north of the sea, and rivers draining this region deposited their sediment along the northern shoreline, forming large, Mississippi-like deltas (**figure 7.22**). These deposits are known as the **Ecca Group** of rocks.

During the time of deposition of the Cape Supergroup, simple land plants were making their first appearance on Earth. These evolved rapidly in the wide range of new habitats the terrestrial environment had to offer. By the time that southern Gondwana emerged from beneath the ice, several large, tree-like plants had already evolved (*see* Chapter 8). These plants, dominated by a genus known as *Glossopteris,* rapidly colonised the large deltas along the northern margin of the Karoo Sea, where they grew in the extensive swamps that flanked the delta distributary channels. In these swamps, dead vegetation accumulated faster than it could decay, and thick accumulations of peat formed, which were ultimately converted to coal (*see* 'The formation of coal', right). Today, these former peat accumulations form the extensive coal deposits of KwaZulu-Natal and the Highveld regions.

During this time, the sedimentary rocks of the Cape Supergroup in the south were further compressed, causing the Cape Fold Belt and its hinterland to the south to rise steadily. The Dwyka Group rocks along the southern Cape now also began to fold. An impressive mountain range formed, comparable to the modern-day Rocky Mountains or the Andes. The crest of this mountain range lay far to the south of the present Cape Mountains, in what is today the Atlantic and Indian Oceans. Like the Andes, this Cape range also had its majestic rhyolite volcanoes, formed as magma rising from the subduction zone to the south erupted at surface (**figure 7.8**). Volcanic ash, produced by periodic violent eruptions of these volcanoes, was carried by wind and deposited with the Karoo Supergroup strata.

The rocks of the Ecca Group in the southern part of the basin, nearer the rising Cape Fold Belt, are different from those in the north of the basin, and do

THE FORMATION OF COAL

In the northern part of the Karoo Basin the glossopterids and cordaitales (a group of now-extinct gymnosperms), ferns, horsetails and clubmosses thrived. Climatic conditions must have been ideal for rapid and luxuriant growth. In fact, the growth rate exceeded the decomposition rate. When an ecosystem is in balance, the decomposition of vegetal matter keeps up with the growth rate and recycling is efficient and complete. When recycling does not occur, which usually happens when plants are growing in water such as in a swamp, the organic matter accumulates to form peat layers. Over time these layers can become buried under sediment. In the case of the Karoo Basin, rivers draining into the basin along its northern margin formed a series of channels and deltas with well-vegetated margins, which formed extensive swamps. These swamps were periodically drowned by subsidence, and the peat layers were buried beneath sediment, only to reform as the water again became shallower. In this way, multiple layers of peat were deposited.

Peat contains about 50% carbon, the rest being made up mainly of oxygen and hydrogen. Once peat is buried beneath sediment, it is compressed and slowly heated. Oxygen and hydrogen are expelled as water, and carbon content increases. Ultimately, the process leads to the conversion of peat to coal. Low heat and pressure results in brown coal or lignite, which contains about 70% carbon. As the temperature and pressure increase, so does the carbon content of the coal, as more oxygen and hydrogen are expelled. The next stage is bituminous coal, which contains about 85% carbon. More extreme heat and pressure results in anthracite, which contains about 93% carbon. Most South African coals are bituminous.

At the time of deposition, peat usually contains some mud, deposited from flood-water that spills from the channels into the swamp. This remains behind with the carbon as the peat is converted to coal. Mud is incombustible so when the coal is burned, the mud is left as a residue, forming ash. Coal often has a banded or layered structure, possessing bright and dull layers. These result from accumulations of different types of vegetal material, such as predominantly leaves or woody material, and also from variations in the amount of decay that has taken place before burial.

Coals, especially those deposited in swamps bordering an ocean, usually contain significant amounts of sulphur. This is released as sulphur dioxide when the coal is burned, and is a major source of acid rain.

Bruce Rubidge

These coal seams are separated by pale-coloured sandstones exposed in the open-cast mine at Rietspruit Colliery south of Witbank.

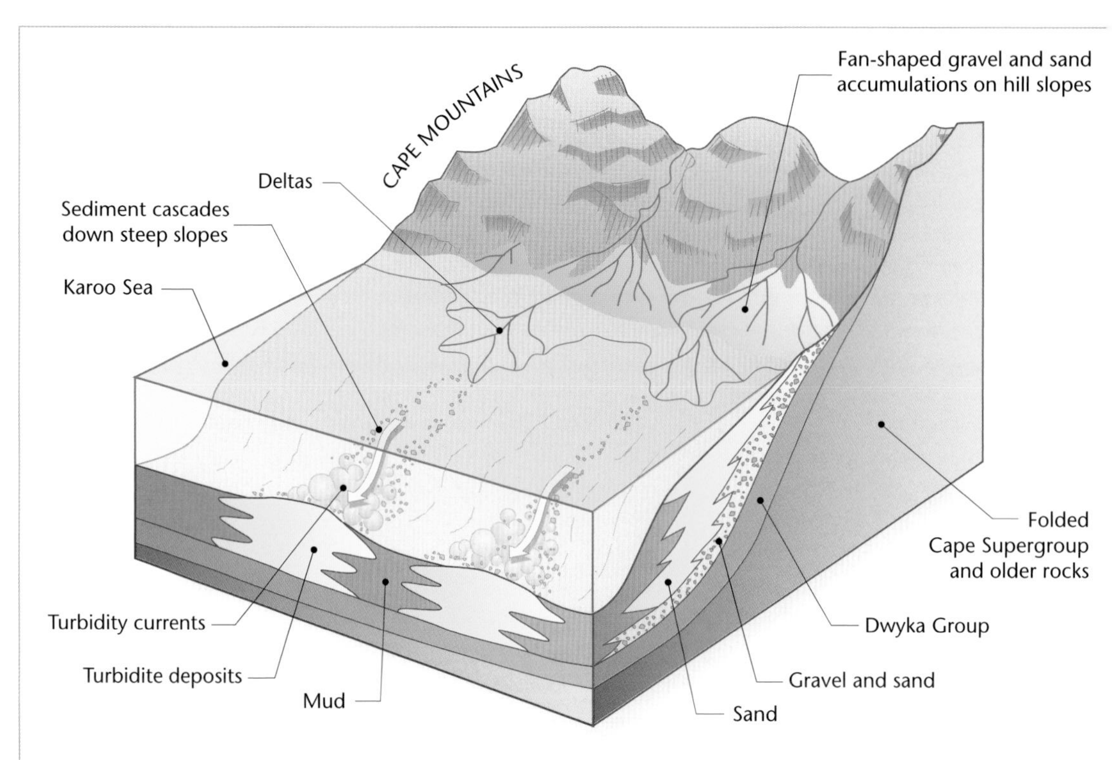

Figure 7.23 ***The Karoo Sea was deep along its southern margin, and numerous small rivers entering the sea from the Cape Mountains built small deltas. Much of the sediment delivered by these rivers cascaded down into the deep as undersea mud-slides to form turbidite deposits.***

not contain coal. In the south the Karoo Sea was deep and the coastal plain narrow. Numerous small rivers draining the Cape Mountains flowed across the coastal plain to the Karoo Sea. Sediment deposited by these rivers formed small deltas that built out to the north from the coastal plain. Fine-grained sediment deposited along the fronts of these deltas periodically cascaded down into the deep trough, producing large, fan-shaped accumulations of turbidites on the sea floor (**figure 7.23**; *see also* 'Sedimentation in ocean trenches' on page 82). These deposits can be viewed today in road cuttings outside the town of Laingsburg. They also form spectacular scenery in the Tanqua Karoo (**figure 7.24**).

The rocks of the southern Ecca Group are important for the fossils they contain. It is within these rocks that the earliest reptile from Gondwana occurs – a small aquatic animal known as *Mesosaurus*. The co-occurrence of *Mesosaurus* in southern Africa and Brazil provided important palaeontological evidence supporting the notion of the supercontinent of Gondwana (*see* Chapter 2).

SILTING UP OF THE KAROO SEA

The Cape Mountains continued to rise, driven by subduction in the south. Consequently, they provided the major source of sediment to the Karoo Sea. This source was augmented towards the end of deposition of the Ecca Group by a second branch of the fold belt to the north of what is now the Cape Peninsula. In contrast, the highlands to the north of the sea became more deeply eroded and contributed progressively less sediment. In fact, these highlands began to be submerged under sediment as the Karoo Sea gradually silted up.

As sediment accumulated in the southern part of the basin, a strip of land was created where there was once only sea. This strip of land, built by coalescing deltas and dominated by delta distributary channels, was originally restricted to the very south of the basin. With time these deltas expanded northwards as the Karoo Sea became increasingly filled with sediment, providing new habitats for a variety of fauna and flora. In time, the sea shrank to a lake (**figure 7.25**). This transition into

Figure 7.24 *Turbidite deposits near Laingsburg were formed by deposition from mud-flows in deep water in the southern portion of the Karoo Sea.*

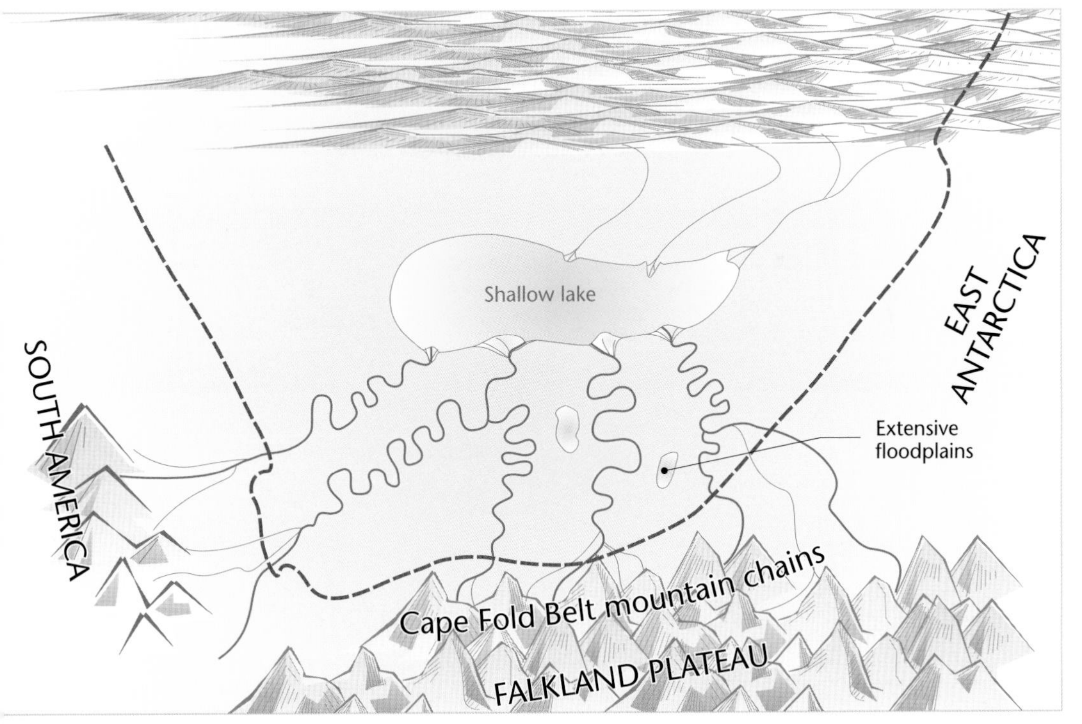

Figure 7.25 *By about 250 million years ago the Karoo Sea was largely silted up and rivers arising in the Cape Mountains meandered across extensive flood plains into an inland lake. Deposits formed during this time are known as the Beaufort Group. The accumulating sediment was steadily submerging the Cargonian Highlands to the north, and rivers rising in this region contributed progressively less sediment to the slowly filling depression.*

FOSSIL PRESERVATION

Only a very small percentage of animals and plants ever become preserved as fossils when they die. Fossil preservation is dependent on at least four circumstances including: rapid burial of the organism, exclusion of oxygen, presence of hard parts such as shells or skeletons, and lack of metamorphism.

Fossils can be preserved in a variety of ways:

Petrified: Hard parts such as bones and shells are very resistant and may survive in rocks. Through time, elements comprising the shell or bone are replaced by minerals from the surrounding rock, which precipitate from groundwater solutions. Via this process the bone or shell is turned into stone.

Moulds and casts: As sediment is deposited around a fossil it takes on the shape of the organism it surrounds. If the organism later decays or dissolves away, a cavity that preserves the surface detail and shape of the organism remains as a mould. If the mould is filled with sediment a cast is formed.

Impressions: These are shallow moulds formed when semi-soft organs such as leaves become compressed between layers of sediment.

Frozen: In geographic regions subjected to freezing temperatures all year round, organisms may be preserved in ice or in frozen ground. Well-known examples include woolly mammoths from Siberia, where even the flesh and hair are sometimes preserved.

Mummified: In very arid environments the desiccated bodies of animals have been preserved with the skin and bone intact.

Trace fossils: These include footprints, trackways, trails and burrows that are made in sediment by organisms. These fossils provide evidence of the behaviour of organisms while they were alive.

Petrified: in this fossil the bone has been replaced by rock.

Moulds and casts: skull has decomposed, leaving a hollow that can be filled to form a replica.

Impression: leaf preserved as an impression in the rock.

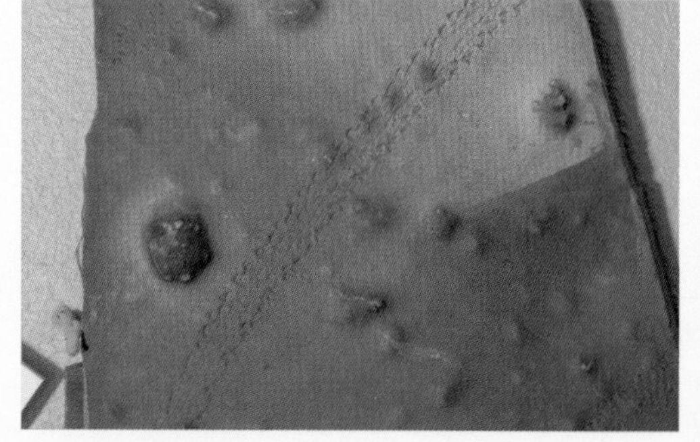

Trace fossil: tracks made by insects crawling over soft mud.

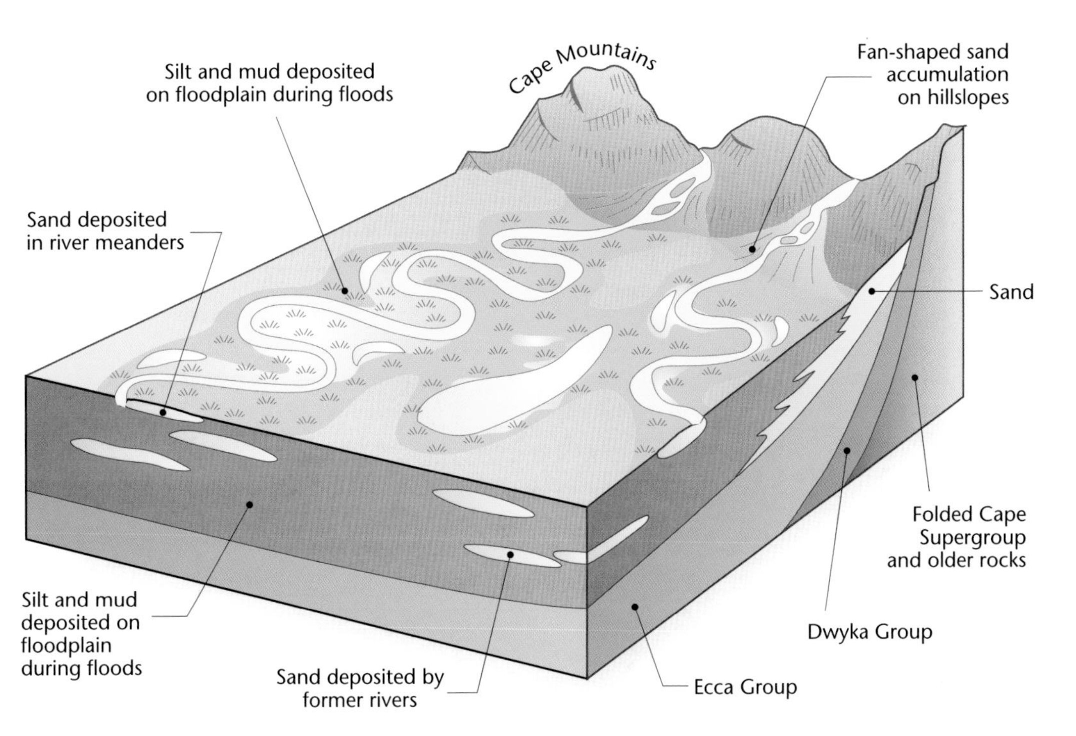

Figure 7.26 ***A schematic illustration of the meandering rivers and floodplains in which sedimentary rocks of the Beaufort Group were deposited.***

terrestrial environments marks the boundary between the Ecca Group and the **Beaufort Group**, and occurred over an interval of time up to about 250 million years ago.

The rocks of the Beaufort Group were deposited by large, northward-flowing, meandering rivers in which sand accumulated (**figure 7.26**), flanked by extensive floodplains where periodic floods deposited mud (**figure 7.27**). It is within the rocks of the lowermost Beaufort Group that the earliest terrestrial reptiles in South Africa are found, and on the floodplains of these river systems that the early evolution of the Permian fauna and flora played itself out.

The rocks of the lower Karoo Supergroup document a period during which climatic conditions were changing in southern Gondwana, when reptiles were beginning their evolutionary diversification and progression towards the forms we see today, including their descendants, the mammals and birds. This evolutionary march changed dramatically 251 million years ago. At this time, for reasons still unknown, life was very nearly extinguished; some 96% of species became extinct. This global event, which has been called the Mother of all Mass Extinctions, brought an end to the Permian Period. The effects of this catastrophe are well preserved in the rocks of the

Figure 7.27 ***Typical sandstone channel deposits and surrounding muddy floodplain deposits (mudstone) of the Beaufort Group.***

Beaufort Group, which provide a window into the aftermath. Unfortunately, however, the rocks are mute as to the cause.

Globally there is considerable research being carried out to identify the cause of this catastrophe. A number of hypotheses have been put forward. These include: changes in the Earth's atmosphere brought on by extensive volcanic activity; a major meteorite impact; and a decline in sea level caused by an ice age that exposed organic-rich sediments on the continental shelves to the atmosphere, causing sudden oxygen depletion and an anoxic ocean as sea level rose again. But to date, little unequivocal evidence as to the cause has been forthcoming.

Following the end-Permian mass extinction, the meandering rivers of the Beaufort Group were replaced by multi-channelled, braided river systems that deposited sand rather than the silts and muds of the earlier meandering rivers. This change may well have been the result of the massive die-off of the vegetation, as vegetation protects soils from erosion and promotes meandering in rivers rather than braiding. Braiding is characteristic of rivers that are overloaded with sediment. The sandstone-dominated strata deposited by these braided rivers, known as the Katberg Formation (**figure 7.28**), can be as much as 1 000 m thick, and extend from Kidd's Beach and Kayser's Beach near East London to the town of Senekal in the northern Free State.

These post-extinction rocks are dominated by a single genus of mammal-like reptile (dicynodont) known as *Lystrosaurus*, one of the few but important survivors of the Permian mass extinction (*see* Chapter 8). It was the discovery of *Lystrosaurus* in

Figure 7.28 *The Earth was affected by an event 251 million years ago, which caused the extinction of 96% of species living at that time – known as the end-Permian mass extinction. The sandstone of the Katberg Formation was deposited in the aftermath of this event. These rocks reflect a change from meandering rivers before the extinction to braided rivers thereafter. The change may have been a result of the loss of vegetation, in the absence of which extensive erosion would have occurred, resulting in rivers overloaded with sediment, which typically braid rather than meander.*

riassic rocks in Antarctica in the ·arly 1970s that provided another mportant piece of palaeontologi- :al evidence linking the southern :ontinents and the idea of drifting ontinents. Since then fossils of his animal have also been found n East Africa, India, China, Russia nd possibly also in Australia.

As time passed, the high-energy, ›raided rivers of the Katberg ormation reverted to a meander- ng form, possibly reflecting ecovery of the vegetation. These edimentary deposits today form he 1 000-m-thick Burgersdorp ormation, the rocks of which nark the end of sedimentation of he Beaufort Group and a change n the depositional style in the ‹aroo depression. The rocks of the ‹atberg and Burgersdorp Forma- ions are particularly important ›ecause they record the recovery ›f life following the devastating ›ermian mass extinction and pro- ›ide a record of subsequent diver- ification of the surviving species, which gave rise o dinosaurs, birds, tortoises and mammals (*see* ›hapter 8).

John Hancox

Figure 7.29 ***Rocks of the Molteno Formation are preserved in the Bamboesberg Mountains of the Eastern Cape. These rocks were deposited mainly by large braided rivers. Sporadic coal seams formed in localised swamp environments.***

ARIDIFICATION OF SOUTHERN AFRICA

›ediment deposition which gave rise to the ›eaufort Group was terminated by a brief period of ıplift of the Karoo strata. During this time, some ›rosion of previously deposited sediment occurred. Jplift was probably caused by changes in the sub- luction environment in the south.

This was short-lived, however, and sedimen- ation was renewed, forming the rocks of the **›tormberg Group** on top of the slightly eroded ›eaufort Group rocks. The rocks of the Stormberg ›roup reflect a gradual change to increasingly nore arid conditions – a change sequentially ecorded in the sequence of rocks that make up he Stormberg Group, that is, the Molteno, Elliot ınd Clarens Formations.

The Molteno Formation rocks were deposited mainly by large braided rivers. Today they form 600 m of sandstones that can be seen in the cliff faces of the Bamboesberg and the Stormberg Mountains in the Eastern Cape (**figure 7.29**). Sporadic coal seams are preserved in these rocks, which formed in localised swamp environments. Alhough of very poor quality, they are historically important as they were the first coals ever to be mined in South Africa.

Even though the rocks of the Molteno Formation do not preserve fossil reptiles, they do contain a truly remarkable assemblage of plant and insect fossils (*see* Chapter 8). The Triassic Period (251 to 203 million years ago), when dep- osition of the Molteno Formation occurred, was globally a warm period, and this is well reflected in the rocks of the Karoo Supergroup. Continued global warming, the relentless northward drift of Gondwana to warmer latitudes, and the position of South Africa within the interior of Gondwana – far from coastlines and possibly in the rain shadow

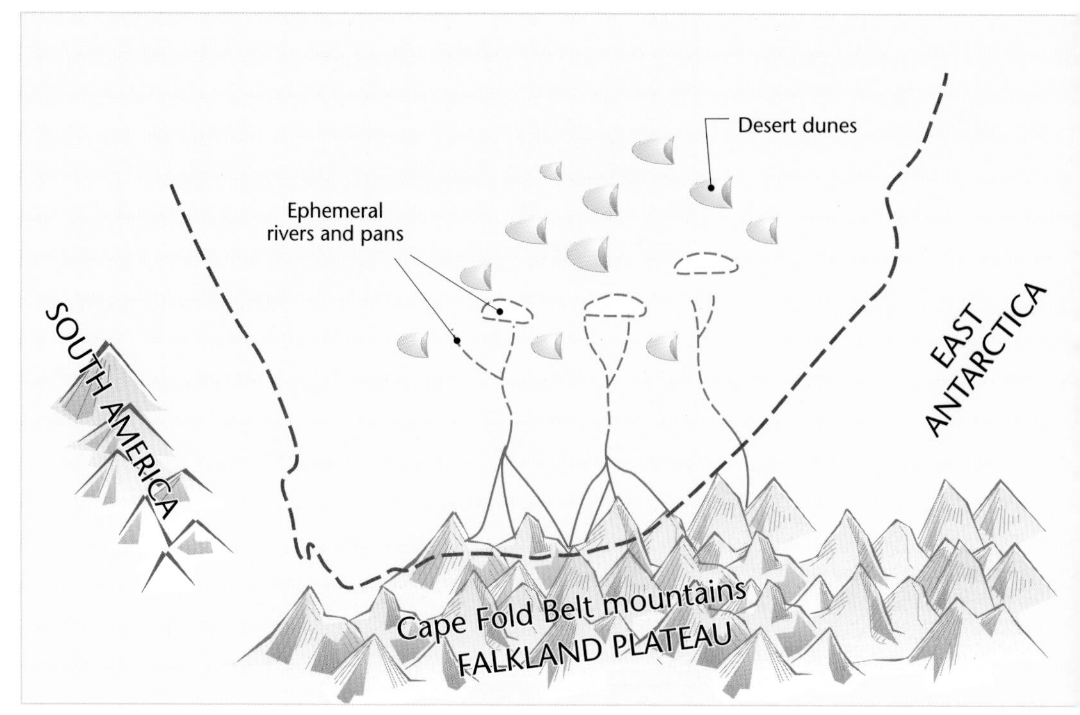

Figure 7.30 ***By 190 million years ago, conditions in the Karoo depression were extremely arid, and a large sand sea similar to today's Namib Desert formed over most of southern Africa to the north of the Cape Mountains. By this stage, the Cargonian Highlands had become completely submerged under sediment, and the sand sea extended unbroken for hundreds, perhaps thousands, of kilometres. These desert sand deposits form the Clarens Formation.***

of the Cape Mountains – all led to the increasing aridification of the Karoo depression.

This change in climate is reflected in the flood-plain sediments of the Elliot Formation (formerly known as the Red Beds because of their red colour, imparted by the presence of iron oxide), which overlie the Molteno Formation. Here there is a return to a meandering river style of sedimentation. Coupled with the oxidised nature of the sediments, this allowed once more for the preservation of fossil bone. The lower strata of the Elliot Formation encapsulate the earliest body fossils of dinosaurs in South Africa.

In addition to meandering river deposits, salt-pan deposits are found, containing fossilised lungfish, as well as fossilised, thick, arid-zone soil layers. These attest to the temporary nature of water bodies during the deposition of the Elliot Formation, as well as to the general warming of the climate. This was an environment of ephemeral pans, possibly seasonally flooded, perhaps akin to the Kalahari Desert that we know today.

The apparent contradiction of large meandering rivers in a generally arid environment is not uncommon today. The Orange River in the Northern Cape, the Nile River in Egypt and the Okavango River in Botswana are modern-day examples of this phenomenon. The rocks of the Elliot Formation also bear witness to another global mass extinction event which occurred around 206 million years ago at the boundary between the Triassic and Jurassic Periods (*see* Chapter 8).

Warming and aridity increased towards the end of the deposition of the Elliot Formation. The overlying rocks of the Clarens Formation (formerly known as the Cave Sandstone, on account of their tendency to weather to form caves, such as in the Golden Gate area) attest to desert conditions. At the base of the Clarens Formation there is still evidence of ephemeral salt pans as well as river activity (**figure 7.30**).

but by the time of deposition of the rocks of the upper Clarens Formation, true desert conditions prevailed, with the development of an extensive sand sea much like the Namib Desert.

These desert sands today form the impressive, pink-coloured cliffs in the Golden Gate National Park (**figure 7.31**). This Clarens desert was part of a giant Sahara-sized sand sea that stretched into Zimbabwe, Botswana and Namibia, as well as into other parts of Gondwana, notably South America. By this stage, the eroded Cargonian Highlands had become completely submerged under the accumulated sediment, and the desert sands of the Clarens Formation formed an uninterrupted sand sea extending northwards from the Cape Mountains for many hundreds, perhaps thousands of kilometres.

THE END OF KAROO SEDIMENTATION

Sedimentation in the Karoo depression was terminated abruptly approximately 182 million years ago when the compression that prevailed throughout the deposition of the sediments of the Karoo Supergroup relaxed. The crust ruptured and huge volumes of basaltic lava flowed out over the Clarens desert, covering virtually the whole of southern Africa, as well as portions of other Gondwanan continents. These volcanic rocks are known as the **Drakensberg Group**. The eruptions heralded the beginning of the fragmentation of Gondwana and the dispersion of its fragments, which gave rise to the continents as known today.

The eruptions occurred mainly from long, crack-like fissures through the Earth's crust, up which

Figure 7.31 *Sandstone of the Clarens Formation, which was deposited in an extensive sand desert, forms prominent outcrops in the foothills of the Drakensberg Mountains, such as these at Golden Gate National Park in South Africa's Free State province.*

magma welled. Apart from the occasional spatter cone, the familiar conical volcanoes rarely formed and lava simply fountained from these fissures over tens of kilometres, to flow away as huge sheets of molten rock. Such eruptions are common on Iceland today. The lava flows were typically between 10 and 20 m thick. Flow after flow erupted from the fissures, building up a pile of lava over 1 600 m thick, reaching its greatest thickness in what is today Lesotho.

The lava contained dissolved gases that formed bubbles in the interiors of flows. These cavities have become filled with secondary minerals, which produce characteristic white spheres and rods (called amygdales) in the black- or red-coloured lava. Occasionally, the cavities are filled with the mineral quartz, which produces concentrically banded agates.

Not all of the magma reached the surface. Much was injected under pressure into the horizontal sedimentary layers of the Karoo Supergroup, where it crystallised to form dolerite sills (*see* 'Igneous intrusions' on page 46). These vary in thickness from a few centimetres to hundreds of metres. Magma also solidified in the fissures, producing dolerite dykes. The dykes, and especially the sills, are resistant to erosion – much more so than the Karoo Supergroup sedimentary rocks that host them. Today, the dykes stand out as linear ridges extending across the semi-arid Karoo landscape, whereas the sills protect the underlying sedimentary rocks from erosion and produce the flat-topped hills so characteristic of the Karoo landscape (**figure 7.32**).

In the Lebombo Mountain region along the Mozambique border, the eruption of basalt was followed by a second phase of volcanic activity involving the explosive eruption of rhyolites and huge quantities of volcanic ash. As much as 4 800 m of volcanic material accumulated during this second phase of activity. This mixed basalt and

Figure 7.32 *Some 182 million years ago, a mantle plume rose under southern Africa, causing rupturing of the crust and the outpouring of vast quantities of lava. Some of the molten rock injected between the layers of Karoo Supergroup sedimentary rock, forming dolerite sills. These sills are more resistant to erosion than the host sedimentary rock, and consequently often form caps to flat-topped hills, such as Platberg outside Harrismith in the Free State, shown here.*

Colour Library / IOA

Figure 7.33 *Extensive volcanic eruptions some 182 million years ago resulted in the accumulation of a pile of lava locally more than 1 600 m thick. The lava field was extensive, probably covering most of southern Africa, and completely buried the former sand desert. Most of this lava has been removed by later erosion, but remnants still remain, one of which forms the high mountains of Lesotho.*

rhyolite volcanic activity in the Lebombo area occurred as a consequence of rifting along the eastern margin of southern Africa as the Indian Ocean began to open. Rhyolites are not developed in the Drakensberg Group lavas of the interior. The volcanic activity which gave rise to the very extensive Drakensberg lavas is believed to have been caused by the rise of a mantle plume beneath Gondwana, and it is possible that it was the rise of this plume that finally triggered the fragmentation of the southern portion of Gondwana (*see* Chapter 9).

The volcanic episode that brought the sedimentation of the Karoo Supergroup to a close seems to have been remarkably short-lived, at least in South Africa, probably lasting less than two million years. This marked the close of nearly 300 million years of almost continuous sediment accumulation in the southern African region of Gondwana.

Since then, the interior of South Africa has been dominated by erosion. Most of the formerly extensive lava sheet has been removed, and today the Drakensberg Group volcanic rocks are only preserved in the high Drakensberg and Maluti Mountains in and around Lesotho (**figure 7.33**), in the Springbok Flats area around Bela-Bela/Warmbaths, in the Lebombo Mountains along the Mozambique border, and in scattered localities in the Soutpansberg area of Limpopo Province.

SUGGESTED FURTHER READING

Anderson, JM. 1999. *Towards Gondwana Alive*. Gondwana Alive Society, National Botanical Institute, Pretoria.

Hancox, PJ and Rubidge, BS. 2001. 'Breakthroughs in the Biodiversity, Biogeography, Biostratigraphy and Basin Analysis of the Beaufort Group.' *Journal of African Earth Sciences*, vol 33, pp 563–577.

Johnson, MR, Anhaeusser, CR and Thomas, RJ. 2005. *The Geology of South Africa*. Council for Geoscience, Pretoria.

MacRae, C. 1999. *Life Etched in Stone – the Fossils of South Africa*. Geological Society of South Africa, Johannesburg,

Smith, RMH, Eriksson, PG and Botha, WJ. 1993. 'A Review of the Stratigraphy and Sedimentary Environments of the Karoo-aged Basins of Southern Africa.' *Journal of African Earth Sciences*, vol. 16, pp 143–169.

Viljoen, MJ and Reimold, WU. 1999. *An Introduction to South Africa's Geological and Mining Heritage*. Mintek, Johannesburg.

THE STORY OF
EARTH & LIFE

8
THE LIFE OF GONDWANA

Bruce Rubidge

A 250-million-year-old rippled mudflat, Estcourt, KwaZulu-Natal. In the foreground is the fossilised body impression of a large amphibian, and in the background are footprints of a herd of dicynodonts that crossed when the surface was still under water.

ROUTE MAP TO CHAPTER 8

Age (years before present)	**Event**
400 million	• *Dutoitia*, the earliest land plant from Gondwana, was preserved in the rocks of the Bokkeveld Group.
300 million	• Southern Africa was positioned over the South Pole.
270 million	• This was the heyday of *Glossopteris* vegetation growing along river courses and delta distributaries of the Ecca and Beaufort Groups. Extensive peat swamps developed in the deltaic environment of the Ecca Group, eventually giving rise to most of the coal mined in South Africa.
260 million	• Anapsid, synapsid and diapsid reptiles radiated in Gondwana, and are especially well preserved in the Karoo rocks of South Africa. Synapsids diversified into a variety of forms that, through time, gradually became more mammalian in form – the mammal like-reptiles (therapsids).
251 million	• The Mother of all Mass Extinctions occurred at the end of the Permian Period. In South Africa, of the more than 50 different kinds of tetrapod species, only four survived into the Triassic Period. Of the abundant dicynodont species present in the Permian, only *Lystrosaurus* survived into the Triassic. The dominant carnivores of the Permian, gorgonopsian therapsids, all became extinct, their ecological role being taken over by the archosaurs (forerunners of dinosaurs). *Glossopteris*, the characteristic Permian flora of Gondwana, became extinct and was replaced by the diverse *Dicroidium* flora.
210 million	• Earliest dinosaurs appeared. In South Africa the earliest dinosaur remains are footprints in the Molteno Formation. By Elliot Formation times, dinosaurs were well established in southern Africa.
200 million	• Therapsids decreased in diversity and earliest mammals appeared alongside a wide variety of ornithischian and saurischian dinosaurs, the dominant form being the prosauropod *Massospondylus*.
180 million	• Sedimentation in the Karoo was disrupted by eruption of flood basalt lavas.
140 million	• Gondwana began to fragment and sediment accumulated in isolated basins such as the Algoa Basin, providing a fragmentary record of dinosaur biodiversity in southern Africa at the time of the Late Jurassic-Cretaceous Age of Reptiles.

A CENTRAL PROBLEM IN PALAEONTOLOGY

Ever since multicellular life evolved, organisms within communities have been dependent on each other for survival in relationships that involve mutual dependence, parasitism or predatory behaviour. This interdependence has created an energy chain that is central to all ecosystems, today and in the past.

The African savanna typifies the many components that make up an energy chain. Extensive grasslands, with their scattered trees and shrubs, convert carbon dioxide and water into carbohydrate, providing the foundation of the energy chain. Most visible to us at the next level of the chain are the grazers and browsers – the zebra, impala, wildebeest, kudu – which convert plant carbohydrates into fats and proteins. They, in turn, are exploited by the predators – lions, leopards and cheetahs. Leftovers are exploited by the scavengers – hyaenas, jackals and vultures – which serve to maximise utilisation of the energy resource.

While the plants and mammals are the high-visibility players in the energy chain, other groups of organisms, mostly less visible, are doing just the same – the birds, reptiles, insects, nematodes, fungi and bacteria. In addition, there is an intricate web of crosslinks among these various exploiters. Mammals prey on reptiles, insects and birds, and *vice versa*, while nematodes, fungi and bacteria consume everything, creating a complex intertwined energy web rooted in photosynthesis.

Competition between organisms is strong as they struggle to survive and reproduce. The pressures of finding a meal – and avoiding becoming one – are relentless. Natural selection operates under this relentless competition, culling inhibitive traits. This either favours specialisation into niches where competition is less, or enhances competitiveness and at the same time sharpens defences against predators. Over time this process, popularly known as survival of the fittest, has produced immense diversity among living organisms.

Africa does not consist only of savanna country; it supports rainforests, grasslands, semi-deserts and desert areas as well. Like the savannas, these other regions also sustain intricate energy webs, but they involve organisms different from those of the savannas. Each is characterised by its own assemblages of primary plant producers, carbohydrate harvesters and predators.

A key question in palaeontology is: how much of this diversity will actually be preserved to be reflected in the fossil record? Will palaeontologists 100 million years from now be able to reconstruct the complexity of today's African ecosystems from the preserved fossil remains of the organisms around us today?

Because organisms decay after death, their preservation as fossils is dependent on a number of special circumstances. The organism must be covered by sediment soon after its death to prevent destruction by such agents as scavengers, or by slower physical destruction through differential heating and cooling, wetting and drying, erosion by wind and water, or bacterial attack. The environment must be neither too acid nor too alkaline. The remains must stay relatively undisturbed, with minimal change during the chemical processes that occur when loose sediment is converted into rock. Moreover, different types of organisms (e.g. animals and plants) require rather different conditions for their preservation. All these conditions are rarely met, with the result that only very few of the organisms that die actually become preserved as fossils (*see* 'Fossil preservation' on page 204).

Much of southern Africa today is experiencing uplift and erosion (*see* Chapter 9), and environments conducive to the long-term preservation of sediments are rare and confined to isolated river and cave deposits. Within a region of active sedimentation, certain local environments are more suitable than others. Particularly important are wet areas such as floodplains, marshes and pans, where rapid burial in an oxygen-deficient environment is possible.

The limited range of environments in which fossils can be preserved creates a natural bias in the types of animals and plants that are preserved – species that live away from such places simply will not become fossilised. The stature, structure and composition of an organism also influences its chances of becoming fossilised. Plants with leathery leaves and hardwood stems are more likely to become fossils than delicate or fleshy varieties. Animals with big, strong bones and which frequent water-holes where they are preyed on will also be favoured in the fossil record. Many other species, though, will simply never be preserved.

Superimposed on these constraints are the effects of changing local and global climate. A region of sedimentation may for a time be occupied by tropical rainforest, which may slowly evolve to savanna and then to desert conditions, as happened in the ancient Karoo (*see* Chapter 7). Different animals and plants are adapted to these different conditions, so a climate change will inevitably result in different types of animals and plants being preserved in the fossil record over time. Animals that prefer wetter climates will no longer be present, but that does not mean that they have become extinct. For these reasons, reconstructing the past history of life from the fossil record is no simple task, as much of the past history of life is not preserved in the fossil record.

The fossil record can be likened to a damaged strip of movie film where the vast majority of the frames are missing, and the story has to be pieced together from the remaining frames. New fossil finds, however, continue to fill in the gaps.

There are other difficulties as well. The further back in time we go, the less familiar the animals and plants become. However, by careful examination of their characteristics it is possible to identify life forms that are related. This, together with knowledge of their relative ages, enables palaeontologists to construct evolutionary histories of organisms. From an animal's body structure, it is possible to get some idea of its diet and lifestyle. From the nature of the rocks in which fossils occur, geologists are able to reconstruct the environmental conditions under which the organism lived and died.

Thus, by careful study of the fossils of animals and plants found together, and the rocks in which they occur, and by piecing together information obtained from widely scattered sites, palaeontologists can reconstruct ecosystems of past eras and track how they changed through time. So, while it is unlikely that future palaeontologists could reconstruct the full diversity of African life from fossils being formed today, they would nevertheless be able to gain a fairly good idea of what life was like.

EARLY LIFE IN GONDWANA

South Africa is uniquely fortunate in having the most complete record of land-living vertebrate animals for a period of more than 80 million years, stretching from the Middle Permian to the Middle Jurassic (260 to about 180 million years ago). This record is preserved in the rocks of the Karoo Supergroup.

Before this, when the rocks of the predominantly marine Cape Supergroup were deposited, the earliest animals with backbones had already appeared in the form of jawless fish (*see* 'How vertebrates evolved from invertebrates', right). Lampreys and hagfish are modern representatives of this ancient group, which subsequently gave rise to a wide variety of fish with jaws. These early fish are well represented in the rocks of the Bokkeveld Group (*see* Chapter 7). They include a wide variety of forms such as the armour-plated predatory **placoderms**, fish with cartilage skeletons, which are possibly the forerunners of skates and rays, and predatory spiny sharks characterised by large spines supporting the fins. More advanced fish with bony skeletons, including both lobe-finned forms like coelacanths and the more common ray-finned forms, are known from the Witteberg Group. Among the latter are the ancestors of the bony fish of today. At that stage fish were the only animals that possessed backbones.

While those early fish were diversifying in the oceans alongside a rich variety of invertebrate animals and algae, land was being colonised for the first time by plants.

PLANTS TAKE HOLD ON LAND

We tend to take our vegetation for granted: the grasslands of the Highveld, the forests of Knysna and Mpumalanga, the short bushes and succulents of the Karoo, and the thorn trees almost everywhere. It has not always been like this. For most of Earth's history land surfaces were bare and such plants as existed then lived in water, where they were protected. Only when plants evolved the ability to absorb enough water and retain it to avoid drying out were they able to colonise dry land. As biologist Lynn Margulis observed:

> *plants made their move to land by recreating their wet environment and sealing it within themselves.*

Some of the earliest **terrestrial** plants are preserved as fossils in fine mudstones of the 400-million-year-old Bokkeveld Group. They are small –

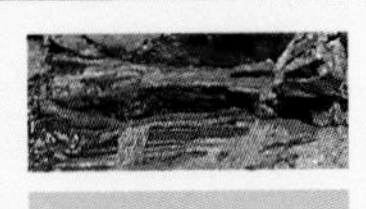

HOW VERTEBRATES EVOLVED FROM INVERTEBRATES

Vertebrates, which belong in the Phylum Chordata, appear to have evolved from an invertebrate ancestor, but the exact ancestral group is uncertain. The fossil record provides few clues to their origin, our knowledge deriving mainly from some present-day soft-bodied marine animals (sea squirts such as red bait, and the lancelet), which appear to be related to the ancestral vertebrates. These animals have, at some stage of their life cycle, a backbone-like structure consisting of a solid rod known as a notochord. A similar structure is present during the embryonic development of all modern-day vertebrates, but is later replaced by the backbone.

Animals belonging to the Chordata have these characteristics at some stage of their life cycle:

- Notochord (rod-like backbone structure)
- Hollow dorsal nerve chord (spinal chord which is hollow)
- Branchial slits (openings into the pharynx from outside; in fish these are gill slits)

When trying to assess which group of invertebrates could have given rise to vertebrates we should determine which invertebrates possess these chordate characteristics. It has been proposed that fossils of carpoids, an extinct group of echinoderms (the phylum to which starfish and sea urchins belong) appear to have had gill slits and a notochord. Biochemical and embryological evidence also shows that echinoderms are more closely related to chordates (and hence vertebrates) than any other invertebrate group. In addition, the embryos of echinoderms closely resemble the early embryonic stages of all chordates.

Adult sea squirts (eg. red bait) look like blobs of nothing in particular, and their bodies have a leathery protective covering. They spend their life attached to rocks in the sea, drawing in a current of water through a hole on one side of the body and passing it out through another while filter-feeding nutrients. Although the adults are sessile they have a free-living tadpole-like larva, which has all three characteristics of chordates.

The lancelet is the most revealing invertebrate chordate. This is a small fish-like animal that inhabits warm seas. It has a notochord, dorsal nerve chord and gill slits, but its anatomy is much simpler than that of the simplest fish. The lancelet thus gives us an idea of what the first 'pre-fish' may have looked like. Palaeontologists believe that the earliest fossil fish, which did not have jaws, would have evolved from an animal similar in morphology to the lancelet.

It is thus reasonable to assume that chordates share a common ancestor with echinoderms, and may even have evolved from the echinoderms themselves.

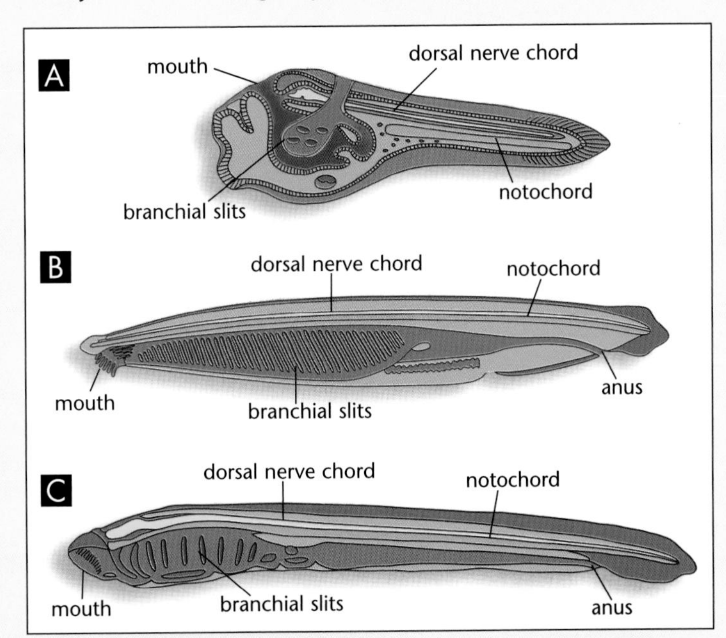

Red bait (A) and the lancelet (B) are primitive chordate animals with the characteristic notocord, hollow dorsal nerve and gill slits. The larva of a lamprey (C) – a vertebrate – also has these features.

less than 10 cm high – and very simple in structure. Their diminutive stems branch in two, and end in cup-shaped structures containing their reproductive spores. These simple plants (called *Dutoitia* after the renowned South African geologist Alex du Toit) are very similar to rhyniopsids from the northern hemisphere, which have **stomata** (tiny pores) in the cuticles of their stems for gas exchange, and primitive cells inside the stem for transporting water from their roots to the aerial parts of the plant (**figure 8.1**).

Features that might be regarded as typical of most plants today – such as cuticle, roots and a vascular system for water and nutrient transport and support – had first to evolve so that plants from an aquatic environment could survive and flourish in a dry atmosphere. These adaptations to living on land were the most important steps in plant evolution and opened up a variety of new niches for colonisation and diversification – so much so that plant biomass on land is now much greater than that at sea, even though land makes up only about one-third of Earth's surface. The successful colonisation of land by plants created a food source and an opportunity for animals to follow.

Plants rapidly became more complex in their branching patterns, and vascular systems and reproductive parts became innovative. Nea Grahamstown, in rocks younger than those bearing *Dutoitia*, is a diverse fossil flora of **clubmosses**, **lycopods**, **ancestral gymnosperms** and algae Many of these Late Devonian plants (from abou 350 million years ago) still have relatives living today. In the fossil record, clubmosses have narrow leaf-like structures attached to stems but, a happens with many plants, these tended to fall of during the life of the plant and are usually preserved separately in the fossil record. The fossi impressions of the stems of these plants show a pattern of diamond-shaped scars where the leave detached, sometimes crowded together when the growth of the stem was slow, or widely spaced when the growth was rapid.

In the humid environment of the northern hemisphere in the Late Carboniferous (about 300 million years ago), early terrestrial floras flourished, eventually giving rise to the enormous European Carboniferous coal deposits. Here huge clubmosses evolved to become the world's first trees, growing up to 20 m tall. By contrast, in southern Africa, then situated in the polar region the climate was more hostile and supported only small clubmosses.

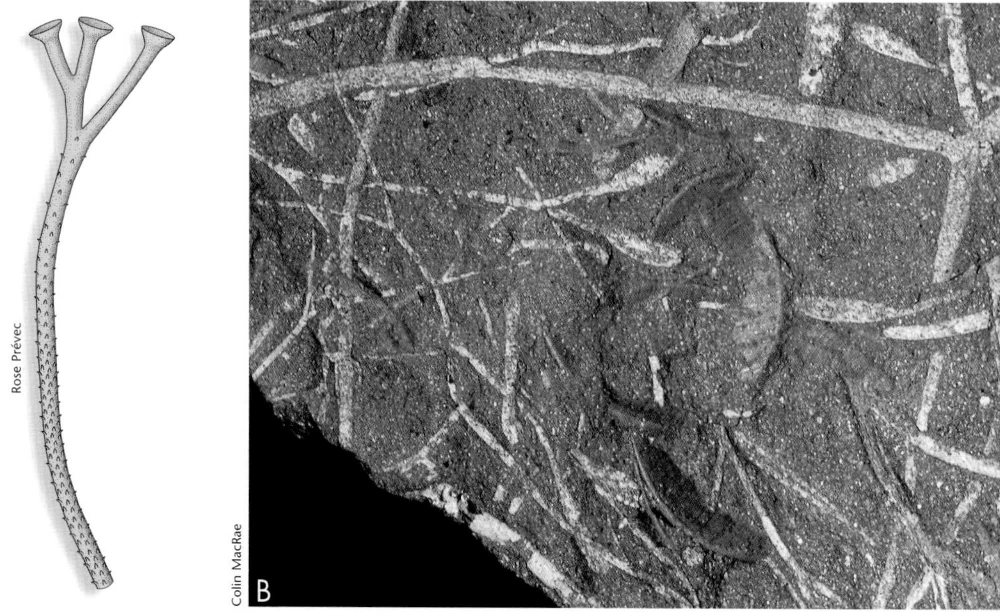

Figure 8.1 Dutoitia, *one of the earliest plants to colonise land, is known from rocks of the 400-million-year-old Bokkeveld Group.* **A** *shows a reconstruction of the plant and* **B** *is a photograph of the fossil.*

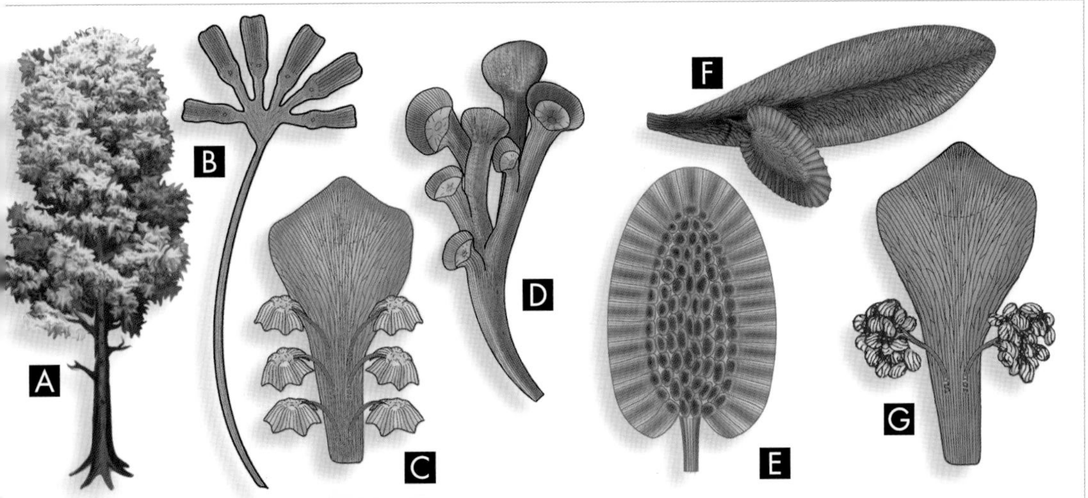

Figure 8.2 *Reconstruction of* Glossopteris *(**A**), and a photograph of its fossilised leaves. These trees dominated the Permian floras of Gondwana. They produced a diversity of seed-bearing organs (**B–F**) and pollen-producing organs comprising clusters of sacs attached to scale leaves (**G**).*

GONDWANA EMERGES FROM THE ICE

Some 300 million years ago, what is now South Africa lay over the South Pole and was slowly drifting northwards (*see* Chapter 7). As the ice sheets began to melt and the glaciers retreated, plants colonised the newly exposed land. Clubmosses, ferns and horsetails were present in these newly exposed lands. A large part of their success can be ascribed to their highly evolved means of reproduction and dispersal.

Clubmosses had their spore sacs clustered together in cones that produced vast quantities of spores. Many of these cones were set high up on tall plants so that the spores could be blown great distances by the wind to germinate in new, distant areas. Another important change was the differentiation of sporangia to produce fewer large female spores (megaspores) and many small male spores (microspores). With this innovation, sexual reproduction in land plants was born and cross-pollination, with all its genetic advantages, became possible.

The evolution of cones was not the only improvement in plant reproduction; some also evolved fertile structures consisting of megaspores or ovules, which had developed a **nucellus** to provide nourishment for the new plants when they germinated. These structures were the forerunners of the seeds of higher plants. Co-incident with these improvements in reproductive structures was a growing threat from the increasing numbers of plant-eating reptiles and insects. While the rapidly diversifying reptiles feasted on vast quantities of leaves, the newly evolved fertile structures were more nutritious because of their nucellus, so they became a much-sought-after prize by herbivores.

Groups of plants developed a variety of strategies for improving their chances of dispersal and germination against threats of being eaten, periodic drying out, or extended droughts. One of these, the **glossopterids** (**figure 8.2**), was so successful that it dominated Gondwana for 60 million years. Glossopterids appeared in the Early Permian (about 280 million years ago) and quickly diversified, but became extinct in the Early Triassic. It was these plants that were responsible for producing southern Africa's vast coal deposits (*see* 'The formation of coal' on page 201).

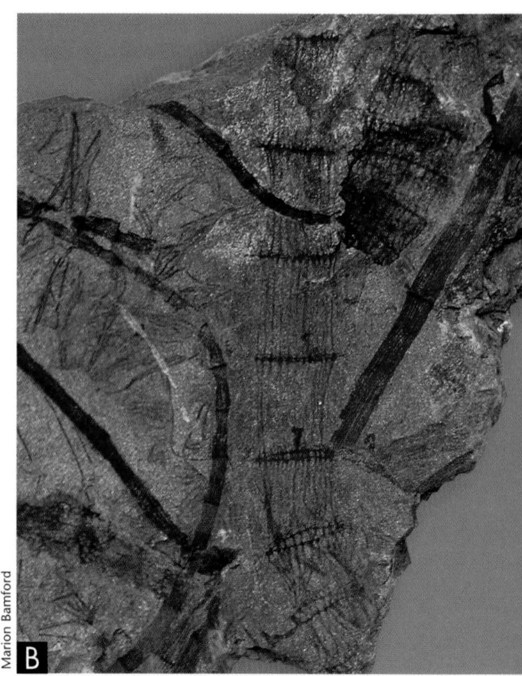

Figure 8.3 Phyllotheca, *a characteristic horsetail of the Permian that is well preserved in rocks of the Beaufort Group, provided a rich food source for vertebrates and insects of the time.* **A** *is a reconstruction and* **B** *shows a fossilised plant.*

Study of fossil plant communities is complicated by the fact that, unlike animals, plants continually shed parts of themselves during their life: leaves drop sporadically throughout the life of the plant and in many cases are shed *en masse* in the autumn; fruits mature and drop off; branches and twigs break and fall continuously. Eventually the plant itself dies and the stem falls over. In this way different parts of the plant end up widely separated from each other, and this is how they are generally preserved in the fossil record. When they are found, there is often nothing to show which parts were associated together on the same plant. As a result, fossils of different parts from the same type of plant often end up with different names: in the case of the glossopterids, the leaves are named as different species of *Glossopteris* (the name refers to the tongue-shape of the leaves); the roots are known as *Vertebraria*; the trunk as some species of *Agathoxylon*; while the male and female reproductive organs have many generic names.

To solve this puzzle the palaeobotanist must seek elusive clues amid the chaos to try to match the separate parts together in order to reconstruct the original whole plant accurately. Without painstaking work it would be all too easy to put the fruit of an orange and the leaves of rhubarb on the trunk of a palm tree, so until the parts are found actually joined together as fossils, palaeobotanists consider it wiser to treat them separately.

PLANTS OF PERMIAN TIMES

During the Mid- to Late Permian (about 260 million years ago), when the Beaufort Group was being deposited (*see* Chapter 7), South Africa was a lowland area traversed by kilometre-wide rivers. They meandered towards the north and were flanked by vast floodplains that were inundated during periods of catastrophic flooding after heavy rains. These resulted in the deposition of thick layers of fine mud and silt over the floodplains and entrapped a variety of animals and plants living there.

The vegetation at that time was vastly different from that we know today, with no grasses or flowering plants. The dominant plant types were the now-extinct glossopterids (today found as fossils in all countries that once formed part of Gondwana). South Africa's long fossil record of glossopterids is particularly well preserved in KwaZulu-Natal. The leaves are variable in shape: we may have as many as 100 species of *Glossopteris*, or as few as 10, depending on how one defines a species. The male reproductive structures are clusters of sporangia on stalks, attached to a small scale leaf. Seeds are borne on the surface of a variety of flat structures attached to the leaves.

Another plant of the Permian times was the horsetail, *Phyllotheca* (**figure 8.3**). Superficially resembling bamboo, the above-ground and underground stems were probably an important food source for the ever-increasing number of plant-eating reptiles. Other plants forming part of this flora were ferns, clubmosses, liverworts and true mosses (**figure 8.4**). As with modern floras, different ecological niches had different proportions of the various groups. The fossil plants are preserved mostly as impressions of leaves, which can be seen when layers of fine mudstone or siltstone are split open. Sometimes staining by minerals highlights the

***Figure 8.4** Photographs and reconstructions of sphenophytes (**A**) and lycopods (**B**), which form part of the famous glossopterid flora of the southern hemisphere.*

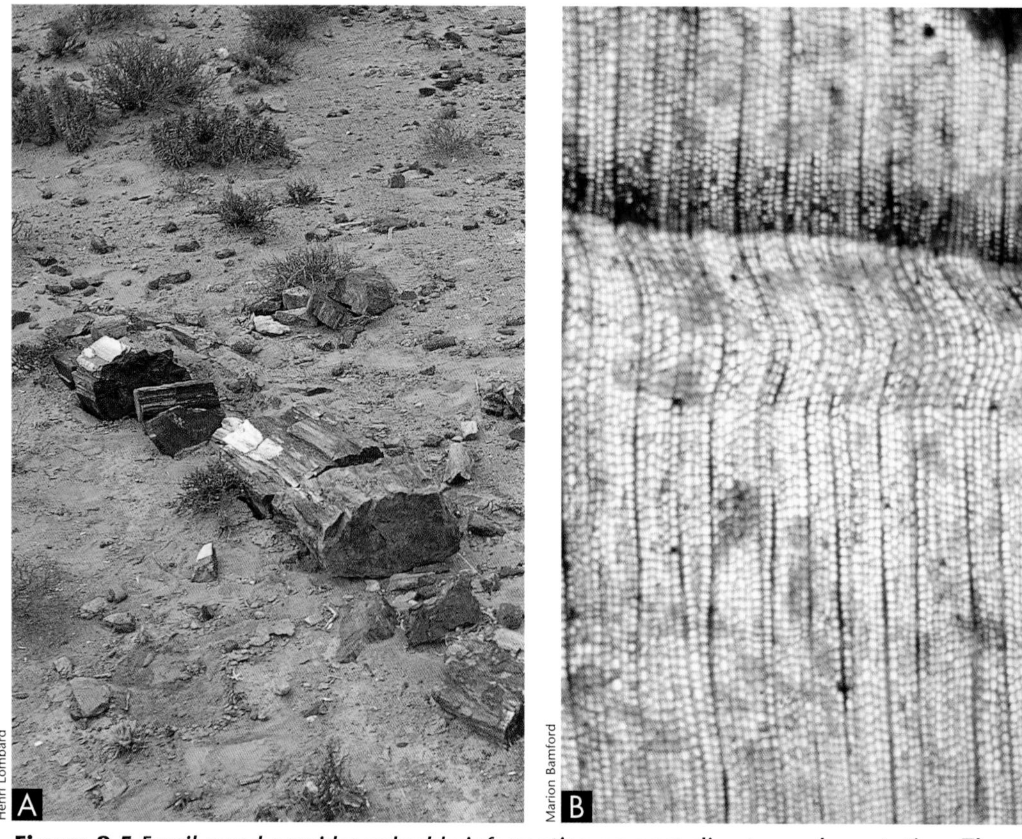

Figure 8.5 *Fossil wood provides valuable information on past climates and vegetation. The morphology of the cells, which appear different in different sections, enables us to identify the wood, as well as to determine the environment in which the tree lived.* **A** *shows a Permian fossil log and* **B** *is a microscope photograph of a piece of fossilised wood. Note the perfect preservation of the individual cells.*

patterns the leaves have left on the rocks, and in some cases the original cuticle is preserved. With good preservation microscopic details of the cells, cell walls, stomata with their guard cells, and papillae (tiny hair-like structures) can be seen.

Petrified wood is an easily recognised type of plant fossil. It forms when tree trunks are buried rapidly in mineral-rich waters, as a result of which they become saturated. Slowly the minerals dissolved in the water replace the organic or woody matter in the original pattern of the cell walls, pits, spiral thickening, tissues and growth rings; eventually, a mineralised detailed replica of the trunk results. Because it is hard and resistant to erosion, it is usually left behind when the softer enclosing sediments erode. When petrified wood is sectioned and studied under the microscope, all fine details of individual cell structure can be seen and the wood can be identified. The presence or absence of growth rings, and their pattern, show whether the tree grew quickly or slowly, whether the growth was seasonal or continuous, and whether or not it experienced hardships such as frost or fire (**figure 8.5**).

Pollens and spores, which have a distinctive morphology depending on the species to which they belong, are abundantly present in the air during spring and summer. As they fall to the ground they become entrapped in accumulating sediment and may be preserved as fossils. Ancient fossilised pollen grains and spores can today be recovered from rocks by chemical processes (**figure 8.6**). This study of the microscopic parts of fossil plants is called palynology and is an important tool in exploration for coal, gas and oil. By comparing the pollens and spores from different borehole cores, equivalent-aged levels can be recognised. Certain pollen grains come from plants that existed for only a relatively short time so they are very useful for matching up or correlating levels in exploration boreholes or coal seams. Other plants that existed for very long time periods are less useful for correlation, but some are good indicators of particular climates at particular times.

THE REPTILES FOLLOW ...

Today southern Africa is rich in reptiles: from turtles and their kin in our oceans, rivers and lakes; tortoises of several kinds plodding through the veld; a huge variety of lizards and snakes of all sizes inhabiting almost every kind of habitat on land; to the monarch of them all – the crocodile, lurking in many a rural river, pond or lake. This seemingly huge variety of reptiles is in fact the reduced remnant of two great reptilian lines that have had a long and much richer history in the fossil record of southern Africa.

The period from the Carboniferous to the Jurassic (350 to 140 million years ago) is often referred to as the Age of Reptiles because they were then the dominant vertebrates. These animals arose from amphibian ancestors in the northern hemisphere during the Carboniferous Period, a time when southern Africa was still situated in the south polar region and covered by ice sheets – certainly not an environment conducive to the development of cold-blooded animals that are dependent on the environment for the maintenance of their body heat.

The first amphibians are considered to have evolved during the Devonian Period from lobe-finned fishes (the group of fishes to which coelacanths belong), which already had many advanced features that were inherited by amphibians. These included bones in the fins/limbs and nostrils opening directly into the mouth. In recent years numerous fossils of transitional forms between lobe-finned fish and amphibians have been discovered in Australia, Greenland, Latvia, North America, Scotland and Russia. So far these forms elude us in the South African record, but may still be discovered in the rocks of the Bokkeveld and Witteberg Groups.

Amphibians themselves are a diverse group that were most abundant during the Permian and Triassic, with a line leading to modern frogs, salamanders and caecilians (worm-like amphibians). Reptiles evolved from an ancient group of crocodile-like amphibians known as the anthracosaurs.

Reptiles are generally classified into three large groups based on the nature of their temporal openings (openings on the skull roof for the attachment of the jaw muscles) (**figure 8.7**). The **Anapsida**, the most primitive reptiles, evolved from amphibians, and have no temporal openings. Today they are represented only by turtles, terrapins and tortoises, but during Permo-Triassic times a variety of these forms flourished and are well represented in the rocks of the South African lower Beaufort Group.

The second major reptilian group, known as the **Synapsida**, have only a single temporal opening. This line eventually gave rise to mammals but in Permian times – in other words, before mammals had appeared on Earth – the synapsids were the dominant land-living vertebrates.

The **Diapsida** have two openings in the skull behind the eye socket on either side, instead of just the one of the Synapsida. Diapsids include dinosaurs, as well as all living reptiles except tortoises. Some scientists today believe that even the tortoise line is not truly anapsid, but rather a modified branch

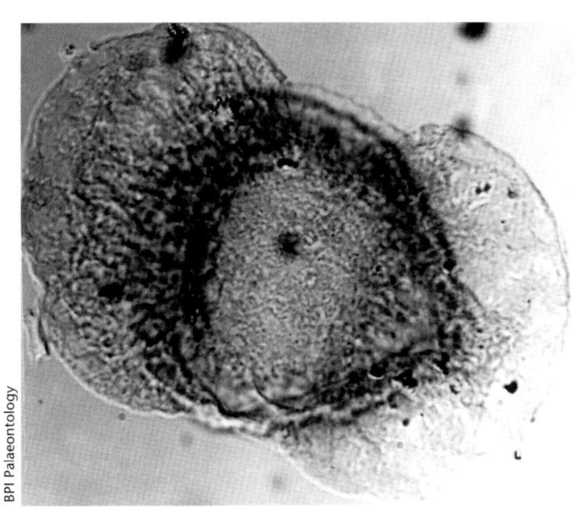

BPI Palaeontology

Figure 8.6 ***Microscopically small pollens and spores also fossilise. Because each type of plant has its own characteristic type of pollen, they provide information on the plant species present. As pollens and spores are very abundant they are used to correlate rock layers that were deposited at the same time.***

of the diapsid line. By any measure, the diapsid line has clearly always been extremely important in the evolution of reptiles. Yet its clear success and variety today is but a shadow of its former glory, for this line also produced the dinosaurs, the flying reptiles (**pterosaurs**) and the birds (**figure 8.8**).

During the Permian (260 to 251 million years ago), representatives of all three groups of reptiles thrived on the vast floodplains of the meandering Beaufort rivers (*see* Chapter 7). As in the case of the African savanna biome today, they formed a balanced ecosystem with some forms eating plants while others fed on the plant eaters. By careful study of the structure of the skull and teeth, stomach contents and fossilised dung of these prehistoric animals palaeontologists can determine what they fed on and thus build up a picture of them as living animals.

Given the wide variety of plant species ther available for food, herbivores were abundant durin the Middle to Late Permian, and were present in variety of forms and sizes. As far as can be ascer tained most of the anapsids were plant-eaters. The rat-sized *Eunotosaurus* from the lower Beaufor

Figure 8.7 ***Reptiles are divided into three main groups, dependent on openings in the skull roof (temporal openings) behind the eye.***

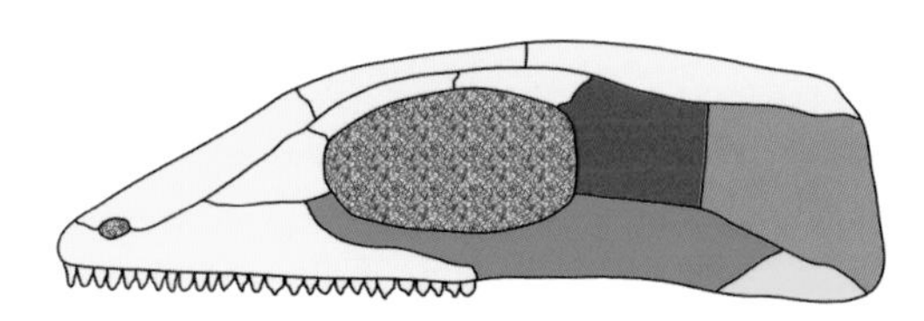

Anapsid: *no openings (the group includes the tortoises and their ancestors).*

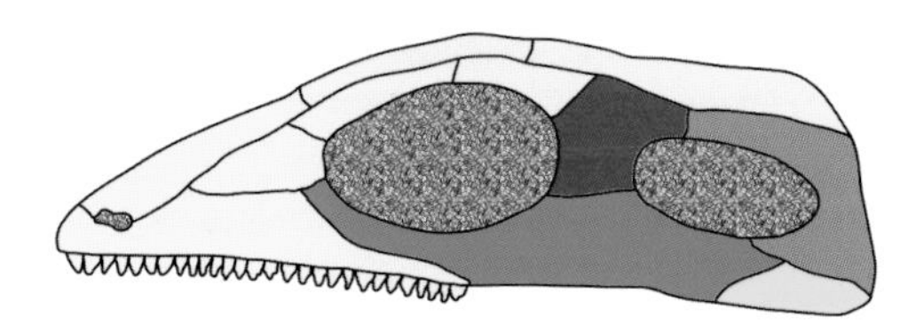

Synapsid: *one opening (the group includes mammals and their ancestors, the therapsids).*

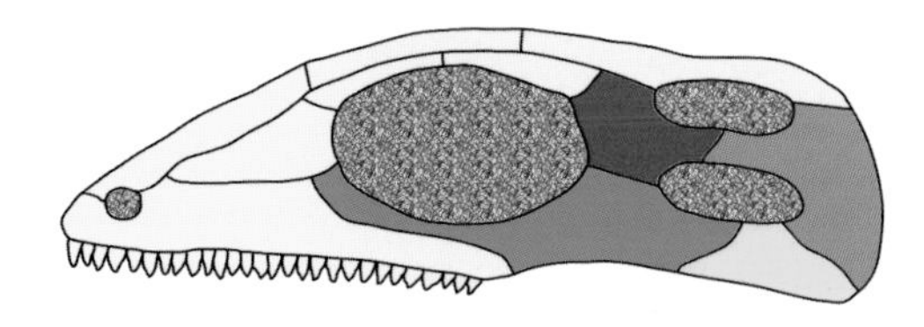

Diapsid: *two openings (the group includes lizards, snakes, crocodiles, birds and dinosaurs).*

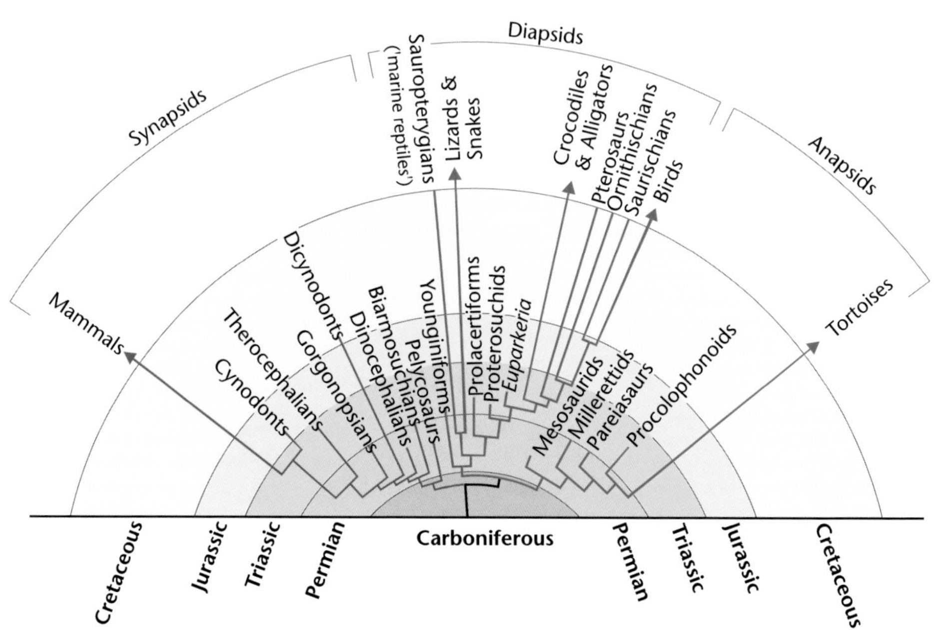

Figure 8.8 *Family tree showing the relationship of different groups of reptiles and their descendants.*

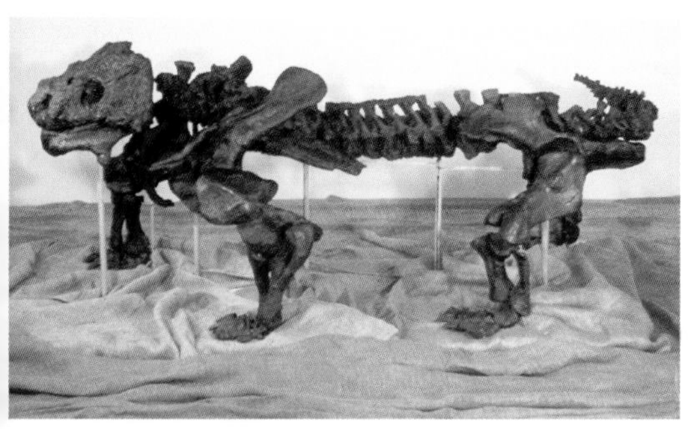

Pareiasaurid skeleton

Billy de Klerk, Albany Museum, Artist: Gerhard Marx

Pareiasaurid reconstruction

Eunotosaurus *skeleton*

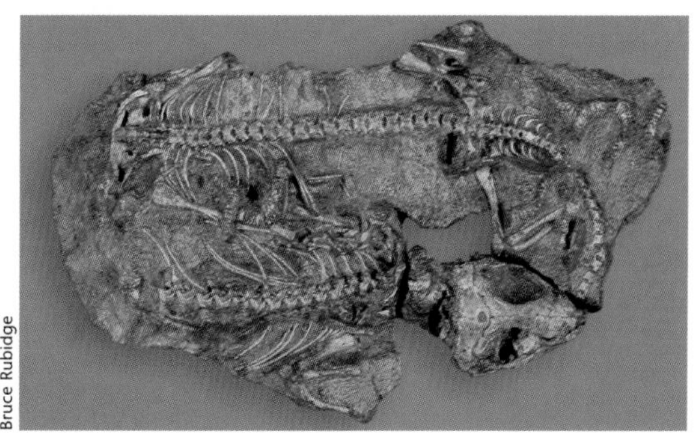

Bruce Rubidge

Owenetta *skeletons*

Figure 8.9 ***Anapsid reptiles from the rocks of the Beaufort Group of South Africa that are considered to be distant ancestors of tortoises.***

rocks, with its bizarre broadened ribs, has recently received renewed attention as possibly the most distant ancestor of tortoises (**figure 8.9**).

Another South African form that is a possible ancestor of tortoises is the 2.5 m long **pareiasaurid** (cheek lizard) *Bradysaurus* (**figure 8.9**). The fact that *Bradysaurus* is often found preserved as complete skeletons has been attributed to their becoming bogged down in the mud of drying floodplains, but it is possibly also the result of their thick armour of bony scales that protected them from scavengers and prevented the decomposing bodies from falling apart. These scales have also been considered as the forerunner of the tortoise shell.

Yet another possible tortoise ancestor from the rocks of the Beaufort Group of South Africa is a diverse group of meerkat-sized reptiles that are collectively known as the **procolophonoids**, of which the oldest form is from the Lower Beaufort. This group diversified in the Triassic (251 to 203 million years ago) into some quite specialised burrowing forms with transversely widened teeth and large horn-like protuberances from the skull.

From this diverse record of early anapsids it has recently been suggested that the group initially diversified in southern African Gondwana during the Permian to give rise eventually to tortoises. Which of the groups mentioned is the true ancestor of tortoises, however, awaits further research.

ANCESTORS OF MAMMALS DOMINATE

Early synapsids (**figure 8.8**), the distant ancestors of mammals, included plant- and flesh-eating forms that increased in numbers during the Permian Period. South Africa is internationally famous for its rich record of **therapsid** synapsids (mammal-like reptiles), which record the evolutionary development of mammals from reptiles. The most abundant herbivores of the Permian were members of the group known as the **anomodonts**. The most primitive forms are known from South Africa.

The discovery recently of the new forms *Patranomodon* and *Anomocephalus* (**figure 8.10**) from the earliest terrestrial rocks of Gondwana in the southern Cape has shown that this group of herbivores originated in Gondwana rather than Laurasia, as had previously been supposed.

Bruce Rubidge

Figure 8.10 *Anomodonts, the dominant herbivores of the Permian, originated in South Africa and spread to other Pangaean continents.* **Patranomodon** *from Prince Albert in South Africa is one of the most primitive anomodonts. It was the size of a mongoose.*

Patranomodon and *Anomocephalus* are primitive in that they retain a complete set of teeth in both jaws. By contrast the dentition of their descendants, the **dicynodonts**, is reduced to only a single pair of tusks (and in many cases no teeth at all), with their jaws covered by a horny beak similar to that of the modern tortoise although they are in no way closely related to each other. In addition, the presence of a sliding jaw hinge that allowed for back-and-forth movement of the lower jaw against the upper enabled them to grind up plant food finely.

Modern antelope are the dominant herbivores of the savanna biome, ranging in size from steenbok to eland. Similarly, the Permian dicynodonts, which probably also moved in herds, ranged from small, mongoose-sized forms such as *Pristerodon*, to large forms such as *Platycyclops*, which would have been the size of a large cow (**figure 8.11**).

Roger Smith/Iziko Museums, Artist Cedric Hunter

Figure 8.11 *Dicynodonts, which belong in the Anomodontia, are a diverse group of plant-eating therapsids that lived from the Permian to the Cretaceous. They range in size from species the size of a mongoose to large forms the size of an ox. This group of therapsids all had bony beaks specially adapted for breaking up plant material.*

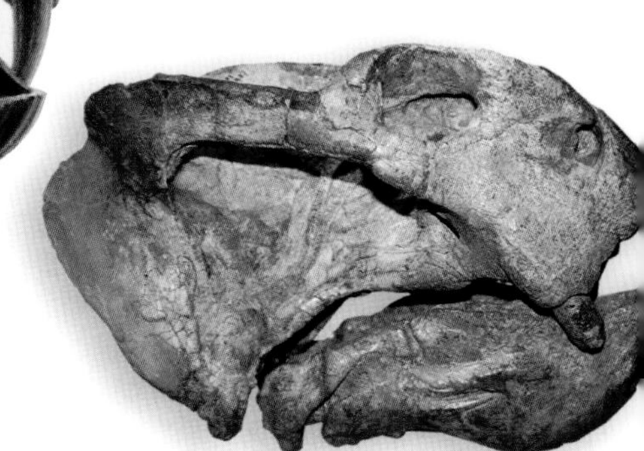

Figure 8.12 *Dinocephalians, the earliest large land-living vertebrates (**A**), are known from mid-Permian rocks of the northern and southern hemispheres, but their fossil remains are most abundant in the lower Beaufort rocks in South Africa.* Tapinocaninus, *a new form recently discovered near Prince Albert in South Africa, has the most complete dinocephalian skeleton (**B**), and has enabled us to reconstruct the animal.*

Dinocephalians (terrible head) (**figure 8.12**) are so named because of their bizarre, thickly boned skulls, which were probably an adaptation for head butting, possibly during courtship displays or in territorial fights. The largest of these strange creatures had a body length of more than 3 m and were the first large animals to live on land. As evidenced by their teeth, both plant- and flesh-eating forms occurred.

These strange beasts had bizarre talon-shaped incisor teeth in both upper and lower jaws that intermeshed with each other and became, in advanced forms, highly specialised for grinding tough plant foods such as horsetails that flourished along the banks of Permian rivers and ponds.

Bruce Rubidge

Bruce Rubidge/BPI Palaeontology; Artist: Marvin Carstens

Figure 8.13 *Gorgonopsians, with their large sabre-like teeth, were ferocious predators of the Permian, typically 2–3 m long. They were victims of the mass extinction that occurred at the end of this period.*

Although dinocephalians were abundant for a short time in the middle and early part of the Late Permian, they inexplicably became extinct, apparently leaving no descendants. Their fossils are known from Russia, China, Zimbabwe and more recently Brazil, but they are most abundantly represented in South Africa. Because they lived only during the Middle Permian and earliest part of the Late Permian, dinocephalians are useful for correlating rocks of this time period in different continents.

The dominant flesh-eating vertebrates of this Late Permian ecosystem were the ferocious **gorgonopsians** (terrible eye) (**figure 8.13**), fearsome predators with large, sabre-like canine teeth. Gorgonopsians varied greatly in size, the largest being the length of an ox, but they did not stand as high off the ground. Apart from having fine serrations on their canines that served to slice through the flesh of their prey, the teeth of these animals were not specialised with cutting or grinding surfaces; they would have fed by ripping chunks of flesh from their prey and swallowing without chewing, much as crocodiles do.

Smaller flesh-eating therapsids were the **therocephalians** (beast head) and **cynodonts** (dog tooth), which include insectivorous and carnivorous forms. They fulfilled the ecological role that jackals, hyaenas and lynxes do in the savanna of today. Members of both fossil groups evolved features typical of mammals, such as a bony secondary palate and the development of cutting and crushing surfaces on the molar teeth in order to grind up food finely (*see* 'Mammals from reptiles' on page 230).

THE MOTHER OF ALL MASS EXTINCTIONS

While the rocks of the Beaufort Group were being deposited 251 million years ago, Earth suffered the greatest extinction event ever – the end-Permian mass extinction. Currently it appears that the extinction extended over possibly 100 000 years, but the reason for it still eludes us. Suggested causes include extraterrestrial impact, volcanism, oceanic anoxia, oceanic overturn, global warming, food-web collapse, and a combination of these.

Given the great numbers and variety of vertebrates that populated the Karoo at the time of this extinction event, the Karoo Basin is one of the best places on Earth to study its effects on the biodiversity of land-living vertebrates. At the end of the Permian there were more than 50 different kinds of tetrapods in existence in South Africa, but the extinction decimated them and only four forms survived: two types of therocephalians, one dicynodont, and the anapsid *Owenetta* made it into the Triassic.

Diapsids and cynodonts were present in South Africa both before and after the extinction, but they were of different types. This means that somewhere they were able to survive the event and then repopulate the Earth from there, but these survivors have not until now been found in the South African rock record. Recent evidence from South Africa suggests that many of the forms that survived the extinction event lived in burrows, and in this way were able to escape the effects of whatever caused the extinction.

After the extinction most therapsids, particularly the cynodonts and therocephalians, began to evolve increasingly mammalian traits, a trend which had already begun in the Permian. These included: a bony palate that permitted breathing to continue while the mouth was full of food; development of grinding and cutting surfaces on the molar teeth to facilitate the breaking up of food while chewing; and reduction of the number of bones in the lower jaw, which enabled the lower and upper jaws to meet in precise occlusion while chewing (*see* 'Mammals from reptiles' on page 230). It is still not known exactly why these structures evolved in this way in the ancestors of mammals, but the fossils document the gradual attainment of mammalness in successive therapsid lineages through time.

Because only the bones and not soft tissues are preserved, it is not known if these animals were warm-blooded or gave live birth to their young. An exciting recent discovery from South Africa has revealed the presence of adult and juvenile specimens of the meerkat-sized cynodont, *Trirachodon,* preserved together in burrows (**figure 8.14**). This is not only evidence of communal living, it also suggests parental care of young and raises the question of warm- or cold-bloodedness in these animals. Perhaps the driving force for the changes that occurred in the evolution of mammals from reptiles might have been the ability to maintain a constant body temperature, and refining the ability to do this.

For all their success in the Permian, only one dicynodont genus, *Lystrosaurus,* survived the extinction and became the dominant terrestrial herbivore of the Early Triassic. Later in the Triassic, dicynodonts diversified and ranged throughout Pangaea, once again becoming a diverse and abundantly represented group. On the other hand the gorgonopsians, the large flesh-eaters of the Permian, were not so lucky and became extinct.

The extinction event also drastically affected plants: glossopterids died out, while clubmosses and

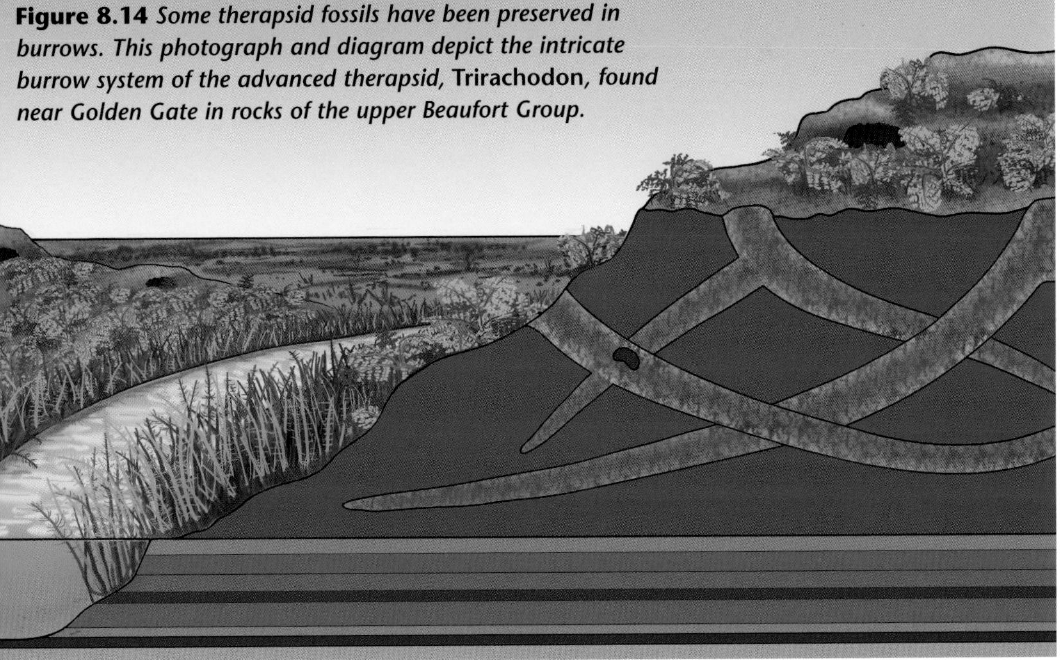

Figure 8.14 *Some therapsid fossils have been preserved in burrows. This photograph and diagram depict the intricate burrow system of the advanced therapsid,* Trirachodon, *found near Golden Gate in rocks of the upper Beaufort Group.*

MAMMALS FROM REPTILES

Therapsid (mammal-like-reptile) fossils from the Karoo portray the evolutionary development of mammals from reptiles. The oldest rocks of the Beaufort Group have yielded the most primitive therapsid fossils. Throughout successively younger Karoo rocks therapsids become progressively more mammal-like in form and the earliest true mammals are found in the uppermost Elliot Formation. In the South African Karoo it is thus possible to take a journey through the evolution of mammals from its most distant roots, starting with early therapsids in the rocks of the lower Beaufort in the vicinity of Laingsburg, and ending with mammals in the foothills of the Drakensberg in the eastern Free State. It is the only place in the world where such an evolutionary journey is possible in a single depositional basin, and in a single country.

Two important structural changes that occurred in the evolution of mammals from their more reptilian therapsid ancestors can be seen in the skulls of therapsids from South Africa – namely, changes in the palate and in the jaw hinge. The same structural changes can be traced during the embryonic development of modern mammals. This is a reflection of Haeckel's Law, which states that the embryonic development of an organism is a speeded-up reflection of its evolutionary history.

ORIGIN OF A BONY PALATE

Reptiles, which are cold-blooded and swallow their food whole without chewing it finely, do not have a bony palate separating the nasal passages from the mouth cavity. By contrast, mammals have a complete bony palate that enables them to chew their food without impairing their breathing. Chewing food increases the rate of digestion and absorption by the body, and thus contributes to faster production of energy and enables endothermic (warm-blooded) metabolism.

Between 200 and 300 million years ago the forerunner of a bony palate evolved in the reptilian ancestors of mammals and its development can be traced through the succession of therapsid fossils from the Karoo. The oldest and most primitive therapsids from the lower Beaufort Group had no bony palate, while the most recent therapsids in the rocks of the upper Karoo possessed a fully developed bony palate as is found in mammals. In the most primitive therapsids such as dinocephalians and biarmosuchians there is no vestige of a bony palate. However, in the slightly more advanced therocephalians and cynodonts, which lived at the very end of the Permian Period, the beginnings of a bony palate are present: the tooth-bearing maxillary bone is expanded towards the centre of the mouth to form a shelf. Although a complete

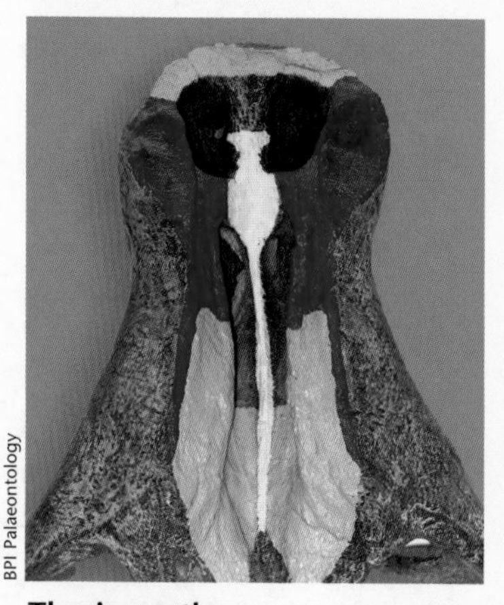

BPI Palaeontology

Theriognathus

Procynosuchus

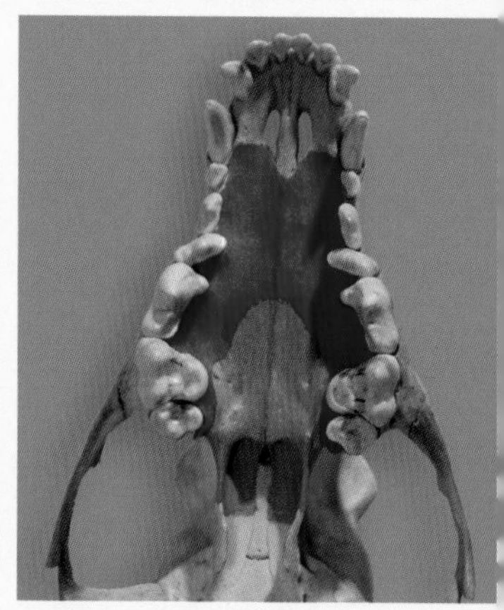

Dog

Skulls showing stages in the evolution of the bony palate in mammals. Notice how the red (maxilla) and pink (palatine) bones expand towards the midline.

bony palate is not yet present, recent research has demonstrated that they did have a soft fleshy palate. By the middle Triassic (upper Beaufort) a fully developed bone palate – where the maxillary and palatine bones meet in the midline of the roof of the mouth – is found in forms such as *Trirachodon* and *Cynognathus*. The embryos of modern mammals show essentially the same sequence of palate growth during their own embryonic development.

DEVELOPMENT OF MAMMALIAN EAR BONES FROM JAW BONES OF REPTILES

The lower jaw of a reptile comprises seven bones, while that of a mammal has only one. Conversely, a reptile has only one ear bone (the equivalent of the mammalian stirrup), while a mammal has three (the hammer, anvil and stirrup). By studying the jaw structure of a succession of therapsids from the Karoo, palaeontologists have discovered that the extra two ear bones of mammals (hammer and anvil) developed from the two bones that form the jaw joint (quadrate and articular bones) in reptiles. The reptilian bones inherited by the mammals no longer take part in jaw articulation, but they have been co-opted into the middle ear mechanism instead. Essentially the same sequence of events takes place in the embryonic development of modern mammals.

The fossil record of the Karoo reflects this story in detail. Therapsids from the lower Beaufort have, as in reptiles generally, a single middle-ear bone – the stapes – and the lower jaw is made up of seven relatively large bones on either side. By Triassic times, in forms such as *Thrinaxodon* and *Cynognathus*, the foremost bone of the lower jaw (dentary) had become larger at the expense of the bones situated at the back of the jaw, which had become smaller. However, the reptilian jaw joint between the quadrate and articular bones was still present and functional. In Jurassic forms such as *Probainognathus*, there are actually two jaw hinges on either side of the jaw – one between the quadrate and articular (the reptilian condition), and a newly acquired one between the dentary of the lower jaw and the squamosal of the upper jaw (the condition seen in mammals); these animals thus had a double jaw joint, neatly intermediate between the reptilian and the mammalian states.

In the earliest true mammals of the Elliot Formation, there is once more a single jaw articulation, but it is between the dentary and squamosal, not the quadrate and articular; these two reptilian bones have dropped out of the mammalian jaw joint and joined the original stirrup bone to become part of the sound-amplifying chain of middle-ear ossicles.

Skulls showing stages in the evolution of the mammalian single jaw bone from the seven jaw bones of their reptilian ancestors. The remainder of the lower jaw bones became modified to form the ear bones (hammer and anvil) of mammals.

horsetails decreased sharply in diversity. New types of seed ferns evolved to fill the vacant niches. These plants, known collectively as the *Dicroidium* flora (**figure 8.15**), were a strange mixture, with fern-like fronds, but protected seeds instead of sporangia. Although they were common and successful for some time they, too, eventually died out.

At their peak in diversity during the Late Triassic, these seed ferns inhabited a variety of ecological niches along with other species: riverine forest was colonised by seed ferns, cycads, and ginkgos; the wetlands were inhabited by seed ferns, clubmosses and horsetails; and the open woodland was inhabited by **cycadeoids** (cycad-like plants), ginkgos and **conifers** (relatives of pines). This environmental information is gleaned from the rocks in which the plant fossils are preserved, the type of vegetation from the impressions and cuticles preserved, and the fauna from the abundant insects and very rare bones preserved in these rocks.

John Hancox

Figure 8.15 Dicroidium *was the characteristic seed fern of the Triassic that diversified when* Glossopteris *flora became extinct at the end of the Permian Period (actual size).*

Also in these rocks of the Molteno Formation, which are beautifully preserved at Little Switzerland, Kannaskop, Umkomaas and Birds River, is a plethora of early **gymnosperms**. Some have ventured the opinion that this was one of the richest times in all plant history and that the Molteno Formation of South Africa represents the clearest window into Late Triassic plant and insect communities. This includes the most explosive of plant radiations, leading to the heyday of gymnosperm diversity.

ARRIVAL OF THE RULING REPTILES

While the end-Permian catastrophe dealt a major blow to the diversity of therapsids, a significant new type of reptile now made its appearance. These were unmistakably members of the diapsid line (possessing the tell-tale twin openings in the skull behind the eye), but of an advanced kind compared with their Permian forebears. Their influence was to be felt throughout the world in the ages to come: they were the forerunners of the aptly named **Archosauria** (ruling reptiles), one of the most spectacularly successful reptilian lines the world has ever known.

They bore the innovation of yet another opening on the side of the face, this time between the eye socket and the nostril on each side. The function of this new opening (named the *antorbital fenestra*, or window in front of the eye) is still debated, some arguing that it housed a gland – possibly a salt-excreting gland such as found in many sea-birds – while others suggest it provided better anchorage for jaw muscles, or it may have acted as a resonator to amplify sound. Still others suggest that it served no function at all other than to lighten the skull.

Once the end-Permian event had obliterated the gorgonopsian mammal-like reptiles, archosaurs became the largest predators, whereas the flesh-eating therapsids became progressively smaller. The first archosaur-like reptiles in South Africa were unremarkable, giving scarcely any hint of what their descendants were to become millions of years down the evolutionary line. They started off like their immediate diapsid ancestors, small and lizard-like, the earliest ones even lacking the tell-tale window in front of the eye that was to characterise nearly all their later descendants. Although the archosaurs probably originated elsewhere and migrated to the position of what is currently southern Africa, it is clear that their first representatives reached this part of Gondwana very soon after the group's origin.

Prolacerta from the earliest Triassic of South Africa is a typical archosaur-precursor. These little carnivores, which probably fed mainly on insects and perhaps on their own smaller diapsid forebears, already showed signs of one of the most important attributes of all but a few of the later archosaurs: speed and agility. Their back legs were significantly longer and stronger than their forelegs, giving them a tendency to lift the front part of the body when running at speed – a tendency that was to be perfected by their descendants, especially the predatory dinosaurs.

During the Early Triassic the fledgling archosaur line diversified to some extent and a larger and more impressive predatory form appeared in the shape of the 2-m-long crocodile-like *Proterosuchus* (**figure 8.16**). This had a curiously mobile snout: the tip of the snout around the nostrils was capable of some degree of forward-and-backward movement, dipping down when extended and lifting up when pulled back by special muscles, probably as part of a feeding specialisation. *Proterosuchus*' teeth indicate that it probably fed on fish and other small prey it could surprise along the edges of the rivers where it lived. Its long, deep tail and powerful hind limbs with broad feet suggest that it was a good swimmer.

By the Mid-Triassic (about 220 million years ago), as the climate warmed and became drier, archosaurs still made up only a small fraction of the total fauna of Gondwana, with plant-eating therapsids being the most abundant. New forms had taken over from the early archosaurian trail-blazers, including two impressive predators. The first of these, *Erythrosuchus* (**figure 8.16**), was 4–5 m long and built much like a deep-bodied crocodile. Its tall, relatively lightly built narrow head consisted largely of huge jaws, each armed with a row of vicious teeth, finely serrated like steak-knives. It was undoubtedly the top predator of its time – an ambush-predator that lay in wait for unsuspecting prey, rushing out to kill and devour with its great slashing teeth.

Figure 8.16 *Archosaurs, the group to which dinosaurs belong, radiated after the end-Permian extinction. These examples are archosaurs of the South African Triassic.*

The other new predator of the Upper Beaufort, *Euparkeria*, was very much smaller – the size of a hen. Its small but powerful hands show it was capable of grabbing hold of prey animals and slashing them with its sharp claws and even sharper teeth. Its large eyes, supported by a ring of thin bony plates identical to the **sclerotic ring** (ring of bony plates surrounding the pupil of the eye) found in the eyes of many birds and crocodiles today, indicate that its sense of sight was well developed.

When the Molteno Formation was deposited during the Late Triassic (210 million years ago), the first true dinosaurs emerged in Gondwana. We know little about these earliest Gondwana dinosaurs because so far their fossil bones have eluded us. They have, however, left behind abundant fossil footprints at many localities in southern Africa, especially in Lesotho, the northeastern Free State, KwaZulu-Natal and the northeastern parts of the Eastern Cape. The Molteno dinosaur-track sites are dominated by tracks of small to medium-sized, fully bipedal dinosaurs that evidently moved together in groups.

The only piece of unquestionable dinosaur bone that has so far been found in Molteno-age rocks in southern Africa is a small piece of a femur (thigh bone) of a **prosauropod** dinosaur (the ancestral group of the gigantic sauropod dinosaurs), which was found together with the bones of a **rhynchosaur** (bizarre pig-like, plant-eating second cousins of the dinosaurs, which are found in Late Triassic rocks at several places around the world, including the Zambezi Valley). Although we have yet to find these late rhynchosaurs in South African deposits, their earlier ancestral forms are known in South African rocks of Early- to Mid-Triassic age.

DINOSAURS DOMINATE

By the Late Triassic both carnivorous and herbivorous dinosaurs occupied niches previously dominated by therapsids. Therapsids and their descendants, the earliest mammals of the Jurassic, were unable to compete and remained small because they lived in the shadow of the dinosaurs until the latter died out 65 million years ago at the end of the Cretaceous Period.

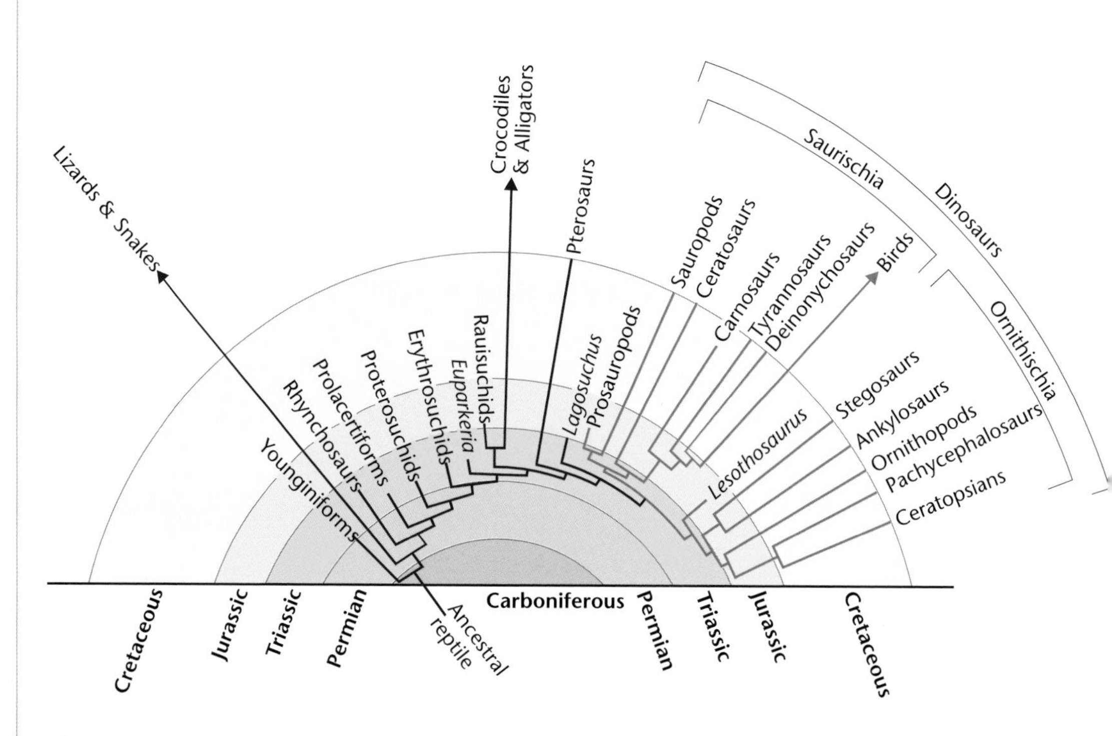

Figure 8.17 *Family tree showing archosaur diversity and interrelationships, and giving an indication of the period when different groups lived. Dinosaurs became extinct at the end of the Cretaceous Period.*

The history of dinosaurs might well have begun within Gondwana, although probably not in the African part of the supercontinent. The earliest known dinosaurs – that is to say, *Herrerasaurus, Eoraptor, Staurikosaurus, Saturnalia* and *Pisanosaurus* – come from rocks reported to be of Middle Triassic age in northwestern Argentina and southeastern Brazil. In these places they are found together with rhynchosaur fossils, just as in the case of the Zambezi Valley finds, but the southern African deposits are thought to be a few million years younger.

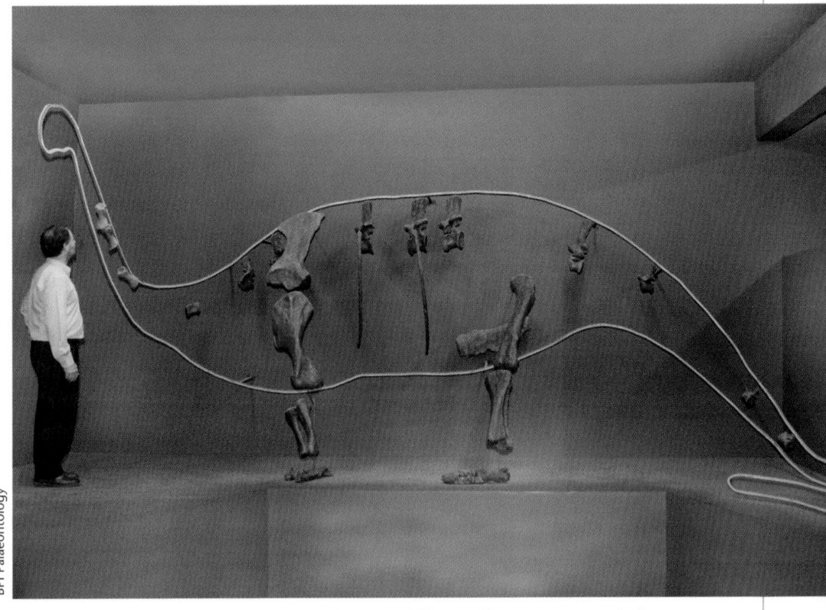

Figure 8.18 Antetonitrus *is the world's earliest sauropod dinosaur from the 200-million-year-old Elliot Formation.*

By the time of deposition of the succeeding Elliot Formation, the dinosaurs had already become well established in the South African part of Gondwana; their bones are relatively common in the red mudstones below the sandstone cliffs and basalt crags of the Maluti and Drakensberg mountain ranges of southern Africa.

Reptile-hipped dinosaurs

The most commonly found dinosaur in the beds of the Lower Elliot Formation was until recently considered to be the large prosauropod dinosaur, *Euskelosaurus*. However, recent research indicates that both prosauropod and sauropod dinosaurs were already present in the Lower Elliot Formation, but that sauropods were more abundant than prosauropods at this time.

The **Prosauropoda** were primitive members of the group of dinosaurs known as the **Saurischia** (**Figure 8.17**) or reptile-hipped dinosaurs. This group included all of the large meat-eaters of later times, such as *Allosaurus* and *Tyrannosaurus*, but also the later gigantic plant-eaters known as the **Sauropoda**. The group name of these early members, Prosauropoda, hints at their ancestral relationship to the giant sauropods such as *Apatosaurus, Brachiosaurus, Diplodocus* and the other huge, long-necked, long-tailed, barrel-bodied gargantuans that so dominated the Late Jurassic landscapes of the world. Current research recognises *Blikanasaurus, Antetonitrus* (**figure 8.18**), and also possibly *Melanorosaurus*, from South Africa as sauropods, while the only true prosauropod has not yet been named, but is closely related to *Riojasaurus* from Argentina.

These early dinosaurs were quite large, reaching up to 6–7 m in total length and weighing a few tonnes. Most of the length was made up of neck and tail, and they had small heads for their body size, equipped with flat, leaf-shaped teeth for cropping soft plants. Although their forelimbs were in many cases significantly shorter than their hind limbs (a trademark of dinosaurs), there is no doubt that they walked on all fours with legs drawn in under the body, so that they left a relatively narrow track.

Towards the end of the Triassic Period (about 210 million years ago) the climate in southern Africa became more arid. Ferns survived, but the seed ferns did not. Not much fossil plant evidence is present for the succeeding Jurassic Period in South Africa, other than a few cycad fronds. In other parts of the world cycads, cycadeoids and conifers flourished at this time. And as the Triassic gave way to the Jurassic, the large early prosauropod

Skeleton of **Massospondylus**

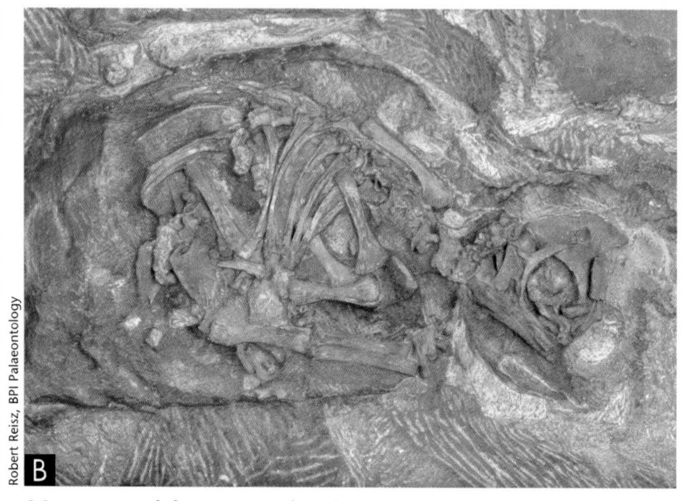

Massospondylus ***egg and embryo (actual size)***

Figure 8.19 **Massospondylus*****, a well-known early Jurassic dinosaur from the Drakensberg foothills (*****A*****). They grew to 4 m in length. Fossilised eggs (8 cm) of this dinosaur from South Africa still contain embryos (*****B*****) and are the oldest dinosaur eggs known.***

dinosaurs disappeared to be replaced by smaller relatives typified by the medium-sized plant-eater *Massospondylus* (**figure 8.19**) and its relatives.

These animals, up to 4 m long as adults, had a number of interesting innovations in their skeletons. Their leaf-shaped delicate teeth indicate that they were mainly herbivorous. This is also borne out by the fact that they had a gizzard of pebble-sized grinding-stones in the stomach to break up the tough leafy plant food to aid digestion. But they probably were not strict herbivores. They had a huge hooked claw on the first finger of the front foot (the equivalent of our thumb), which suggests it played an important role in their feeding. Some believe that the Early Jurassic prosauropods may have been generalised feeders that, although they may have eaten mainly plant foods, they also ate whatever else they could find in the harsh, semi-arid environments of the Elliot Formation. Perhaps their diet even included some carrion, where a large claw would have been useful for ripping open decaying carcases to get at the softer and more digestible inner bits.

An exciting discovery was made during the 1970s in the eastern Free State by the legendary fossil-hunter James Kitching. A number of fossilised eggs of *Massospondylus* were found in blocks of scree below a road cutting. Some of the eggs had already hatched by the time they were buried, and only the broken empty shells remained; others were unhatched. In some of these unhatched eggs tiny remains of fossilised dinosaur embryos lie tightly curled up as mute testimony to individual tragedies that happened almost 200 million years ago (**figure 8.19**). These fossil eggs are among the oldest reptile eggs known, and are certainly the oldest known dinosaur eggs.

Also sharing the Upper Elliot floodplains were flocks of small, bipedal, group-hunting, meat-eaters formerly known as *Syntarsus*, but now considered to be a local species of the American genus *Coelophysis* (**figure 8.20**). These small dinosaurs, about the size and weight of a modern secretary bird, are thought by some to have been covered in primitive feathers mainly to protect them from the rapid, severe fluctuations in temperature in the harsh desert-like conditions of the late Elliot and especially the Clarens Formation. They are early members of the group of dinosaurs known as the **Theropoda**, which were to rise to great prominence in later times. It is to this group that the world's largest ever land-dwelling meat-eaters, such as *Carcharodontosaurus, Spinosaurus* and *Tyrannosaurus* belong. One reason why the theropod dinosaurs are considered important today is because it is from this group that most scientists now believe modern birds descended (*see* 'The origin of birds' on page 240).

Coelophysis (*Syntarsus*) lived in fairly large social groups of several dozen individuals. They almost certainly hunted together as a group, enabling them to tackle prey that was several times larger than themselves. In southwestern America, one quarry has yielded over 1 000 individuals of one species of this genus, all killed and preserved together in a single catastrophic event – probably a flash flood. The small, sharp, serrated teeth of *Coelophysis* show clearly that they were meat-eaters, and one specimen found in Zimbabwe in the 1960s even had the fossilised bony remains of its 200-million-year-old last meal preserved within its body cavity.

Figure 8.20 ***This reconstruction of a Karoo scene during the Early Jurassic shows*** **Coelophysis (Syntarsus)** ***attacking*** **Massospondylus.**

specialised for cutting up tough plant foods. *Heterodontosaurus* (varied teeth lizard), as its name implies, had a variable set of teeth which included a toothless beak-like section at the front of the jaw, followed by a pair of sharp canine teeth. These, in turn, were followed by a battery of cutting-chopping teeth.

Figure 8.21 *These fossilised dinosaur footprints from the eastern Free State are from the Triassic Period.*

Bird-hipped dinosaurs

Living in the Upper Elliot Formation with the relatively abundant saurischian (reptile-hipped) dinosaurs were the earliest two-legged representatives of the great line of herbivorous dinosaurs known as the **Ornithischia** (bird-hipped), most of which later abandoned bipedalism and reverted to all fours. These early ornithischians (**figure 8.17**), represented by forms such as the turkey-sized *Heterodontosaurus* and the smaller *Lesothosaurus*, were all small bipedal runners that had closely packed teeth

Other fauna

Not much is known about the rest of the fauna that shared the latest Triassic Gondwana environments with these dinosaurs. There were large (2-m-long) crocodile-like amphibians inhabiting the rivers and lakes; some of the straggling remnants of the mammal-like therapsids of the Beaufort Group still clung on tenaciously; and large, ferocious archosaurs, known as **rauisuchids**, filled the top predator niche. These animals are known mainly by their isolated dagger-like teeth preserved in South African deposits. Because their shed or broken teeth are so often found associated with the bones of dinosaurs, it was at one time thought that the earliest large dinosaurs from South Africa were in fact carnivorous. It is now known that the teeth actually belong to large and ferocious rauisuchid archosaurs that were feeding on the dead bodies of the dinosaurs, perhaps as scavengers.

During the late 1970s James Kitching discovered the oldest tortoise from Africa in the rocks of the Elliot Formation near the Free State town of Clocolan. The oldest true tortoise is *Proganochelys* from the Triassic of Germany, but the slightly younger early Jurassic Clocolan representative, known as *Australochelys*, is almost as primitive as its German relative.

Despite the relative abundance of fossils in the Elliot and Clarens Formations, sedimentological evidence points to these rocks having accumulated in an arid environment. During favourable seasons this desert was covered by large ephemeral lakes that dried up during periods of drought.

Dinosaurs watered at these places and left their spoor in the sediment (**figure 8.21**). In good seasons these lakes were colonised by crocodile-sized amphibians and metre-long lungfish, as is evidenced by their fossilised bones.

THE FIRST MAMMALS APPEAR

From a human point of view the most significant animals of the Jurassic desert were advanced meerkat-sized therapsids whose bodies were probably covered by hair and were mammalian in many respects. But, from a biological point of view, they were not yet mammals. They had not yet developed the three small bones of the middle ear – the hammer, anvil and stirrup – which are characteristic of mammals; and they still retained the reptilian jaw joint (*see* 'Mammals from reptiles' on page 230).

Towards the close of Elliot Formation times, during the Early Jurassic (about 200 million years ago), the world's first true mammals appeared, living alongside the last survivors of their therapsid ancestors and the dinosaurs. Mammals and dinosaurs evolved alongside each other, but anatomically they are very different. The two separate evolutionary lines continued side by side until the end of the Cretaceous Period 65 million years ago, when dinosaurs became extinct while mammals survived.

The first true mammals were small shrew-sized creatures, and in southern Africa are known from only two species represented by three specimens (*Megazostrodon* (**figure 8.22**) and *Erythrotherium*). It was from these tiny animals that the many hundreds of different kinds of mammals – and ultimately humans – evolved (*see* Chapter 10).

THE END OF KAROO TIMES

The tough desert life of late Karoo times was violently disrupted about 180 million years ago by the eruption of the flood lavas that cap the KwaZulu-Natal Drakensberg and Maluti Mountains, marking the end of Karoo sedimentation and the beginning of the breakup of Gondwana (*see* Chapter 7). Almost no record of life was preserved in southern Africa during these turbulent times. Molten lava does not favour the preservation of fossils. Fortuitously, a glimpse of what was happening is preserved on a tiny island in the middle of what is now Lake Kariba in Zimbabwe.

About 180 million years ago a pause in the lava eruptions allowed sediment to accumulate in a large lake that existed at the time, preserving the bones of a variety of animals, including a large dinosaur. These remains were discovered in 1969 by a fisherman out for a weekend's sport on Lake Kariba. Little did he know what he had stumbled across. It turned out to be a new kind of relatively large dinosaur neatly intermediate both in time and in structure between the prosauropods of the Early Jurassic Elliot Formation and the large sauropod giants of the Middle and Late Jurassic.

The new dinosaur was named *Vulcanodon* and at the time of its discovery was generally considered the earliest and most primitive representative of the group of dinosaurian giants, the Sauropoda (**figure 8.17**). Recently Adam Yates and James Kitching described an even older and more primitive sauropod from the Triassic Elliot Formation of South Africa, which they named *Antetonitrus*, meaning before the thunder, in reference to the fact that it lived before the giant sauropod (thunder) dinosaurs (**figure 8.18**).

Southern Africa's dinosaur record is minimal after the lava eruptions at the end of Karoo sedimentation, mainly because there are very few deposits of Late Jurassic to Cretaceous age preserved on land in this part of the world. Two places in Zimbabwe have yielded some dinosaur bones representing this time, which in other parts of the

Figure 8.22 *The shrew-sized* **Megazostrodon** *is the oldest mammal in Africa and is known from the rocks of the Elliot Formation in South Africa and Lesotho.*

THE ORIGIN OF BIRDS

For more than a century palaeontologists have asked themselves the question: How did birds originate?

South Africa's famous palaeontologist, Robert Broom, suggested that they came from an ancestor like *Euparkeria*, the little archosaur of Triassic age from Aliwal North in the Karoo. He noted that it was an unspecialised small bipedal hunter and therefore ideal as an ancestor for many different kinds of later diapsid descendants from crocodiles to dinosaurs and birds. Broom's idea became the virtually unchallenged ruling theory for nearly a century. However, in the 1970s an idea first put forward a century earlier by Darwin's great champion, Thomas Huxley – that birds are descendants of the dinosaurs – was resurrected by John Ostrom of Yale University.

Ostrom was studying the small dinosaur *Compsognathus*, which had been found in a Jurassic limestone in Bavaria, southern Germany, in the early 1860s. From much the same locality two other skeletons of a different animal were also found in the 1860s, both bearing remarkable impressions of feather-like structures on the wing-like forelimbs and the long tail. This little animal, about the size of a crow, was named *Archaeopteryx* (ancient wing) and quickly became accepted as the first bird: it had wings and it had feathers, two of the features that effectively define birds today. Since those initial discoveries a handful of further specimens have been found, some with feather impressions and some without.

During his study of small dinosaurs, Ostrom noted that a couple of the specimens he was examining were actually misidentified specimens of *Archaeopteryx*; they had not been recognised because they lacked the tell-tale feather impressions. He noted that if *Archaeopteryx* is stripped of its feathers, it becomes classified as a dinosaur. Ostrom demonstrated that in details of its skull structure – its possession of teeth, its wing skeleton with a three-fingered hand in which each finger ends in a sharp claw, its shoulder girdle with a slender collar bone (rather than a wish-bone as in birds), its hind limb and pelvis, and its long bony tail – *Archaeopteryx* was very dinosaurian. It was suggested that even the possession of feathers was something that birds had inherited from dinosaurian ancestors, initially for temperature regulation as in the ancestral stock, but that the earliest birds later co-opted some of the feathers and converted them into wing surfaces.

Ostrom was at the time also researching a new dinosaur his team had found in beds of early Cretaceous age in Montana, United States. He named his new dinosaur *Deinonychus* (terrible claw), in reference to a large sickle-like slashing claw it bore on its first toe (the equivalent of our big toe). *Deinonychus* showed many similarities in its skeleton to that of *Archaeopteryx* and Ostrom suggested

world is often called the Golden Age of Dinosaurs. The Cabora Bassa basin hosts the best Late Jurassic deposits in Zimbabwe with a variety of fossils, including the gigantic *Brachiosaurus* unearthed in the 1970s. This fauna is very similar to the dinosaur fauna from the Tendaguru region of southern Tanzania, excavated by German geologists in the first decade of the 20th century. It is also similar in several ways to the much more diverse and spectacular fauna of the Morrison Formation of the same age in the southwestern United States.

The final chapters of dinosaur history in South Africa are preserved patchily in the rocks of the Kirkwood Formation of the Early to Mid-Cretaceous Age in the Algoa Basin near Port Elizabeth, in volcanic crater-lake deposits of Mid-Cretaceous age in Namaqualand, and in Late Cretaceous coastal plain rocks in Zululand on the east coast of KwaZulu-Natal. Discussion of these occurrences will be deferred until Chapter 9, however, as we first need to consider an important event in our journey through time, namely the break-up of Gondwana.

that it was from this group of dinosaurs – commonly referred to as the Deinonychosaurs or sometimes the Raptors – that birds originated. This view quickly gained wide acceptance among palaeontologists, but one stumbling block was that no non-bird dinosaurs had ever been found with evidence of feathers. The skeptics noted that feathers had only ever been found in *Archaeopteryx* and other undoubted birds, so feathers were therefore a trademark only of birds.

However, in the 1980s and 1990s, new spectacular finds of fossils in remote regions of Mongolia, China, showed that there were indeed feathered dinosaurs living during the Cretaceous. Here, as in the case of *Archaeopteryx*, freak conditions of preservation had ensured the preservation of delicate feather impressions along with the bones of several different types of dinosaurs without any hint of wings or any other unique bird-like structures. The best known of these feathered Mongolian dinosaurs are the forms known as *Sinosauropteryx* (Chinese feathered reptile) and *Caudipteryx* (tail feather), which featured prominently in press and other media reports for some time after their announcement.

Although there are still some sceptics, the overwhelming majority of palaeontologists now accept that small bipedal carnivorous dinosaurs belonging to the group known as the Coelurosauria (which includes the Deinonychosaurs) were the ancestors of today's birds.

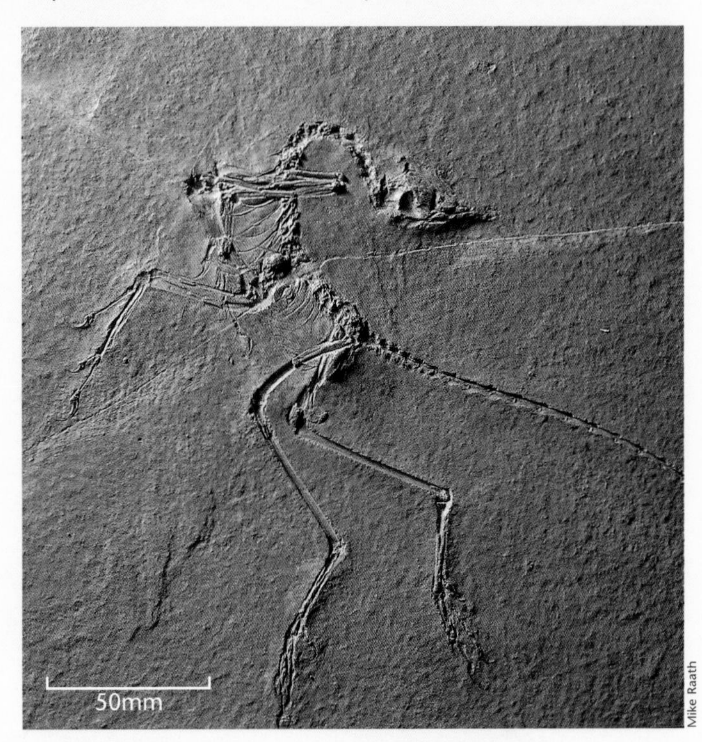

Archaeopteryx, ***the first bird, dates back to the Jurassic Period.***

SUGGESTED FURTHER READING

Cluver, MA. 1991. ***Fossil Reptiles of the South African Karoo.*** South African Museum, Cape Town.

Johnson, MR, Anhaeusser, CR and Thomas, RJ. 2005. ***The Geology of South Africa.*** Council for Geoscience, Pretoria.

Long, JA. 1997. ***The Rise of the Fishes, 500 Million Years of Evolution.*** University of New South Wales Press, Sydney.

MacRae, C. 1999. ***Life Etched in Stone: Fossils of South Africa.*** Geological Society of South Africa, Johannesburg.

Viljoen, MJ and Reimold, WU. 1999. ***An Introduction to South Africa's Geological and Mining Heritage.*** Mintek, Johannesburg.

White, ME and White, W. 1986. ***The Greening of Gondwana.*** Reed Books, Frenchs Forest, New South Wales.

THE STORY OF
EARTH & LIFE

9

THE MODERN WORLD TAKES SHAPE

IOA

The opening and cooling of the Atlantic Ocean, as well as the uplift of southern Africa, created arid conditions along the west coast that formed the Namib Desert.

ROUTE MAP TO CHAPTER 9

Age (years before present)	Event
180 million	• South Africa formed part of Gondwana, nestled between South America, the Falkland Plateau and Antarctica. A mantle plume appears to have risen beneath what is today southern Mozambique, and was responsible for vast out-pouring of basalt lava, which buried the southern African region of Gondwana. This mantle plume appears to have initiated the break-up of Gondwana.
140 million	• Gondwana began to separate into two fragments, initiating the formation of the Indian Ocean. They were West Gondwana (South America and Africa) and East Gondwana (Antarctica, India, Madagascar and Australia). By 140 million years ago oceanic crust had begun to form between them.
120 million	• West Gondwana began to fragment, possibly also initiated by a mantle plume. By 120 million years ago oceanic crust began to form between them – the start of the Atlantic Ocean. As the two landmasses separated, continental crust along the southern Cape (Falkland Plateau) began to detach along a major fracture (Agulhas Falkland Fracture Zone) and moved westwards with South America. Smaller fractures developed along the southern Cape as the Falkland Plateau slid by, creating depressions in which sedimentary deposits began to accumulate.
90 million	• The Falkland Plateau finally separated from the Cape on its westward journey, and southern Africa was surrounded by warm seas. The interior was elevated and was experiencing erosion. Three main river systems existed: the ancestral Limpopo River, which drained most of central southern Africa; the Kalahari River, which drained what is now the central Kalahari; and the Karoo River, which drained the southern interior. Sedimentary deposits formed in volcanic craters atop kimberlites indicate that the climate was warm and humid. Deep, tropical soils developed across southern Africa.
65 million	• A major meteorite impact caused collapse of the global ecosystem and 70–80% of species (including the dinosaurs) became extinct. Little is known about the effects of this event in southern Africa, as rocks of this specific age are rare.
55 million	• The warm Earth climate peaked, then began a steady decline towards lower temperature. The carbon dioxide concentration also declined steadily. The cause is uncertain, but may have been due to a combination of factors, including changes in ocean circulation brought about by plate motion, the rise of the Himalaya and Andes Mountains, or perhaps to the spread of grasslands. The interior of Africa began to experience warping, disrupting drainage patterns. The Limpopo River was cut from its upper tributaries and the ancestral Lake Makgadikgadi formed. Upwarping also caused the Kalahari River to capture the Karoo River, forming the Orange River. The Southern Ocean began to cool because of the opening of the Drake Passage.

20 million	• Uplift of southern Africa occurred, mostly in the east. The south Atlantic cooled further and marine upwelling commenced along the west coast, leading to arid conditions in the west.
5 million	• Further uplift occurred, especially in the east. The eastern escarpment began to trap moisture, increasing the rainfall gradient across southern Africa and causing more arid conditions to develop in the interior, resulting in the formation of the Kalahari Desert.
2 million	• The Earth's climate continued to cool, perhaps due in part to the formation of the Panama isthmus, which isolated the Atlantic from the Pacific Ocean. Ice ages commenced in the northern hemisphere, during which extremely arid conditions developed in the interior of southern Africa, and desert sands spread as far north as the Congo. The last ice age ended about 10 000 years ago, and the local climate has been comparatively moist since.

THE BREAK-UP AND DISPERSAL OF GONDWANA

Major reorganisation of the world's continents characterised the Jurassic and Cretaceous Periods. The world 170 million years ago was geographically very different from today. All of the continents formed the single landmass, Pangaea, extending almost from the South to the North Pole. A period of rifting commenced around this time that led to the opening of vast new oceans. The next 170 million years saw radical transformation of the planet. The break-up of Pangaea was an important event, heralding the destruction of the most recent supercontinent and the beginning of the cycle of plate movements that produced the present configuration of the continents. This break-up and dispersal has had a profound effect on the current surface topography of the continents, on the distribution of oceans, and on ocean and atmospheric circulation, global climate and the biodiversity of the planet.

This chapter focuses on events affecting Gondwana, the southern portion of Pangaea. South Africa has only a fragmentary record of this break-up history preserved in the rocks on land, but the record is more complete in rocks lying below sea level on the continental shelf around southern Africa.

At the time break-up of Gondwana commenced, southern Africa nestled between the southern tip of South America, the Falkland Plateau and Antarctica (*see* **figure 7.2**). There was no Atlantic or Indian Ocean, and South America, southern Africa, the Falkland Plateau and Antarctica formed a continuous landmass. A great mountain range extended from South America across the Falkland Plateau into East Antarctica. The foothills of this range lay across the southern Cape. To the north of this mountain range lay the great Drakensberg lava field, extending across the whole of South Africa, most of Zimbabwe and Namibia, and into parts of South America and Antarctica.

The elevation of South Africa may have ranged from about 1 000 m above sea level along what is now the west coast to over 2 000 m in the region that is today Lesotho, although some believe these estimates to be too high. South Africa lay about 15° further south than it does today, but the Earth was warmer then and the climate was probably still hot and arid. The early dinosaurs were eking out a living in this harsh landscape.

Birth of the Indian Ocean

The break-up of Gondwana began with the opening of the Indian Ocean along the African east coast, heralded by the eruption of basalts and rhyolites of the Lebombo region. Gondwana began separating into two fragments: West Gondwana, consisting of South America and Africa; and East Gondwana, made up of Madagascar, India, Antarctica and Australia. What caused this split is still a matter of conjecture. A possible candidate is the rise of a mantle plume beneath Gondwana, with its centre situated in Mozambique, to the east of Pafuri where the Limpopo River crosses the Lebombo Mountains (**figure 9.1a**). A similar plume , currently centred beneath northern Ethiopia, is believed to be responsible for the recent opening of the Red Sea, the Gulf of Aden and the East African Rift Valley (*see* Chapter 2).

The Mozambique plume is believed to have caused the crust to rise, and at the same time perhaps stretched and weakened the underlying mantle lithosphere. The vast outpouring of lavas that marked the end of the Karoo period of sediment accumulation has also been attributed to this plume (*see* Chapter 7). It has been estimated that about 3 million km^3 of lava was erupted during this short event.

Plumes typically create three rift arms arranged at angles of about 120° to each other, a pattern clearly evident at the southern end of the Red Sea. The Mozambique plume also started to develop in this way about 180 million years ago, with extension taking place along what is now the Mocambique-Zimbabwe border, along the Lebombo range and along the Limpopo valley. Then, major fractures formed in the crust along the eastern edge of the Lebombo Mountains and to the northeast, and East Gondwana began to slide southward relative to West Gondwana, opening rifts in what is today the Mozambique Channel (**figure 9.1a**). A seaway began to develop between the two land masses and oceanic crust began to form between them. Rifting continued along the Limpopo valley in a west-northwest direction, extending as far as the present position of the Okavango Delta, and dykes were intruded into the crust along this rift. However, this particular rift branch failed to develop further.

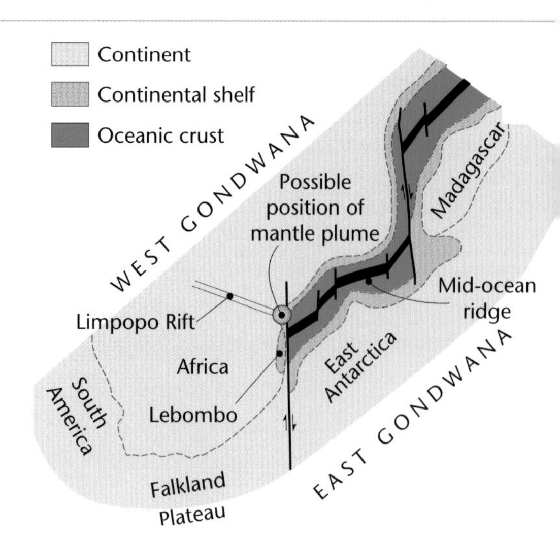

Figure 9.1a ***Southern Africa once formed part of Gondwana, a single landmass that included South America, Africa and Antarctica. About 180 million years ago, a mantle plume is believed to have risen beneath what is today southern Mozambique, generating extensive volcanic activity over southern Africa and neighbouring parts of Gondwana. This mantle plume is believed to have initiated the opening of the Indian Ocean. A rift opened between West Gondwana (Africa and South America) and East Gondwana (Antarctica, Madagascar, India and Australia), the sea invaded and ocean floor began to form. A rift also began to form in the interior of Africa (along what is now the Limpopo valley) but this failed to develop further. (The outline of South Africa is included in this diagram only for reference; it did not exist at that time.)***

Birth of the Atlantic Ocean

Some 120 million years ago, South America began to detach from Africa, opening rifts along the southern African west coast. This thinned the continental crust (**figure 9.1b**): the start of the Atlantic Ocean. Again, plume activity may have contributed to this, producing voluminous basalt lavas that today form the Parana basalts of South America and the Etendeka basalts of Namibia. Crustal stretching along the west coast detached a large slab of continental crust from the African continent off the southern Cape coast, now known as the Falkland Plateau. This began to move westwards as South America separated.

The fracture along which this detachment took place is today known as the Agulhas Falkland Fracture Zone. It extends across the width of the Atlantic Ocean, one of the longest fracture zones on Earth. In its early history, it probably generated huge earthquakes along the KwaZulu-Natal and southern Cape coasts, much as the San Andreas Fault in the United States does today. The still-active segment of the fracture zone is today situated in the middle of the southern Atlantic Ocean where its earthquakes cannot do any damage.

As the Falkland Plateau was tugged away from southern Africa by the receding South American

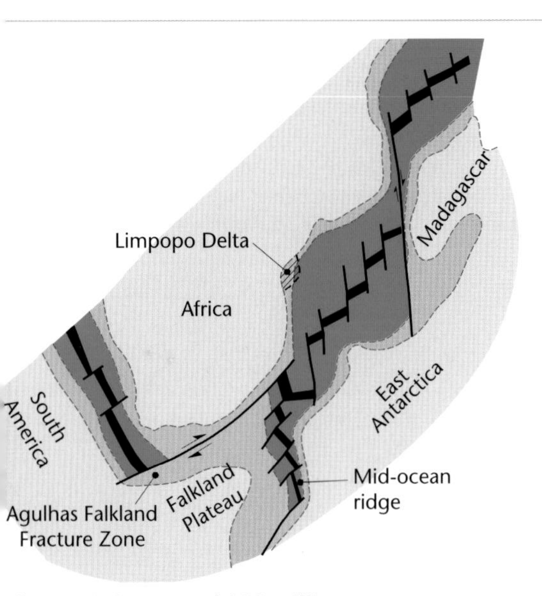

Figure 9.1B ***Around 120 million years ago, South America and Africa began to separate, opening the Atlantic Ocean. This may also have been initiated by a mantle plume that rose along what is today the west coast of Namibia. A large fracture developed along what is today the southern Cape coast (the Agulhas Falkland Fracture Zone). As the Atlantic Ocean widened and South America separated, the Falkland Plateau (consisting of continental crust) remained attached to South America and slid away from the Cape towards the west. A mid-ocean ridge developed along its eastern side, and oceanic crust formed to fill the void created by westward separation of the Plateau.***

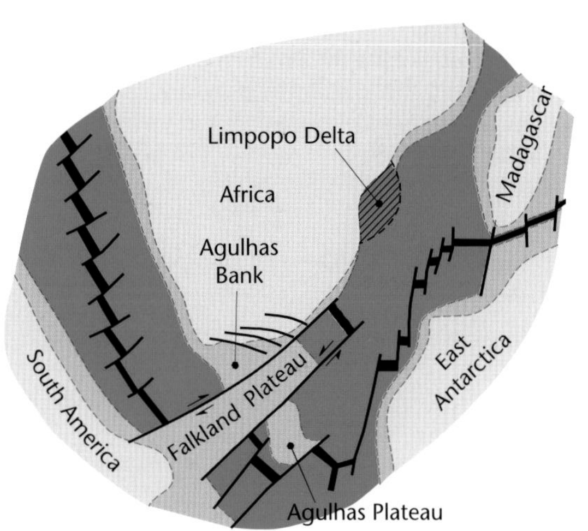

Figure 9.1c ***The Atlantic continued to widen. The westerly slide of the Falkland Plateau caused tears to develop along the southern Cape coast, and depressions formed in which sediment began to accumulate. A river system had developed in the interior (the ancestral Limpopo River) and a large delta formed at its mouth in what is today the Mozambique Channel. Some reorganisation of the mid-ocean ridge system had occurred by about 110 million years ago, and a portion of the Falkland Plateau became detached. This remained stranded as the Falkland Plateau receded, and today forms the Agulhas Plateau. South America and the Falkland Plateau continued on their westward journey and by about 90 million years ago the Falkland Plateau had cleared the Cape.***

continental landmass, it stretched and tore the continental margin of southern Africa, creating a number of depressions along what is now the southern Cape region and the Agulhas Bank.

This stretching – as well as that taking place on the Falkland Plateau itself – thinned the crust, and as a result the mountain range that formerly lay on the Falkland Plateau gradually sank beneath the ocean. The sea spread northwards into the depression that was created by rifting between Africa and South America.

Meanwhile, East Gondwana continued its southerly drift. India and Madagascar separated from Antarctica and Australia, starting approximately 120 million years ago. India was about to start its northward journey, which was to culminate in a collision with Asia – an event that began about 50 million years ago.

By about 90 million years ago, the Falkland Plateau had cleared the southern Cape coast on its westward journey with South America, and oceanic crust had formed in its place. A small sliver of the plateau remained stranded (**figure 9.1c**) and today forms the Agulhas Plateau off the southern Cape coast. While the Atlantic and Indian Oceans continued to open, the Antarctic Peninsula remained close to the southern tip of South America. This was achieved by the simultaneous westward movement of South America and the southward movement of Antarctica (plate motions are actually rotations on a sphere). Africa drifted slowly northwards, reaching its present position about 30 million years ago. Meanwhile, Australia separated from Antarctica some 50 million years ago, and gradually began moving northwards to its present location.

THE END OF THE JURASSIC: METEORITE IMPACT IN SOUTH AFRICA?

The geological time scale is divided primarily on the basis of the nature of the fossil remains of organisms which lived at various times. Such subdivisions are only possible for the period following the appearance of organisms with hard body parts that could become fossilised – the Cambrian Explosion, which happened about 540 million years ago (*see* Chapter 6). The period after this, the Phanerozoic Eon, is subdivided into the well-known divisions called Periods, such as the Triassic, Jurassic or Cretaceous (*see* 'The geological time scale' on page 71).

The terminations of these periods are usually based on sudden changes in prevailing life forms – generally considered to represent mass extinctions of species. Five big mass extinctions have been recorded in the Phanerozoic: the end-Ordovician (about 443 million years ago); the end-Devonian (about 354 million years ago); the end-Permian, Mother of all Mass Extinctions (about 251 million years ago); the end-Triassic (206 million years ago); and the end-Cretaceous (65 million years ago).

The end of the Jurassic Period (142 million years ago) is also marked by a mass extinction, which is comparable to the end-Devonian event in terms of the number of families – zoologically speaking – that became extinct. As with most other such events, the cause is unknown. A meteorite impact may well have played a role, and the impact may have happened in South Africa while the Atlantic and Indian Oceans were just beginning to open.

The evidence comes from the region north of Kuruman in the North West Province. In this area, a vaguely circular structure about 160 km in diameter

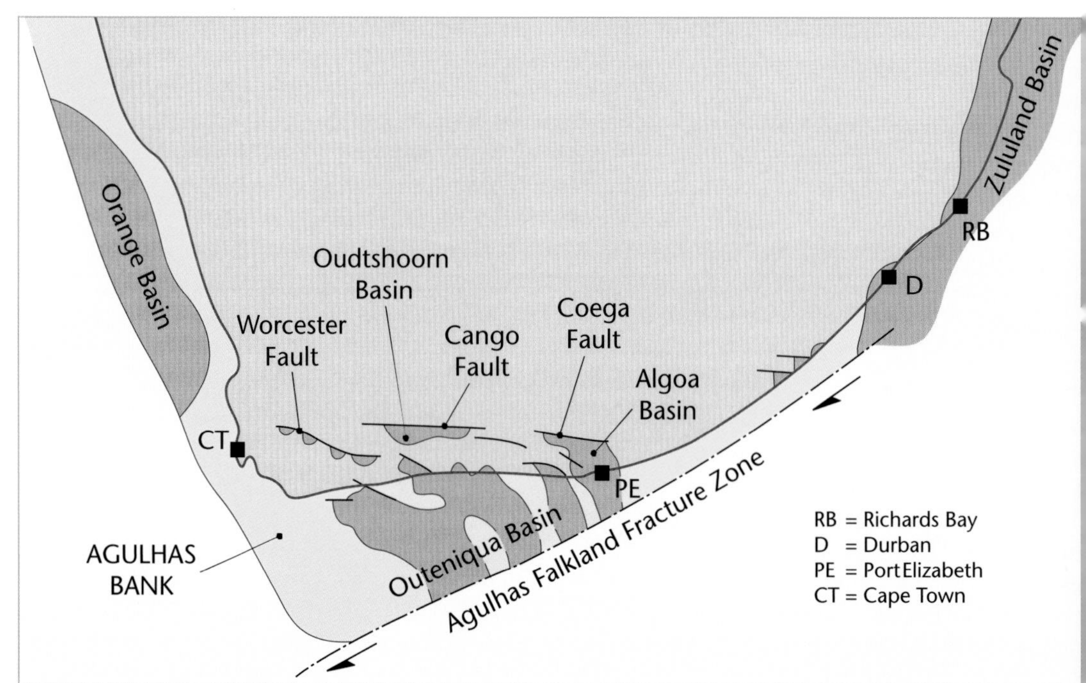

Figure 9.2 ***The westward slide of the Falkland Plateau caused fractures to develop along the southern Cape coast. The crust on the southern side of these fractures dropped downwards, creating depressions in which sediment, eroded from the higher-lying side, was deposited. Sediment eroded from the interior was also deposited off-shore in the Zululand, Outeniqua and Orange Basins. (The locations of major cities are shown for reference.)***

is defined by the ancient rocks of the Transvaal Supergroup. The little village of **Morokweng** in the region has given its name to the structure. Unfortunately, most of it lies buried beneath the sands of the Kalahari Desert. Geophysical investigation recently confirmed the existence of the circular feature beneath the sand, and a borehole was drilled to investigate. This revealed the presence of a layer of frozen impact-like melt about 900 m thick lying on top of fragmented rock, or breccia, which confirmed an impact origin for the structure. The impact melt was even found to contain small fragments of meteorite, a remarkable discovery because normally a sizeable meteorite is completely vaporised on impact (*see* 'Impact craters' on page 134).

The age of the melt has been measured to be 145 million years, very close to the end of the Jurassic. Could this be the event that precipitated the Jurassic extinction? Research has only just begun on the newly discovered structure, and it will be several years before its importance can be adequately assessed.

SEDIMENTATION AND MARINE LIFE DURING GONDWANA'S BREAK-UP

Unlike the Karoo – where sedimentation occurred for nearly 120 million years, much of it on land, providing a complete record of terrestrial life during the Permian, Triassic and early Jurassic Periods – the later Jurassic and Cretaceous Periods during which the break-up of Gondwana occurred are poorly documented in the rocks of South Africa. During this time, it seems that southern Africa was elevated and the interior was experiencing erosion. Sediment deposition was taking place mainly in the developing Indian and Atlantic Oceans, now all off-shore areas.

There were a few exceptions in the form of local depressions created by stretching and tearing of the crust as the Falkland Plateau slid past the southern Cape, as well as on the northern KwaZulu-Natal coastal plain (**figure 9.2**). The on-shore southern Cape basins were mostly developed along large faults, tears that were created as the Falkland Plateau slid past in a westerly direction, stretching the crust. Collapse towards the south occurred along these tears, creating depressions in which sediment, eroded from the higher ground to the north, accumulated (**figure 9.3**). The largest of these are the Oudtshoorn and Algoa Basins (**figure 9.2**). There are also extensive off-shore sediment accumulations, which are known as a result of oil and gas exploration off the coast (*see* 'Formation of oil and gas' on page 250).

Figure 9.3 ***Sedimentary rocks that form the Robberg peninsula at Plettenberg Bay were deposited during the break-up of Gondwana.***

During the early stages of break-up new, shallow seaways were created around South Africa – the proto-Indian and Atlantic Oceans. These warm, shallow seas were host to an abundant variety of life forms that would not have looked too different from a modern reef ecosystem, although the mix of species present differed from today. Deposits of this age in southern Africa occur from northern Mozambique (near the port of Nacala), through southern Mozambique and northern KwaZulu-Natal, to the coasts of the Eastern and Western Cape. Rivers entering the sea along the northern KwaZulu-Natal coastline gave rise to a variety of settings of sedimentation, including estuaries, beaches and shallow-marine conditions. These deposits today form much of the flat coastal belt of northern KwaZulu-Natal, and are well exposed in the Lake St Lucia area such as at Hell's Gate.

In contrast, sedimentary deposits formed on land are small and very localised, and are mostly related to active faults. The largest of these occurs in the Algoa Basin in the Port Elizabeth area (**figure 9.2**). Here as much as 14 km thickness of

FORMATION OF OIL AND GAS

Oil (or petroleum) is one of the main energy resources we use today. It is also used in a range of products from tar for road-building to cosmetics and wax wrappings. Its presence underground has major socio-economic and political implications, sometimes even causing wars. Oil therefore touches our lives every day.

Oil is biological in origin, formed mainly from the alteration of organic matter deposited on the floors of the oceans. Organic productivity is particularly high in oceans that are warm and shallow (much like algal blooms that occur in swimming pools in summer). For this reason, one of the main periods of organic production and accumulation was during the warm Jurassic Period of Earth's history, when the supercontinent of Pangaea was splitting and creating new shallow seas. Numerous other factors also play a role, however, and not all suitable settings preserve oil fields.

Oil fields occur in areas where thick piles of marine sediments have been deposited and preserved. Such piles are often found in rapidly subsiding basins in rift and foreland settings (*see* 'Sedimentation in continental rifts' on page 98 and 'Sedimentation in foreland basins' on page 162). These sediments in turn must contain significant amounts of organic matter (predominantly phytoplankton) that forms the raw material for petroleum.

Petroleum formation takes place when such organic-rich sediments are slowly heated during deep burial to between 100 and 150°C (the so-called **oil window**) – too little heat and petroleum is not created, too much and the organic material is changed to gas. Once created, the oil (or gas) is expelled from the **source rock** in which it formed and migrates upwards via fractures and joints until it encounters rocks with a high proportion of open spaces between the grains, where it accumulates (a **reservoir** – like a giant sponge). The reservoir rocks in turn need to be covered by a fine-grained **cap** or **seal**, such as a salt or mudstone layer, to prevent the oil simply leaking out to the surface and decomposing. Such **oil traps** may be created by normal sedimentary layering or by faulting or folding. They are common in areas where ancient rifted continental margins have been subjected to later compression.

Because sediments have to be buried to be heated to the temperatures required for oil formation, but not beyond, most of the world's oil originates from Jurassic-aged rocks, and is typically stored in reservoirs of Cretaceous age.

SOUTH AFRICAN OIL POTENTIAL

The coastline of South Africa has nearly all the correct parameters for the formation of oil. There are abundant Jurassic-aged marine sediments deposited during the break-up of Gondwana; there are Cretaceous sandstones of good reservoir potential; and there are suitable traps. Furthermore, extensive drilling of these rocks has been undertaken by SOEKOR (now PetroSA) since 1965 that has identified many petroleum plays.

Why then do we have so few producing oil fields? The answer may lie in the fact that most of the Jurassic rocks have been heated through the oil window and are therefore gas prone, or more likely that insufficient detailed exploration has been undertaken. Exploration continues today, however, with a number of companies positive about South Africa's off-shore oil potential.

Photo Access

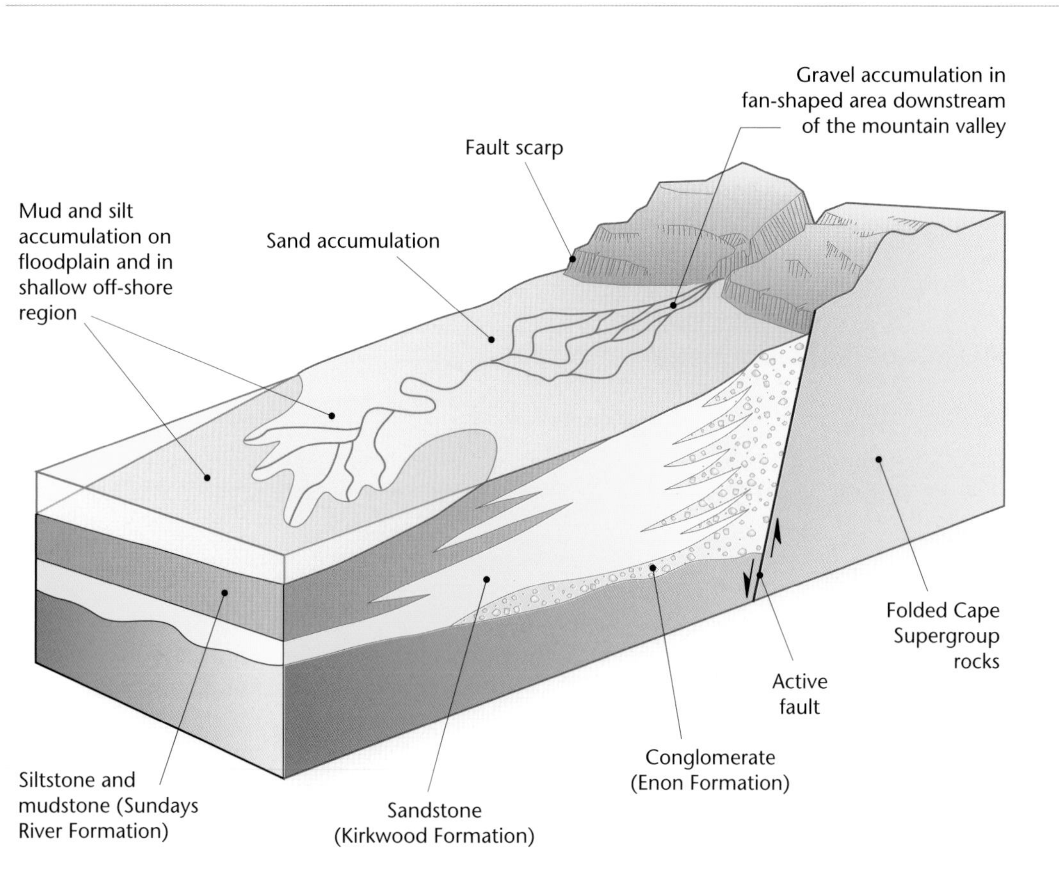

Figure 9.4 *This schematic illustration shows how fractures created by the westward movement of the Falkland Plateau created depressions, in this case the Algoa Basin in the Port Elizabeth area (see* **figure 9.2***). Rivers flowing from the higher ground to the north of the fracture deposited first gravel, then sand and finally silt and mud in deltas and on the bed of lakes and a shallow sea that filled the deeper parts of the depression. Sediment accumulation kept pace with subsidence along the fault and the resulting piles of sediment became extremely thick, as much as 14 km in one instance. Fossils from these sedimentary deposits provide an indication of marine and terrestrial life at the time.*

sediment accumulated, collectively known as the **Uitenhage Group**. The sedimentary processes involved in this deposition are illustrated in **figure 9.4**, which is a reconstruction of this environment about 120 million years ago. Active faulting related to the westward drift of the Falkland Plateau tore the crust, creating local fault scarps, and rivers arising on the high ground crossed the scarps to deposit coarse gravel and sand in gently sloping, fan-shaped deposits in the slowly subsiding depression. Down river, these gave way to extensive floodplains and ultimately deltas, which debouched into the shallow seaways around the coast.

Today, these gravel deposits form impressive conglomerates of the Enon Formation, which are well exposed around Knysna and at Plettenberg Bay. The sand deposits, on the other hand, form the Kirkwood Formation, while the silt and mud form the finer-grained sedimentary rocks of the Sundays River Formation.

During the Cretaceous (142 to 65 million years ago) there were three major episodes in which the world's sea levels rose, covering the coastal plain with shallow marine deposits. The marine deposits of this age are rich in fossils, containing numerous invertebrates, including ammonites, as well as

Figure 9.5 ***Ammonites, with their characteristic spiral shells, as seen here, were a common marine animal in the Cretaceous seas that developed around the South African coast as Gondwana broke up. They became extinct along with the dinosaurs at the end of the Cretaceous Period.***

marine microfossils. Ammonites were animals that built coiled shells, like the nautilus of today, and were among the most common life forms in the Cretaceous seas. Many different species, ranging in size from less than a centimetre to more than a metre, inhabited the shallow, warm oceans at that time (**figure 9.5**).

A single skull and partial skeleton of an approximately 3-m-long plesiosaur (**figure 9.6**), a carnivorous fish-eating marine reptile, is also known from the Sundays River Formation. This specimen, called *Leptocleidus capensis*, was collected at Picnic Bush in the Zwartkops River Valley south of Uitenhage. It is the same genus as the famous fossils found in England and Australia, indicating that this taxon probably had a global distribution during the Cretaceous.

Based on the limited amount of fossil data, one could imagine the seas of the Cretaceous around South Africa as warm and shallow, with ample penetration of sunlight. Small and large ammonites would have been either floating about in the upper water column looking for their next meal of small crustaceans, or jet-propelling themselves away by means of their siphons, like the modern-day nautilus, from predators such as *Leptocleidus*, the denizen of the deep. Various corals would have formed atolls and barrier reefs on which sea urchins, snails and marine segmented worms would have lived (much like in the tropical seas of today). Between the reefs, numerous bivalves and gastropods would have thrived on the sandy sea floor.

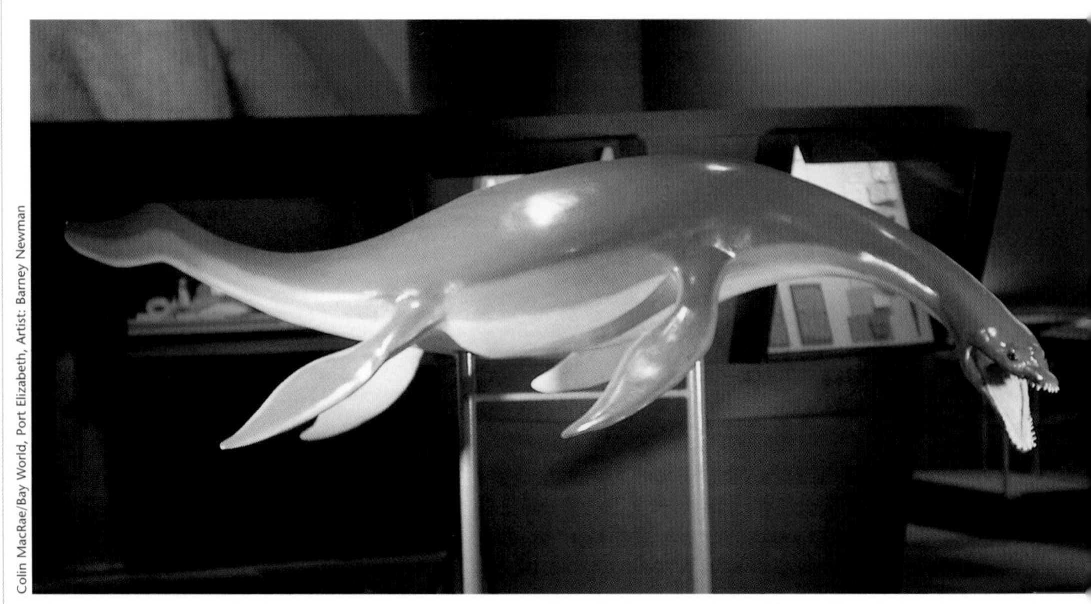

Figure 9.6 **Leptocleidus capensis,** ***a marine reptile (plesiosaur), found in rocks of the Algoa Basin.***

LIFE ON THE COASTAL PLAIN

Due to the limited width of the coastal plain, rocks deposited in terrestrial settings (such as river deposits) are sparsely developed. Where present, they often indicate high energy conditions such as are formed by steep, gravelly, fast-flowing rivers (for example, in gravel and sand deposits of the Enon Formation). Such conditions are not amenable to fossil preservation. For this reason, although terrestrial fossils are known from these coastline deposits, they are rare, and the remains are more often than not fragmentary, often consisting of little more than resistant teeth. Rivers also carried terrestrial animals out to sea, and their scattered and fragmentary remains are known from some near-shore deposits.

Billy de Klerk/Albany Museum, Artist: Gerhard Marx

Figure 9.7 **Paranthodon africanus**, ***a relative of* Stegosaurus, *was the first dinosaur that was discovered in South Africa.***

Although very rare, animal fossils include a number of dinosaurs, among which are the first dinosaur remains found in South Africa (and one of the first world-wide), which were discovered in 1845 along the Bushman's River. This fossil – originally considered by its discoverer William Atherstone in 1856 to be Cape *Iguanodon*, similar to the well-known plant-eating dinosaur from Europe - has recently been shown to be a large stegosaur (similar to the well-known North American *Stegosaurus*) called *Paranthodon africanus* (**figure 9.7**).

Other dinosaur remains from this period in South Africa are the vertebrae of a large sauropod dinosaur called *Algoasaurus* (similar to but smaller than its North American cousin, *Apatosaurus*), which were discovered in near-shore marine deposits of the Algoa Basin. A life-size reconstruction is on display at Bay World, Port Elizabeth (**figure 9.8**). This plant-eating sauropod probably browsed on the leaves and cones of the tall pine-like trees growing on the coastal floodplains of the Algoa Basin.

Colin MacRae/Bay World, Port Elizabeth, Artist: Barney Newman

Figure 9.8 **Algoasaurus**, ***a relative of the giant* Apatosaurus *of the northern hemisphere, lived in the southern Cape in the Cretaceous Period.***

Whereas the Karoo was dominated by gymnosperms (naked seed plants, *see* Chapter 8), the Cretaceous marks the period when the first angiosperm (flowering plant) bloomed. This early diversification of the angiosperms gave rise to the

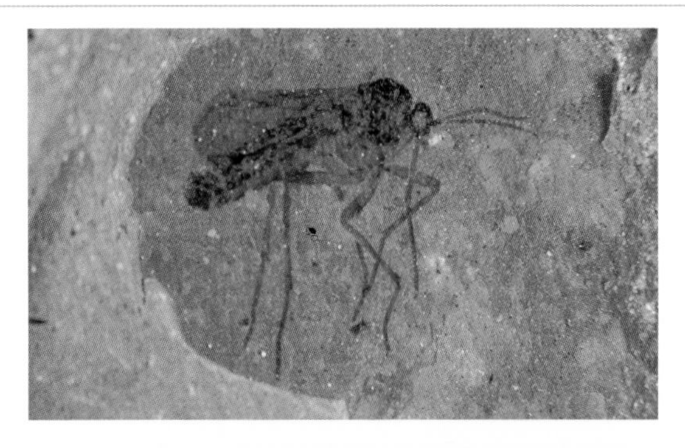

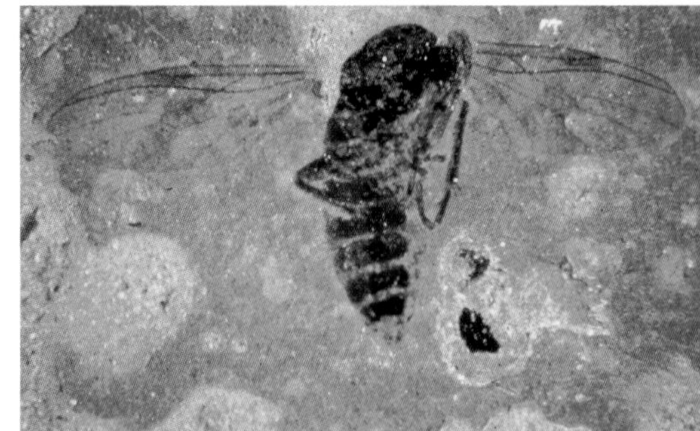

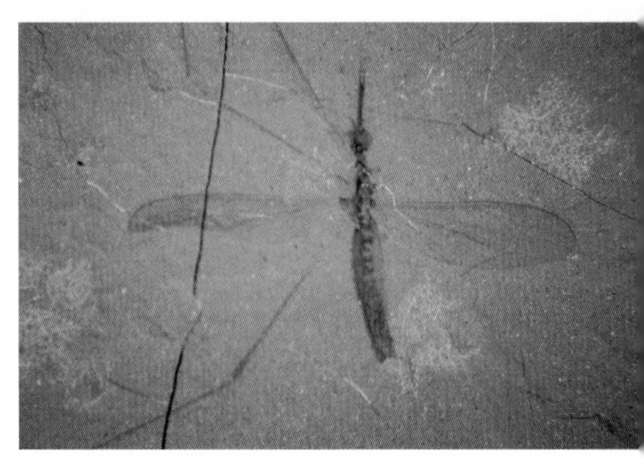

BPI Palaeontology

Figure 9.9 *During the Cretaceous Period, most of the interior of South Africa was elevated and undergoing erosion. Sites where fossils could accumulate are thus rare. An exception was in the volcanic craters that formed atop kimberlite volcanoes. Crater sediments from the 90-million-year-old Orapa kimberlite in Botswana have provided abundant fossil insects, such as those shown here. They also provide a glimpse of life in the continental interior.*

floral diversity seen today in regions like the Cape fynbos biome. Plant fossils are quite well represented in the Algoa Basin and include abundant fossilised wood, as well as leaf remains, and microfossils of the various pollens and spores. A number of well-preserved fern fronds are known from the Kirkwood Formation in particular. The abundant fossil tree trunks suggest that the coastline of southern Africa was well vegetated, probably by forests.

LIFE IN THE INTERIOR

During the Cretaceous, most of South Africa lay above sea level, perhaps lifted by the mantle plumes that initiated the break-up of Gondwana. The elevation may have exceeded 2 000 m above sea level and the landscape was undergoing erosion. Nevertheless, brief glimpses of terrestrial life in the interior of the continent are afforded by the crater deposits associated with the intrusion of **kimberlites**, volcanic rocks that arise from deep within the Earth's mantle. It was these kimberlite intrusions that delivered southern Africa's vast diamond wealth to the surface (*see* 'Kimberlites and diamonds' on page 256). There were two major periods of kimberlite intrusion during the Cretaceous, at 120 million years and 90 million years ago (there are older kimberlites in southern Africa, but these are far fewer in number than those of the Cretaceous).

Kimberlite magmas are rich in gas, and rise from a depth by forcing open cracks to form fissures or dykes. As the magma approaches surface, the gas expands and the magma rapidly bores a circular hole. Finally, it explodes through this to the surface, forming a crater. This crater sits atop a carrot-shaped pipe filled with solidified magma and fragments of the surrounding rock that have been ripped off the walls during magma ascent or have fallen from above into the hole. Lakes form in these craters and gradually become filled with sedimentary detritus collapsed and eroded from the sides, including the fossilised remains of animals and plants that live there. Crater deposits such as these provide the only direct information about the diversity of life and the climatic conditions on the South African Cretaceous land surface in the interior of the country.

In southern Africa, two pipes in particular preserve a remarkable record of intervals in the Cretaceous. These are the crater deposits at Orapa, a very large diamond deposit in central Botswana, and at Stompoor in the Kenhardt district of the Northern Cape. Very rarely in the fossil record does one find a locality where abundant and exceptionally well-preserved animals and plants occur together, but where they do occur, they provide great insights into the conditions of ancient life.

Such deposits are known as fossil *lagerstätten* (a German mining term for an extremely rich body of ore, now used to describe an exceptionally well-preserved fossil assemblage – a jewel of the fossil record). The Orapa site has yielded a unique assemblage of fossils. Before the start of mining operations, some 6 000 specimens were recovered, approximately half of which were insects (**figure 9.9**) and the rest the remains of angiosperms. Studies of these fossils have considerably aided the understanding of Middle Cretaceous ecosystems, and have shown that southern Africa was a centre for the diversification of both lineages.

The Orapa crater formed about 90 million years ago and the sediments were probably deposited fairly rapidly. The nature of the vegetation suggests that the climate of central Botswana was temperate (i.e. seasonal with cool winters) and relatively wet, and the surrounding area was forested. The rapid deposition of clays and silts, derived from the breakdown of the crater walls, quickly buried animal and plant remains and is one of the main factors contributing to the unique nature of the deposit. The presence of insects that require decaying organic matter (leaf mould) at some time in their life cycle suggests that the margins of the crater lake must have been covered by permanent, fairly lush vegetation. The types of flies, beetles and aphids suggest the presence of deciduous forests, requiring fairly high humidity and rainfall – an interpretation supported by the presence of various types of ferns.

The crater lake fill at Stompoor in the Northern Cape is unique in that it provides the only record to date of life in the interior of South Africa during the Late Cretaceous (approximately 70 million years ago). Fossils discovered from this amazing locality include fish, frogs (**figure 9.10**) and tadpoles, mussels, snails, insects and rare land-dwelling animal remains, which have tentatively been identified as dinosaurian. Some of the fish remains are so exceptionally preserved that they still show colour spots on the scales.

Based on extensive collecting and sedimentological analysis, Roger Smith of the South African Museum in Cape Town has reconstructed the environment of the Bushmanland landscape. Towards

Eddie van Dijk

Figure 9.10 ***A fossil frog from sedimentary deposits in the 70-million-year-old Stompoor crater in the Northern Cape. Crater deposits suggest that the climate in the interior at that time was warm and humid.***

KIMBERLITES AND DIAMONDS

The first diamond was discovered in South Africa in 1866 near Hopetown in gravel deposits related to the Orange River. Other discoveries followed along the Orange, the Harts and especially the Vaal Rivers, particularly near Barkly West. At that time, the source of the diamonds was unknown. In 1870, diamonds were discovered at Koffiefontein, remote from any river, suggesting a new type of deposit. Discoveries of similar material were made at Jagersfontein and in the Kimberley area that same year. In 1871, Colesberg Koppie, now the Big Hole at Kimberley, was also found to be of diamondiferous rock.

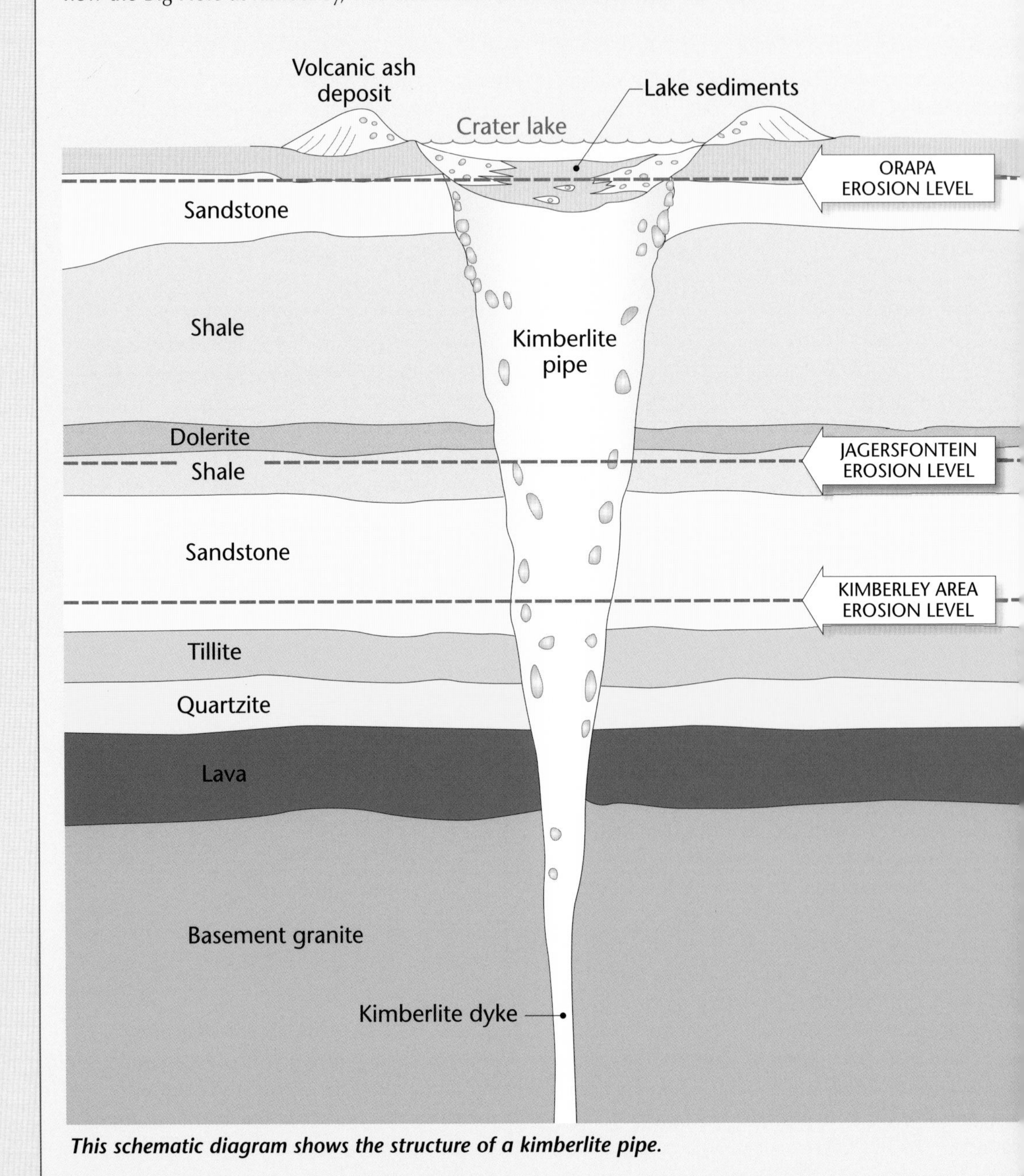

This schematic diagram shows the structure of a kimberlite pipe.

Production from the mines in these areas rapidly made southern Africa the leading diamond-producing region in the world. These new deposits turned out to be volcanic pipes, the carrot-shaped feeders to old volcanoes, and contained a very unusual type of igneous rock rich in magnesium and alkalis. It was given the name **kimberlite** by HC Lewis in 1887. These were the ultimate source of diamonds, and it was the erosion of these pipes that gave rise to the diamond deposits along the rivers in the interior as well as the deposits along the west coast. Not all the kimberlite pipes have been eroded; some still retain the remains of the actual volcanic crater, such as the Orapa kimberlite pipe in Botswana.

A schematic illustration of a typical kimberlite pipe (opposite) also illustrates the extent of erosion of various southern African kimberlite pipes. At depth, the pipes take on the form of a fissure or dyke.

Diamonds are composed totally of carbon, chemically the same as graphite, which is used to make pencil lead. When carbon is subjected to very high pressure it forms diamond. Diamonds can be manufactured synthetically, and the temperature and pressure needed to do this have been determined (below). Diamonds form only at extreme pressure, but they exist at atmospheric pressure. We use them in jewellery and for cutting hard materials. Although they are unstable at atmospheric pressure, they will not change spontaneously to graphite – they need energy to make the change. Heat a diamond, and it will change to graphite.

Also plotted on the diagram is the temperature gradient within the Earth below southern Africa where kimberlite pipes occur (rock temperature rises with depth in the Earth). Diamonds originate from depths below the point where the temperature gradient line crosses the stability line of diamond – below a depth of about 150 km.

Not all kimberlites contain diamonds; it has been observed that only those located on ancient continental nuclei or cratons do so. This is believed to be because the mantle lithosphere beneath these areas is cooler, promoting diamond stability. Studies of carbon isotope abundances in diamonds indicate that there are two sources of carbon that form diamonds: primordial carbon, part of the original mantle of the Earth; and carbon that has had its isotopic abundances changed by photosynthesis – that is, carbon in the form of organic matter that has been recycled from the Earth's surface down into the mantle by subducted oceanic crust.

The famous De Beers slogan 'A diamond is forever' is perhaps more accurate than the company ever realised. Diamonds themselves cannot be dated, but certain types of mineral inclusions occasionally found in diamonds can, and the inclusion age is assumed to be the age of the diamond itself. A wide variety of diamond ages have been measured, many of which exceed 3 000 million years, more than two-thirds the age of the Earth, so have survived almost forever.

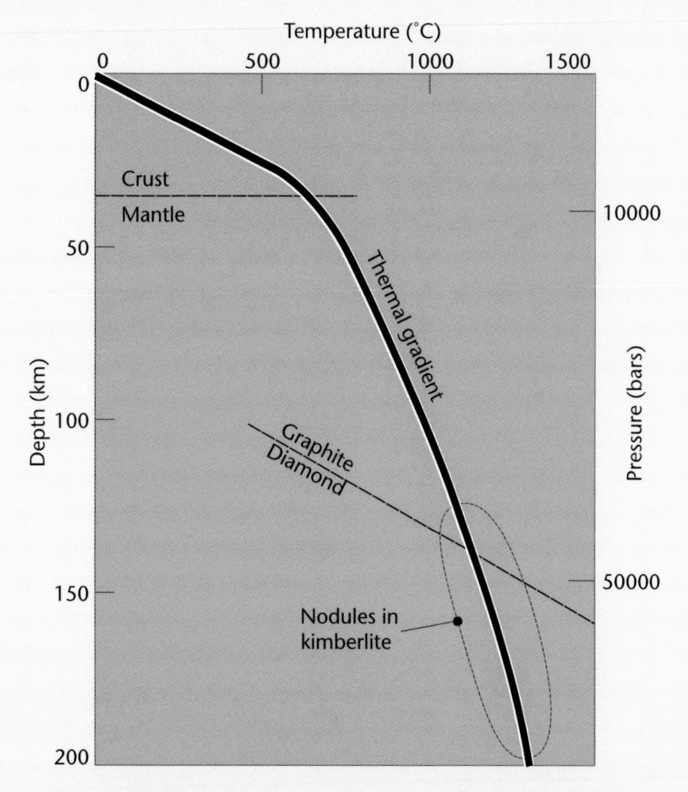

Phase diagram showing diamond stability and the geotherm.

Christian Koeberl

Figure 9.11 *The end of the Cretaceous is marked by a distinctive layer, often only a few centimetres thick, such as this example from Italy.*

the end of the Cretaceous Period, Bushmanland was flat and dotted with clusters of volcanic cones surrounding sunken craters. The climate was cooler and wetter than today, and the landscape vegetated by tall conifers (*Araucaria*). Within these light airy forests, large herbivorous dinosaurs browsed on the lower branches, while smaller dinosaurs, lizards and primitive shrew-like mammals foraged for food on the forest floor. Pterosaurs (flying reptiles) and perhaps early birds may have glided between the trees. Reeds growing around the crater lake itself provided a haven for small frogs, freshwater snails and mussels, while the waters of the lake were home to abundant small fish.

These conditions were very different from the hot, arid environment that prevailed before the break-up of Gondwana. Earth was much warmer during the Jurassic and Cretaceous than today, and the opening of the Atlantic and Indian Oceans created warm seas, which provided abundant moisture in the interior.

THE END OF THE CRETACEOUS

The mass extinction event that ended the Cretaceous, although only the third largest in Earth history, is certainly the most famous, for it was this event which extinguished the dinosaurs that had dominated the landscape for a hundred million years or more. The ammonites, so common in the Cretaceous seas, were also victims, along with 70–80% of all species living at the time (the actual number is still disputed and will probably never be known).

The mass extinction was ultimately caused by the collapse of the global ecosystem. But what brought this about? It is clear that the event was almost instantaneous, as can be seen recorded in areas where sediment was accumulating at the time. In these piles of sediment it is possible to place one's finger on the layer that marks the event (**figure 9.11**). This usually thin layer is invariably enriched in the element iridium, uncommon in the Earth's crust, but relatively abundant in meteorites. Moreover, the layer often contains small spheres of glass, impact melt blown high into the atmosphere or even low Earth orbit by the explosion, subsequently to rain down around the globe. More significantly, the layer contains tiny grains of the mineral quartz that all show damage characteristic of severe shock, such as can only be caused by a hypervelocity impact.

In the Gulf of Mexico, rocks formed at this time suggest deposition under chaotic and violent conditions. This indicates that huge waves or **tsunamis** occurred in this region and that the impact site must have been close by. Geophysical evidence has recently revealed the probable site, at Chicxulub on the Yucatan Peninsula in Mexico. The crater itself is not visible, as it is buried beneath younger sedimentary deposits, but evidence from drilling and other studies has confirmed the presence of an impact crater of the right age. The crater is estimated to be about 180 km in diameter.

It is now widely accepted that this impact blew millions of tonnes of dust high into the atmosphere, creating a globe-encircling cloud that massively reduced solar radiation reaching the ground for months or even years. The impact occurred on rocks that happened to be rich in sulphur, and probably released vast amounts of sulphur dioxide, which returned to Earth as acid rain. Photosynthesis shut down and the world's food web collapsed, taking with it the larger terrestrial and marine animals. These global problems may have been exacerbated by the eruption of huge quantities of basaltic lava at this time in India, forming the so-called **Deccan Traps**, which added further dust and gases to the already burdened atmosphere. The cumulative effect was catastrophic for the world at the time.

The end of the Cretaceous saw the complete destruction of the global ecosystem, and as the dust began to settle, it had to be rebuilt by the survivors. Old hierarchies had been smashed, and a whole range of new environmental niches began to open to the survivors. The insects, birds and the small nocturnal mammals that were to be our ancestors took the world by storm, along with the angiosperms, with their superior reproductive ability. The wide range of emerging habitats favoured rapid evolution of new species, and the flora and fauna of the modern world began to take shape, creating a world very different from that of the dinosaurs.

What South Africa was like that fateful day we do not really know, as terrestrial deposits of this age are not preserved anywhere in the coastal basins or on land. Many boreholes have intersected rocks deposited off-shore at this time in the course of petroleum exploration, and palaeontological evidence suggests that the uppermost part of the Late Cretaceous and lowermost part of the Tertiary are missing from most. Cores that may record the event have not been studied in detail, so we do not really know how the region was affected by the event.

GLOBAL CLIMATE AFTER THE CRETACEOUS

Living things often incorporate subtle imprints of their environment into their tissues and bones while they are alive. After death this information can be extracted by careful analysis of their remains. Isotopes, atoms of the same chemical element with slightly different masses, are particularly useful for this purpose. For example, by analysing a piece of elephant tusk for isotopes of the elements strontium, carbon and nitrogen, you can tell in what region the elephant lived and also something about its diet, whether it fed mainly on grass or shrubs and trees.

Marine organisms similarly encapsulate information about their environment into their shells. Minute surface-dwelling marine organisms known as foraminifera, which build tiny shells of calcium carbonate, are particularly valuable. The relative abundances of oxygen isotopes in the shells gives an indication of the temperature of the water in which the organism grew, while the isotopes of boron are sensitive to the pH of the water, which in turn is related to the amount of carbon dioxide in the atmosphere.

Foraminifera have been around for many millions of years and their shells are common in deep-sea sediments. In the remote regions of the oceans, such as the central Pacific, there is a constant rain of these shells onto the ocean floor. Being far from land, there is very little dilution by eroded material from the continents. By studying shells extracted from sediment cores from these remote regions, it is possible to obtain information about past climates, sea-surface temperature in particular, and atmospheric carbon dioxide abundances. Carbon dioxide abundance is significant because it is one of the more important greenhouse gases and has a strong influence on atmospheric temperature.

Surface temperature of the Pacific Ocean now and in the past has only marginal relevance to the climate of southern Africa. However, trends in Pacific Ocean temperature in the past are far more relevant, as these probably have a more global significance. This is the real value of studying past sea-surface temperatures. Carbon dioxide concentration in the atmosphere is everywhere essentially the same, so the values determined from shells in the Pacific Ocean apply globally.

Trends in the surface temperature of the central Pacific Ocean over the past 65 million years are shown in **figure 9.12**. At the end of the Cretaceous, the sea was warm and became warmer, reaching a maximum some 55 million years ago. Thereafter, temperature began to fall. Major growth of the Antarctic ice cap probably occurred between 35 and 25 million years ago, followed by slight warming. About 14 million years ago, rapid cooling took place, with probable further growth of the Antarctic ice sheet. Finally, dramatic cooling occurred, starting about two million years ago. The last million years has seen lower Pacific Ocean surface temperatures than at any time since the extinction of the dinosaurs.

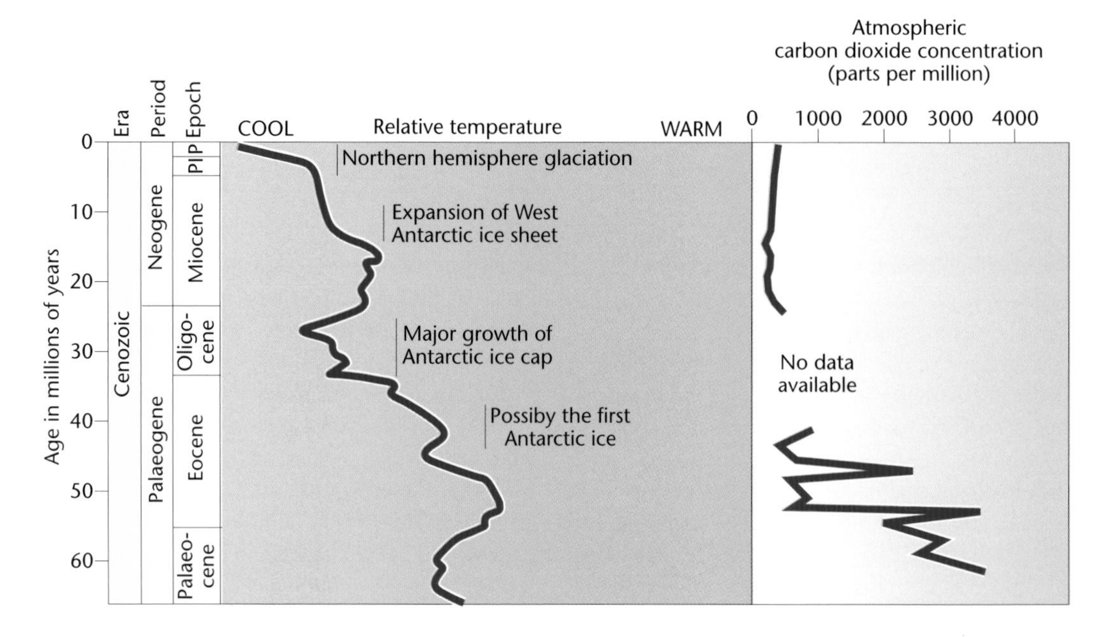

Figure 9.12 ***An indication of sea surface temperature can be obtained from measurements of oxygen isotope abundances in shells of micro-organisms deposited on the sea floor in the past, while isotopes of boron provide information on the carbon dioxide content of the atmosphere. These measurements indicate that over the last 60 million years the Pacific Ocean, and by inference, the entire Earth, has cooled significantly, while the carbon dioxide content of the atmosphere has fallen dramatically.***

Carbon dioxide abundance shows a similar trend to temperature. Some 60 million years ago, the carbon dioxide concentration in the atmosphere was about 3 500 parts per million (it is about 360 parts per million today) and the Earth was warm. As the Earth cooled, carbon dioxide concentration fell, suggesting some link. The patterns of carbon dioxide abundance and temperature are not exactly the same, indicating that whereas the greenhouse effect of carbon dioxide may have played an important part in global cooling, there were probably other factors as well. These may be related to the growth of the Himalayan and Andean mountain chains, and changes in ocean circulation brought on by the movement of the continents.

Three such changes are believed to have been particularly important. Firstly, the severing of the land link between Antarctica and South America opened the Drake Passage, which occurred about 35 million years ago. This allowed circum-Antarctic circulation to take place and it also thermally isolated Antarctica. Secondly, the northward movement of Australia increased the size of the Southern Ocean and reduced the flow of warm, Pacific water into the Indian Ocean. Antarctica began to cool dramatically and an ice sheet formed and expanded. By 23 million years ago the ice had reached the coastline. The Southern Ocean began to cool as well, and the cold water spread northward, cooling the Pacific, Indian and Atlantic Oceans.

Finally, an important change in ocean circulation occurred about 3.5 million years ago, with the formation of the Panama Isthmus, closing the seaway between the Atlantic and Pacific Oceans. This seaway had allowed warm water from the Atlantic to enter the Pacific, raising the temperature especially of the northern Pacific. Once this circulation stopped, the north Pacific cooled, contributing to the rapid fall in temperature that began about three million years ago. The north Atlantic may have warmed, increasing moisture in the atmosphere of the region, and thus the amount of snow in northern high latitudes.

But carbon dioxide was also an important player. What caused it to fall? Several possible explanations have been put forward: global volcanic activity may have declined as India collided with Asia; there may have been more calcium released into the oceans as the Himalaya and Andes rose, and weathering and erosion increased (calcium added to the oceans removes carbon dioxide from the atmosphere – *see* Chapter 6); or there may have been a burst of coal formation in response to the high temperatures just after the Cretaceous. A novel and somewhat radical idea was recently proposed by soil scientist Greg Retallack, who suggests that carbon dioxide was stripped from the atmosphere by grasslands.

THE RISE OF GRASSLAND

Grass is so pervasive that it is difficult to imagine a world without it. But grass is actually a relative newcomer in the plant world – it only emerged some time prior to 30 million years ago. Once grass appeared, so did grazing animals, and these two guilds of organisms have been co-evolving ever since.

So what was there before grass? Let us go back 40 million years. Imagine a land with a gradual increase in rainfall from very low on one side to high on the other – perhaps something like southern Africa today. On the arid side, there would have been sand or rock desert. At slightly higher rainfall, small bushes would have appeared, much like the Karoo or Namaqualand today. These shrubs would have become more abundant and taller as rainfall increased, with some trees. This was dry woodland. Finally, at much higher rainfall, forests would have occurred. No grass – its niche was occupied by dry woodland (**figure 9.13**).

The earliest grasses appear to have been adapted to more arid conditions and they competed with desert scrub. Over time other grass species appeared, adapted to somewhat wetter conditions. They began to compete with and displace the dry woodlands, partly replacing them with short grassland, perhaps something like the central Kalahari today. Finally, by about four million years ago, grassland had almost completely replaced the dry woodland vegetation, producing open savanna (**figure 9.13**).

How could humble grass achieve this? Perhaps the grasses' main weapon was fire. During the dry season, grasses offer copious combustible material in the form of dead leaves and stalks, often covering extensive areas. Grasslands are thus prone to hot wildfires. Most grasses have their growing points below ground, well protected from fire, unlike shrubs and trees, which usually have their growing points exposed on their branches. Grasses therefore easily survive fires, while shrubs and small trees are frequently killed. A succession of fires will eliminate shrub and tree seedlings, and given time, dry woodland will succumb to an invading grass. In the absence of grassland fires due to overgrazing, for example, bush encroachment ensues. Sub-humid woodland is less vulnerable. The taller trees shade out the grass and because conditions are more humid, fires are less common. Once grass appeared on the scene, it steadily eliminated and eventually replaced dry woodlands world-wide (**figure 9.13**).

Retallack believes the expansion of grassland played a major role in cooling the Earth over the past 30 million years or so. Grass evidently contributed to this in several ways. Grasslands store far less carbon than forests in actual plant material, but grassland soils are usually far richer in carbon than forest soils, so grasses effectively remove carbon from the atmosphere and store it in soils. This may in part be due to frequent fires, which produce copious charcoal that is not easily decomposed by bacteria and becomes incorporated in the soil.

Grasslands are much lighter in colour than woodland and they reflect a greater proportion of solar radiation into space, contributing to cooling. The air over grasslands is generally much drier than over woodland because trees tap deep-water sources and pump the water into the atmosphere by transpiration. Water vapour is a powerful greenhouse gas, so more grassland means drier, and thus cooler, air. The combined effect of the rise of grasslands, Retallack believes, was the cooling of the Earth.

The rise of grasslands also had a profound effect on animal life, particularly on mammals. As grasslands expanded, grazers steadily replaced browsers and became the dominant large animals by the Pleistocene. Grasslands are extremely productive. Ecologist Norman Owen-Smith notes:

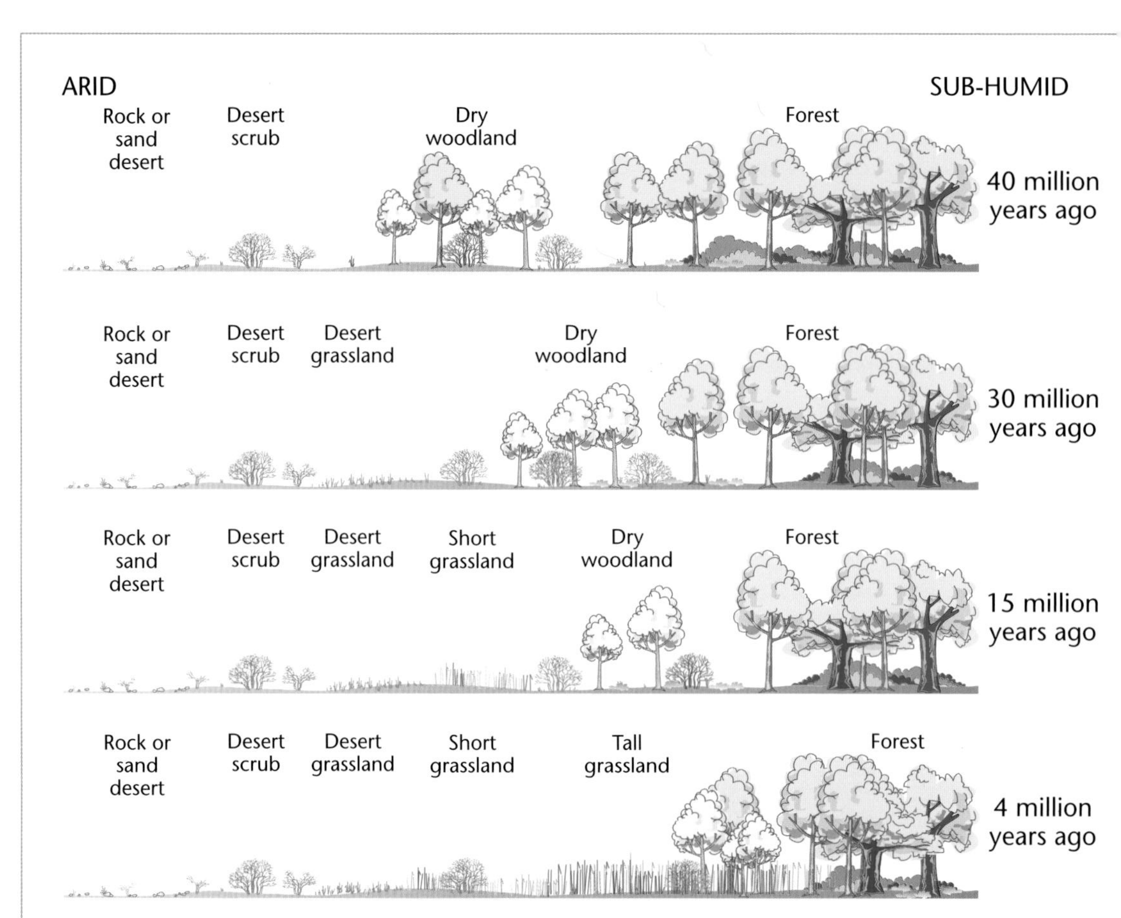

Figure 9.13 ***Grasslands have expanded enormously at the expense of dry woodland over the last 30 million years. Initially, grass competed with desert scrub, but gradually its environmental range expanded and it became the dominant vegetation type in intermediate rainfall regions. The expansion of grassland may have contributed to the cooling of the Earth over the last 40 million years.***

the population abundance attained by grazing ungulates is an order of magnitude greater than that reached by browsing ungulates of similar size, because of the greater year-round accessibility of grass relative to browse.

Diversity among ungulates emerged as grasslands spread, some adapting to eating short grass, others longer grass, and yet others to mixed grazing and browsing. Diversity also developed in respect of forage quality, some species favouring more nutritious grass, others less nutritious types. Carnivores were forced to adapt to grassland too. The typical African savanna – with its vast herds of zebra, wildebeest, buffalo and impala, preyed on by lions, cheetahs and hyaenas – is a product of grassland evolution.

THE AFRICAN LAND SURFACE

At the time of the break-up of Gondwana, southern Africa is believed to have been elevated, the highest terrain lying towards the eastern side of the subcontinent because of the mantle plume activity in that region (**figure 9.14a**). Rifting and the formation of seaways around the subcontinent had resulted in a marginal escarpment, a step in the landscape from the elevated interior down onto the newly formed coastal plains (**figure 9.14b**). This escarpment seems even to have been accentuated by the bulging of the crust that typically precedes rifting, raising it slightly relative to the interior.

The Cretaceous was a relatively warm and humid period. Surrounded by warm seas, southern Africa basked in a moist climate that supported

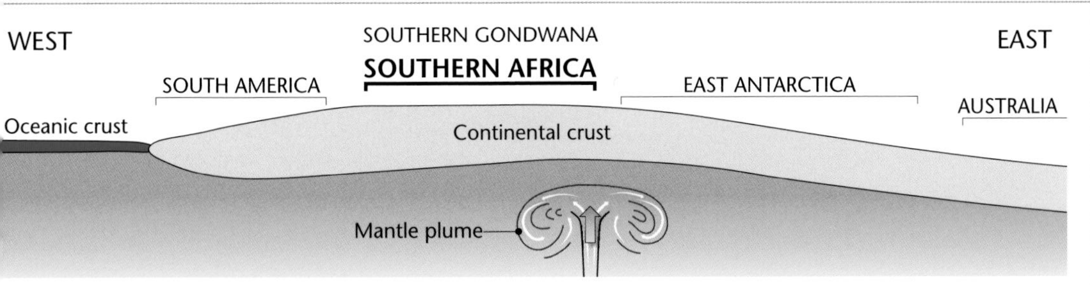

Figure 9.14A ***Schematic diagram showing the bulging of the crust of southern Africa as a result of the rise of a mantle plume about 180 million years ago. The vertical scale has been greatly exaggerated***

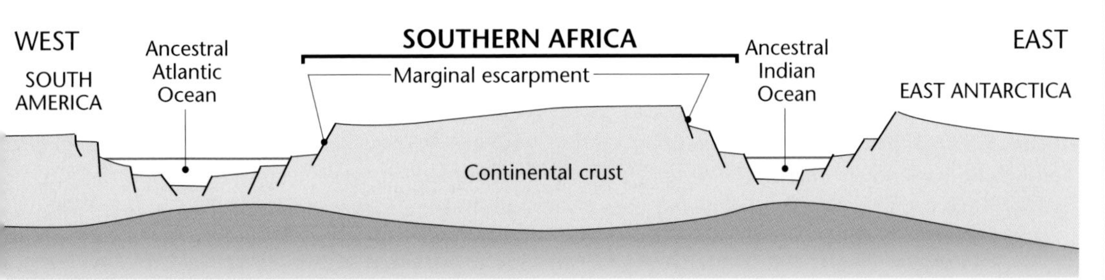

Figure 9.14B ***By 120 million years ago, rifting along the east and west coasts of southern Africa had thinned the continental crust, opening the proto-Indian and Atlantic Oceans. The interior of southern Africa became separated from the coastal plain by an escarpment. The interior remained higher in the east.***

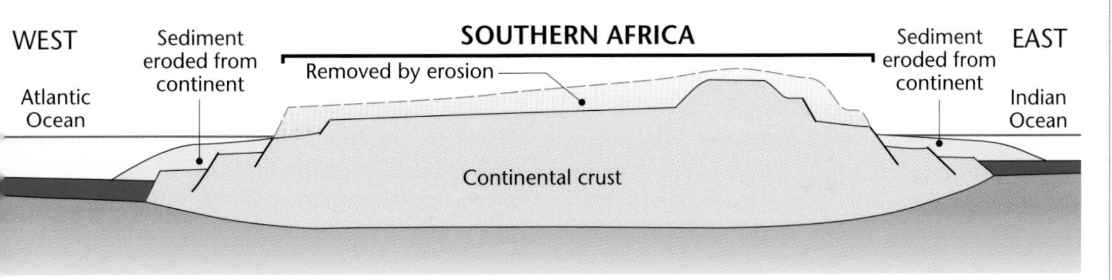

Figure 9.14C ***By 65 million years ago, oceanic crust floored the growing Atlantic and Indian Oceans. The interior of southern Africa had experienced considerable erosion, and the eroded material had been deposited on the continental shelf. The marginal escarpment had been cut back during this process. The Lesotho Highlands remained as a small remnant of elevated ground on the east of the subcontinent.***

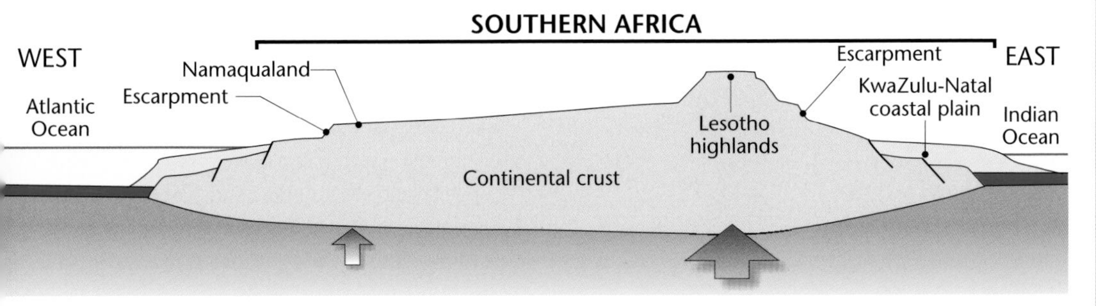

Figure 9.14D ***At 20 million and five million years ago, southern Africa experienced episodes of uplift and tilting towards the west, which created renewed erosion in the interior, but the major features of the ancient landscape remained.***

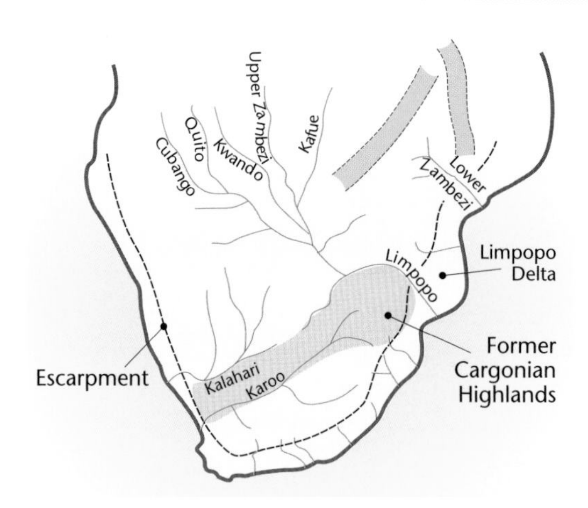

Figure 9.15A ***Shortly after the break-up of Gondwana, three rivers dominated drainage in southern Africa, the most extensive being the Limpopo system. These were responsible for the erosion of Karoo Supergroup strata from the former Cargonian Highlands, exposing underlying older rocks. The interior of the country was relatively moist, and thick, strongly leached soils had developed. The interior formed a plateau, sloping gently to the west, which was separated from the coastal plain by an escarpment. This was drained by short rivers, which can still be seen today along the southeast coast. The land surface that existed at that time is known as the African Surface.***

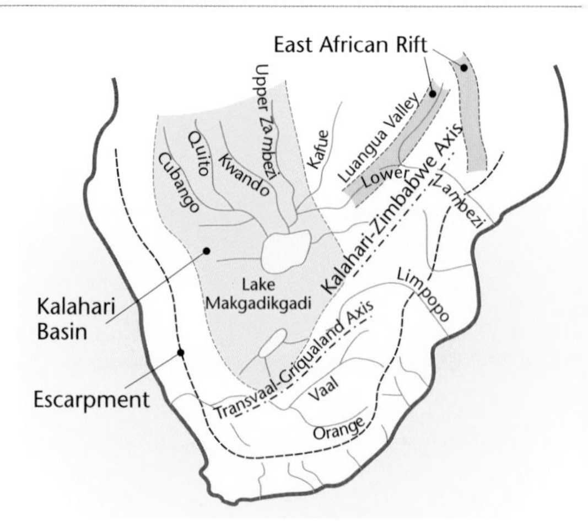

Figure 9.15B ***About 60 million years ago, gentle arches (called 'axes') began to form in the interior of the African continent. Two of these had particular importance for southern Africa, as they resulted in a depression in the interior known as the Kalahari Basin. The Kalahari-Zimbabwe Axis cut off the headwaters of the Limpopo, and large lakes began to form in the interior, notably Lake Makgadikgadi. Uplift of the Transvaal-Griqualand Axis resulted in the capture of the Karoo River by the Kalahari River to form the modern Orange River system. By about 20 million years ago, the East African Rift system was beginning to propagate into southern Africa, influencing the Zambezi River system in particular.***

extensive tropical or sub-tropical forests. Deep, strongly leached soils developed in the interior. A drainage network had established itself in the interior, involving three major river systems: the **Limpopo River**, which drained the vast northern regions of southern Africa; the **Karoo River**, which drained the eastern highlands and flowed to the west; and the **Kalahari River**, which drained the western interior (**figure 9.15A**). The escarpment itself was drained by numerous short rivers still evident today along the KwaZulu-Natal and Cape coasts (**figure 9.15A**).

The peculiar asymmetry of South Africa's drainage, with the major Vaal and Orange Rivers rising close to the east coast and flowing westwards across the entire country, had ancient beginnings – a product of the plume activity that initiated break-up in the east.

The Limpopo River had developed along a rift formed at the time of opening of the Mozambique Channel (*see* **figure 9.1A**). Perhaps subsidence of this failed rift had created a depression which the Limpopo came to occupy. This river originally drained much of the interior of central southern Africa (**figure 9.15A**), and sediment eroded from this huge catchment was deposited at the river mouth, building a large delta on the Mozambique coast (*see* **figure 9.1B, C**). This delta today forms the arc-shaped coastline between Maputo and Beira.

The Limpopo, Kalahari and Karoo river systems were responsible for the erosion and removal of Karoo Supergroup strata in a broad arc extending from the Namaqualand region in the west towards the northeast through the North West, Gauteng, Limpopo and Mpumalanga provinces. This arc originally formed the Cargonian Highlands at the time

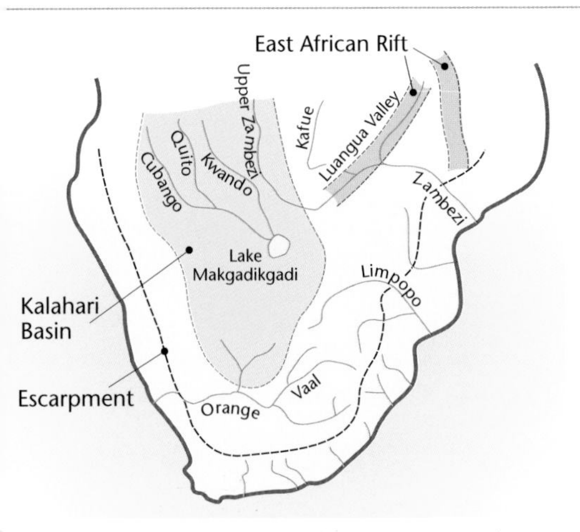

Figure 9.15c ***About 14 million years ago the upwelling of cold water began on the west coast, causing extremely arid conditions to develop in the west of the country. Lakes in the Kalahari Basin began to dry up, a process exacerbated by the loss of inflow caused by the progressive capture by the Zambezi River of major tributaries of the lakes. Lake and river deposits in the Kalahari Basin gave way to desert sand.***

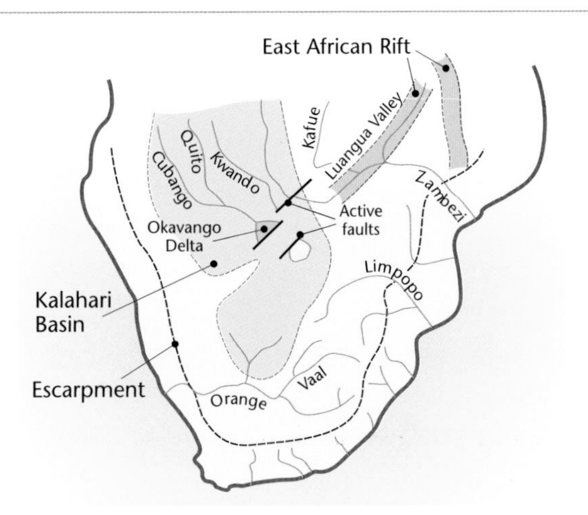

Figure 9.15d ***The East African Rift system continues to propagate into southern Africa, and the associated faulting has led to the diversion of the Kwando River into the Zambezi River and caused the formation of the Okavango Delta in northern Botswana.***

of filling of the Karoo Basin (*see* **figure 7.14**). Along this arc, ancient rocks of the Kaapvaal Craton and its flanking metamorphic belts became exposed. It seems that only the Karoo strata were stripped off and very little of the harder, ancient underlying basement rocks were eroded. Cretaceous erosion has therefore essentially resurrected the old land surface which existed at the time the Karoo strata were being deposited.

Glacial features that formed during the Dwyka period are therefore commonly found in this region (*see* Chapter 7), such as along the Vaal River near Kimberley. In fact, the majority of the mountain ranges in the northern part of the country, such as the Waterberg, Soutpansberg and Magaliesberg, as well as the Witwatersrand ridge, were actually shaped by glaciation during Dwyka times (*see* Chapter 7), and then slightly modified by post-glacial erosion.

The topography we see today along the arc is therefore very ancient. How much of the Karoo strata were eroded and when exactly it occurred is difficult to say. In Namaqualand and central Botswana there has been very little erosion since the Late Cretaceous, 70 million years ago, and this applies to the highlands of Lesotho as well. In contrast, it has been estimated that about 1 400 m of material have been removed in the vicinity of Kimberley since the kimberlite pipes intruded about 90 million years ago. This erosion, variable as it appears to have been, was important; among the debris eroded from the interior were diamonds, which were carried to the Atlantic Ocean by the ancient Cretaceous rivers, where they became concentrated by wave action in gravels along the beaches. Some diamonds never made it to the coast, but were trapped in river gravel deposits *en route*. Today they are still mined along the banks of the Vaal and Orange Rivers and their tributaries.

Some believe that by the end of the Cretaceous, the escarpment around southern Africa had been eroded back approximately to its present position (**figure 9.14c**) and a broad, gently sloping coastal plain had been cut at its base. In the interior, the land surface was probably broadly similar to what we see today, with gently rolling hills and the occasional ranges such as the Magaliesberg, Waterberg, Soutpansberg and parts of the Drakensberg rising above the general terrain. This surface has been termed the **African Erosion Surface**. Soils on this surface were deep and tropical in character, showing

Figure 9.16 *Along the eastern escarpment in Namaqualand 60-million-year-old soils formed during the Cretaceous Period have been exposed by modern erosion. These soil profiles have a hard silica-rich cap that tends to protect the softer underlying material from erosion and can be seen capping these hilltops.*

extreme leaching. Around the end of the Cretaceous, the upper soil layer seems to have become cemented by silica, which has tentatively been attributed to changes in the atmosphere at the time, and possibly also to changes in vegetation cover. Today, remnants of these very ancient soils can still be seen in various parts of the country (**figure 9.16**), which is quite remarkable considering that they are more than 65 million years old.

THE KALAHARI BASIN

Starting at around the end of the Cretaceous, the southern African crust began to flex, forming broad swells and depressions. This phenomenon is widespread on – and apparently unique to – the African continent. The cause is unknown, but it could be due to convection in the mantle below Africa or may be a response to the inward push from the mid-ocean ridges that surround most of Africa. (The African tectonic plate is virtually entirely surrounded by ocean ridges – *see* Chapter 2.)

The first of these swells to form in southern Africa is known as the **Kalahari-Zimbabwe Axis**, which developed across the main headwaters of the Limpopo River, followed by the **Transvaal-Griqualand Axis** further south (**figure 9.15B**). These had a profound effect on the rivers in southern Africa. The Limpopo lost most of its water, and inland lakes began to form behind the ridge in the interior. The largest of these was the ancestor of the present Makgadikgadi pans, which was a body of water not unlike Lake Victoria today (**figure 9.15B**). Smaller lakes formed further south.

The formation of these lakes marks the start of the **Kalahari Basin**, the great depression from which no rivers emerge. Uplift along the Transvaal-Griqualand Axis, aided by tilting of the continent (discussed on page 267), caused tributaries of the Kalahari River to cut southeastward, where they intersected the Karoo River and diverted its course, forming a single river from what were formerly two separate rivers – now called the **Orange River** (**figure 9.15B**).

The Orange River from Prieska to the coast is a remnant of the Kalahari River system, whereas the Orange upstream of Prieska, as well as the Vaal River, are remnants of the Karoo River system. The marked kink in the Orange River near Prieska in the Northern Cape, marks the position where the river capture occurred. It is possible that lakes in the interior spilled over into the Orange River from time to time; geological evidence suggests that a major river formerly entered the Orange from the north near the village of Douglas in the Northern Cape.

ARIDIFICATION OF SOUTHERN AFRICA

Meanwhile events were unfolding elsewhere on the globe that would profoundly influence southern Africa. The opening of the Drake Passage and the resulting circum-polar circulation resulted in expansion of the Antarctic ice sheet and cooling of the Southern Ocean. In the atmosphere, a strong, semi-permanent high pressure system became established over the South Atlantic Ocean, producing off-shore drift of water off the west coast of southern Africa. This gave rise to the Benguela upwelling system about 14 million years ago.

Cooling of the ocean water along the west coast radically changed the climate of southern Africa. Whereas previously, moist air was supplied to the subcontinent from both the Indian and Atlantic Oceans, producing relatively moist conditions on both sides of the continent, the upwelling of cold water on the west coast cut off the moisture supply from the Atlantic. The west coast became very arid, the **Namib Desert** formed, and the rainfall gradient from the east to the west coast was established. Only the Orange River has remained a perennial river in this region since that time.

RISE OF THE CONTINENT

The arid conditions in the west were further exacerbated by events inland. The interior of southern Africa began to rise. How, why and when this occurred are not fully understood. One view is that uplift started around 20 million years ago. The eastern portion of the continent rose more than the west, further exaggerating the already tilted landscape of southern Africa. Two main periods of uplift are proposed: the first, about 20 million years ago, caused a rise of about 250 m in the east and 150 m in the west. The second, about five million years ago, involved a 900 m rise in the east, but only a modest 100 m in the west (**figure 9.17**).

Uplift substantially increased the height of the eastern escarpment, which reduced rainfall in the interior because moist air from the Indian Ocean lost more of its water as it rose against the escarpment. This further increased the east to west rainfall gradient, which today ranges from over 1 000 mm per year on the east coast to less than 100 mm per year on the west coast. As a consequence, there was an expansion of grassland across the interior of southern Africa, as well as an increase in the seasonality of the rainfall. This increase in grassland gave impetus to the diversification of grazers, and most of the African antelope genera evolved during this time, not only in southern Africa, but in East Africa as well, where the start of the East African Rift was also causing uplift.

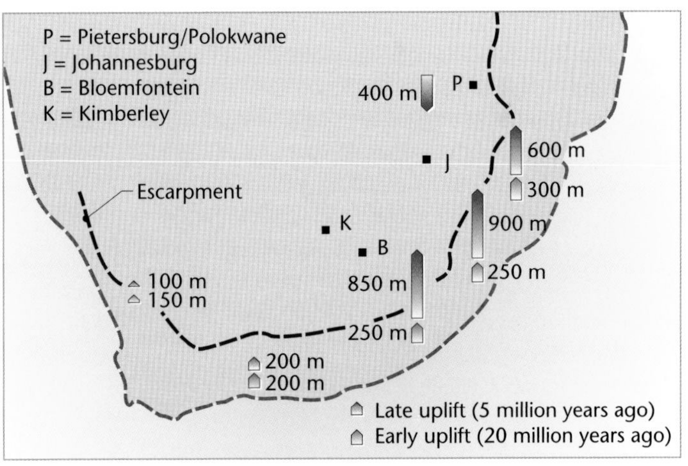

Figure 9.17 ***Southern Africa experienced two major periods of uplift, the first about 20 million years ago and the second about five million years ago. These accentuated the escarpment and the westward tilt of the interior land surface, giving rise to renewed erosion in the interior (forming the Post-African I and II Surfaces) and the erosion of deep valleys and gorges on the coastal plain. (The locations of major cities are shown for reference.)***

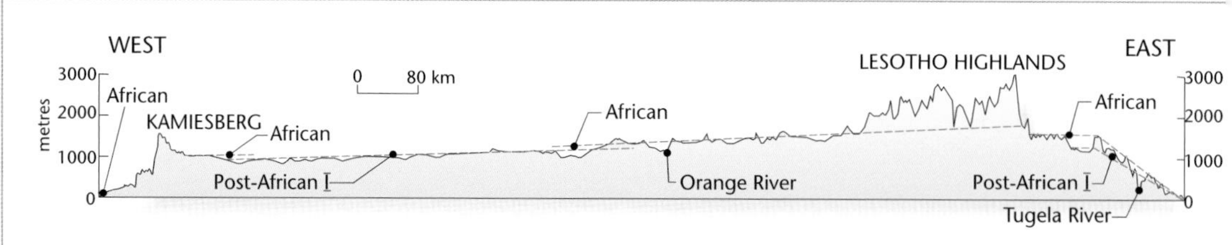

Figure 9.18 *A section across South Africa today showing the marginal escarpment, the westerly sloping interior plateau and the African and Post-African Erosion Surfaces. Many of these features originated with the break-up of Gondwana more than 100 million years ago.*

The uplift events in southern Africa, which involved tilting of the continent to the west, increased the slopes of the major rivers, increasing their energy and thereby creating two pulses of erosion in the interior, producing new land surfaces known as the **Post-African I and II Surfaces**. The first of these (related to the 20-million-year period of uplift) is more extensive. The effect of these in the interior was relatively modest, however. It involved essentially the erosion of most of the old soil cover that had formed on the African Surface. Much of the present landscape of the interior, such as the gently rolling hills and the typical Karoo *koppies*, is due to this erosion period.

The effects of uplift were more dramatic around the coast, especially in the east, where the slopes of the short rivers were greatly increased, particularly during the second, larger period of uplift. The increased slope induced rapid down-cutting by the rivers, producing striking topography such as the Valley of a Thousand Hills in KwaZulu-Natal, and the impressive gorges along the southern Cape coast, such as that of the Storms River.

Rivers carried the eroded material to the coastal plains where it was deposited. Included in the sediment were grains of ilmenite and zircon, respectively titanium- and zirconium-rich minerals. Both are signicantly denser than common sedimentary minerals, and as a result of winnowing by waves and wind they became locally concentrated on the coastal plains. Further concentration occurred during the sea-level fluctuations of the Pleistocene (discussed on page 272), and resulted in the important titanium and zirconium deposits of the Western Cape and KwaZulu-Natal.

Whereas most of South Africa experienced uplift, there was an exception. The Bushveld region actually experienced subsidence of some 400 m (**figure 9.17**). This resulted in preservation o an extensive flat region of the African Erosion Surface, now known as the **Springbok Flats**, which extends from Modimole/Nylstroom to Mokopane/ Potgietersrus. This area is very unusual as it has virtually no drainage network and is prone to extensive flooding during wet periods.

The present scenery of southern Africa was long in the making, but the imprints of events that took place tens of millions of years ago remain. Mos striking is the marginal escarpment, a relic from the break-up of Gondwana that has been accentuated by later uplift (**figure 9.18**), especially the eastern escarpment. The many mountain passes between the interior and the coast – such as the Abe Erasmus, Long Tom and Van Reenen passes in the east and Anenous, Vanrhyns and Middelburg passes in the west – are striking reminders of its presence.

The escarpment tends to become emphasised by the local geology in some areas, particularly where thick layers of very hard rock overlie less resistant types. In these areas, the escarpment becomes very steep, forming a spectacular rampart, such as can be seen in the KwaZulu-Natal Drakensberg in the Royal Natal National Park (harder basalt on soft Karoo sedimentary rocks; **figure 7.33**), or in Mpumalanga in the Blyde River area (hard, resistant Wolkberg Group quartzites on more easily weathered and eroded granite; **figure 4.11**). Where the rocks are of more uniform resistance to erosion, the slope of the escarpment is less extreme.

The interior of the country is essentially a plateau that slopes gently to the west (**figure 9.18**), a land surface which in the main dates back to the Cretaceous – more than 60 million years old – with only minor later modification. Rising above this are local mountain ranges, such as the Maluti of Lesotho, also formed in the Cretaceous Period (**figure 9.14c**), as well as much older ranges including the Cape Fold Mountains and the ranges of the old Cargonian Highlands, such as the Waterberg, Soutpansberg and Magaliesberg, which were carved by ice some 300 million years ago. In the western interior is the Kalahari Basin, also formed during the Cretaceous, which has been slowly accumulating sediment since that time. In all, the South African landscape is ancient – among the oldest in the world.

THE AFRICAN SUPER SWELL

The massive uplift of southern Africa in the last 20 million years is unusual. In fact, it has resulted in a topographic anomaly of global significance (**figure 9.19**). Whereas areas of similar geology and geological history elsewhere on Earth, such as Western Australia or central Canada, today lie at elevations of around 300 m to 400 m above sea level, most of southern Africa lies at elevations above 1 000 m. This topographic anomaly, which has been called the **African Super Swell**, has recently become the focus of intense research.

While the Theory of Plate Tectonics can explain most of the topographic features of the Earth, such as the Himalayan Mountains and the Mariana Trench, the African Super Swell poses a problem, because it lies far from tectonic plate boundaries, where most topographic features occur. Hence the interest from scientists. To investigate this feature, earth scientists have been using a technique called seismic tomography, analogous to a medical CAT (or CT) scan, to image the interior of the Earth. Rather than X-rays, as used in CAT scans, earth scientists use seismic waves from distant earthquakes, which are recorded by sensitive seismometers. By measuring the small differences in the travel times of the energy pulses from the earthquakes to the various seismometers, it is possible to build up images of the Earth's interior. These images highlight areas that are more dense and probably colder (pulses travel faster), distinguishing them from areas that are less dense and probably hotter (pulses travel more slowly).

It turns out that deep in the mantle beneath southern Africa there is a huge blob of hot material, nearly 2 000 km in diameter, which appears to be rising towards the surface. Like a rising bubble in thick syrup, the Earth's surface is being pushed up, creating the Super Swell. Why this is happening is not known. The blob has a tail (like a giant tadpole), which rises closer to surface under the East African Rift Valley, and is undoubtedly responsible for the rifting in that region. Perhaps the blob will express itself by major rifting of southern Africa some time in the future, as it nears the top of the mantle.

THE ZAMBEZI RIVER AND THE MAKGADIKGADI PANS

The East African Rift system began to exert an influence on southern Africa possibly as far back as 20 million years ago, as its various branches began to

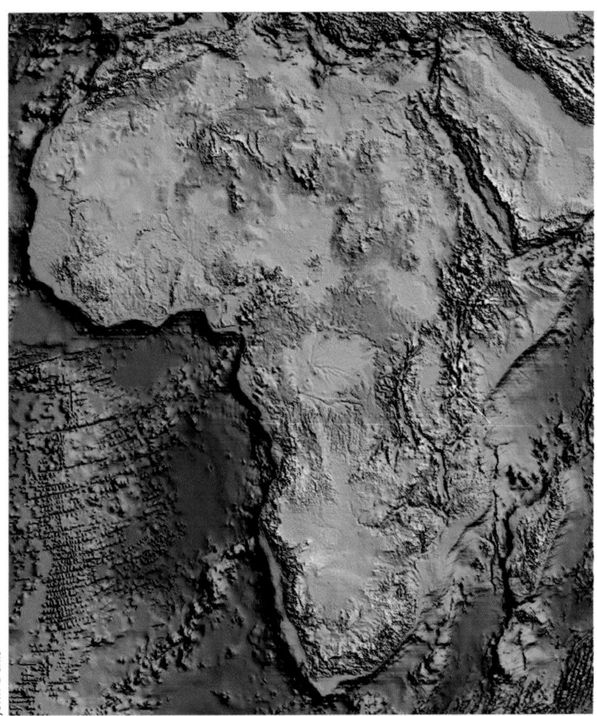

Figure 9.19 ***Southern Africa is unusually elevated. Most regions of the world with similar geology lie at elevations of around 400 m above sea level or less. In contrast, most of southern Africa lies above 1 000 m, forming a topographic anomaly known as the African Super Swell. This is thought to be due to accumulation of heat near the core-mantle boundary beneath southern Africa.***

extend southwards. The Zambezi River in particular has been strongly affected. Originally a small coastal river draining the eastern escarpment, the lower Zambezi cut back into the interior, where its progress was initially assisted by uplift along the Kalahari-Zimbabwe Axis, and later by faults related to the East African Rift along the Luangwa valley (**figure 9.15B**).

As this branch of the rift system extended south-eastwards, it has caused the lower Zambezi to capture rivers arising to the north: first the Kafue, then the upper Zambezi and most recently the Kwando were diverted into the lower Zambezi River (**figure 9.15C, D**). Next in line is the Okavango River, which has yet to be diverted to the Zambezi, but in the interim, the rift-related faults have blocked the Okavango River, creating the spectacular Okavango Delta of northern Botswana (**figure 9.20**). Once the Okavango River has been captured, an entire drainage network that once formed part of the Limpopo River system (**figure 9.15A**) will have shifted to the Zambezi system.

These captures of the Kafue, upper Zambezi and Kwando Rivers by the lower Zambezi River deprived the ancestral Lake Makgadikgadi of its water over time, and it shrank in size. At the same time, the central Kalahari became more arid, because of the cooling of the Benguela current along the west coast and the rise of the eastern escarpment. Lake Makgadikgadi vanished, and the salt pans of today are reminders of what was once a mighty lake whose western shoreline is still visible on satellite images taken from space. More arid conditions began to prevail in the Kalahari, and the lake and river sediments of the early Kalahari Basin gave way to desert sand. It is possible that the link between the Kwando/upper Zambezi river system and the Makgadikgadi pans was periodically re-established during the last 50 000 years due to fault movement, leading to extensive, but intermittent flooding of the pans.

NASA

Figure 9.20 *Southwestward propagation of the East African Rift System has created a large depression that accommodates the Okavango Delta.*

THE PLEISTOCENE EPOCH

The last two million years, the Pleistocene Epoch, have seen further changes in southern Africa, arising from changes in the climate and in sea level around the coast. The Pleistocene was characterised by a succession of ice ages, when the polar ice caps, particularly the Arctic ice cap, expanded greatly. There were several periods of major ice advance,

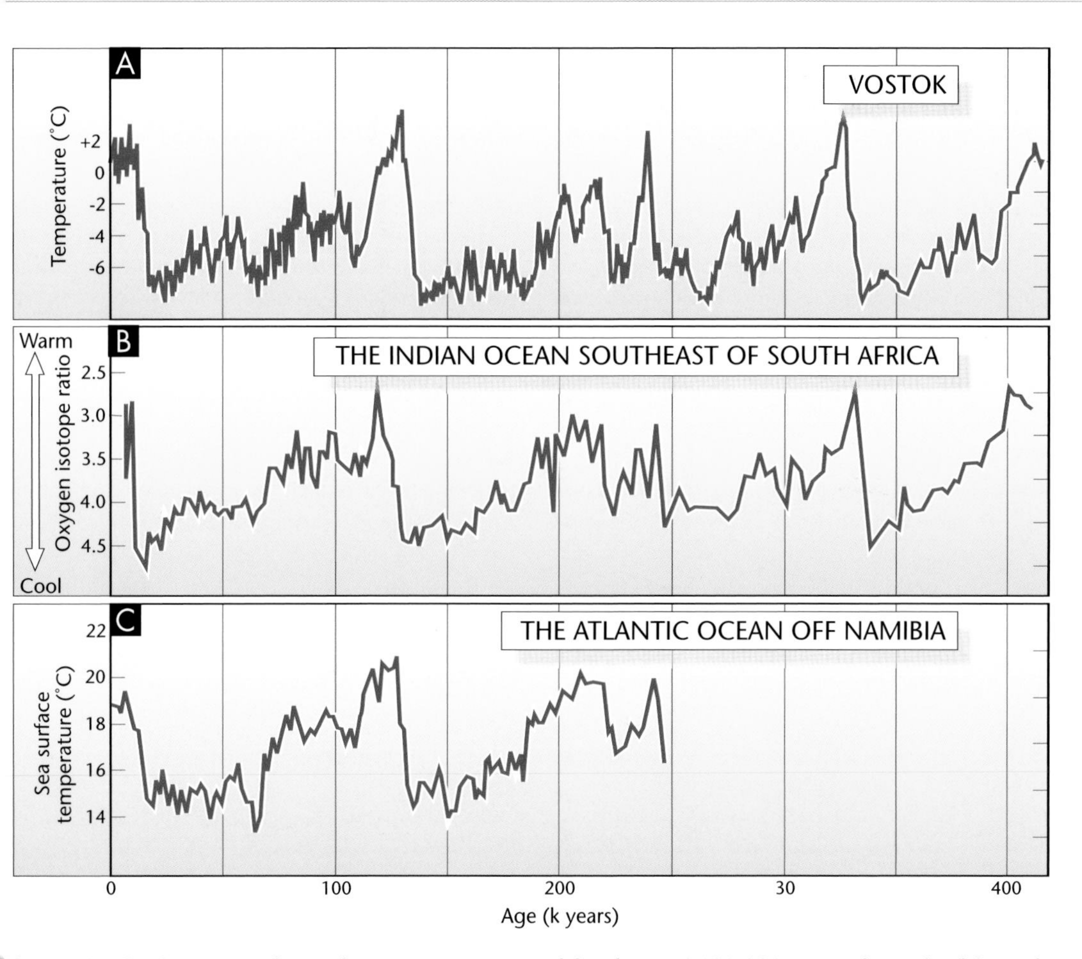

Figure 9.21 *Diagram* **A** *shows the temperature record for the past 420 000 years, determined from the Vostok ice core from Antarctica. Ice cores from Antarctica and Greenland provide a means of measuring past temperatures, since the relative abundances of certain isotopes change with changing temperature of formation of the ice. Ice is compacted snow, so the isotopes tell the temperature at the time the snow fell. For most of the last 420 000 years the Antarctic has been relatively cold, with a few brief warmer periods. The present warm period is the longest in the entire record. The temperature records from sediments on the ocean floor to the east and west (***B,C***) of southern Africa reveal a similar temperature history to Antarctica, affirming the global significance of the Vostok record.*

during which large portions of the continents of the northern hemisphere were buried beneath ice sheets kilometres thick. The landscape of the northern hemisphere continents has to a large degree been shaped by these ice ages and is very young compared to that of southern Africa.

Records of global temperature changes during the later Pleistocene have been obtained from a variety of sources, particularly from cores extracted from the ice sheets of Antarctica and Greenland. These ice sheets are made of compacted snow. The relative abundances of different isotopes of oxygen in the ice is a measure of the temperature at the time the ice formed, while the age can be determined from the seasonal layering in the ice.

One of these cores, known as the Vostok core from East Antarctica, has been particularly well studied and its record extends back about 420 000 years. The Vostok temperature record is illustrated in **figure 9.21A**. It shows four distinct cycles, each characterised by initially warmer conditions, followed by slow and erratic cooling. Each of these cooling episodes coincided with a major glacial cycle during which the polar ice caps, especially the Arctic, expanded. These glacial periods ended abruptly as the temperature rose.

Cores of ocean floor sediment collected off the east and west coasts of South Africa reveal a very similar temperature record to Vostok, providing confidence that the Vostok record truly reflects wider conditions (**figure 9.21B, C**). The Vostok record tells us that over the last 400 000 years cold conditions have been the norm, interspersed with a few warmer periods.

Initial results from a new Antarctic ice core, Dome C, have recently been released, and extend the climatic record back to 740 000 years. Generally colder conditions with brief warmer interludes (although generally colder than today) prevailed back to this time (**figure 9.22A**).

The most recent cold period culminated about 18 000 years ago and ended only 10 000 years ago. During this period ice advance in the north caused a global fall in sea level as sea water became locked up in the ice, and the sea level around the southern African coast fell by 130 m. During this and earlier cold periods, rivers on the coastal plain eroded down into the bedrock to form deep valleys, especially in those regions where the bedrock consisted of soft, Cretaceous sediment, such as on the coastal plain of northern KwaZulu-Natal.

During the subsequent rise in sea level as the ice melted, these valleys drowned, forming estuaries and coastal lakes such as Lake St Lucia and Lake Sibayi. During the ice ages, icebergs may have been a common sight off the southern Cape coast, as rock fragments (which seem to have come from Antarctica) dropped from melting icebergs lie scattered on the sea bed off Cape Town.

Ice advance in the northern hemisphere also seems to have caused more arid conditions in southern Africa. During these ice ages, rainfall in the interior decreased markedly and the vegetation became much more sparse. A vast desert formed in the inte-

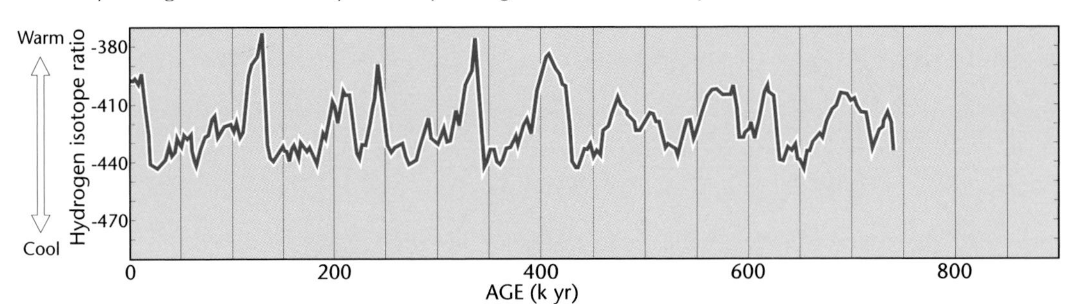

Figure 9.22A ***The temperature record (as reflected by hydrogen isotope abundance) for the southern hemisphere has been extended back to 740 000 years by the Dome C core from Antarctica. Colder conditions than today have prevailed over most of that time.***

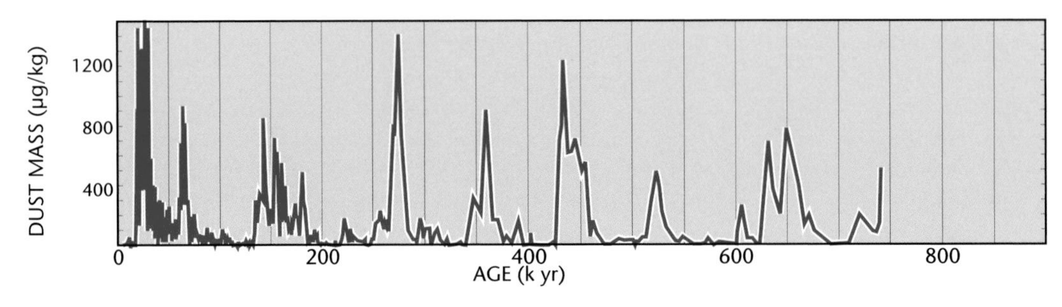

Figure 9.22B ***The dust content of the ice in the Dome C core is higher during colder periods, partly a result of generally drier conditions in the southern hemisphere at these times and possibly also exposure of the Antarctic continental shelf.***

rior of southern Africa, of which the Kalahari Desert is a small remnant. The sands of this former desert (the so-called Kalahari sands) are still present over large areas of the Northwest and Northern Cape provinces, Botswana, Namibia, Angola, eastern Zambia and Zimbabwe and the southern Congo, forming the largest expanse of sand on Earth. Regions in Angola and the Congo that today receive rainfall in excess of 1 000 mm per annum probably received less than 200 mm during the last glacial maximum. During glacial advance, the sands, which today are anchored by vegetation, were mobilised by the prevailing winds and whipped into dune systems tens to hundreds of kilometres long.

Evidence for drier conditions on southern hemisphere continents during glacial maxima is also provided by the Dome C core. The dust content of the ice rose dramatically during glacial maxima, indicating generally dustier conditions in the southern hemisphere at these times (**figure 19.22B**). This is partly a consequence of more arid conditions on the continents, and perhaps also exposure of the Antarctic continental shelf as a result of sea-level fall.

THE HOLOCENE EPOCH

Today we are experiencing the relatively mild climate of an interglacial period (known as the **Holocene Epoch**), and it is in this interval that all of the world's great civilisations developed. The prospects that another ice age would pose for humanity are indeed awesome.

Although the Holocene has been characterised by generally mild climates globally, the climate has nevertheless been changing, and southern Africa has experienced both drier and wetter periods. What causes these and how much the climate has changed are still matters of intense interest and concern. As the concentrations of greenhouse gases in the atmosphere continue to rise due to human activity, we need a comprehensive understanding of the climate in order to separate natural influences from those we induce, so we can predict the climates we are likely to experience in the future. In the past, when human beings were hunter-gatherers, the effects of climate change such as ice ages were mitigated by migration, but that option is no longer available. In order to prepare for the future, no matter what it might hold, we need to understand climate change.

The dramatic changes in climate in southern and central Africa that took place in the past few million years, driven by forces within the Earth, in the oceans and in the atmosphere, had a profound effect on the environment and consequently on mammalian evolution and distribution. Primates in particular underwent diversification, leading to the evolution of *Homo sapiens*.

SUGGESTED FURTHER READING

Alvarez, W. 1998. *T. rex and the Crater of Doom*. Vintage Books, New York.

Anderson, JM. 2001 (2nd Edition). *Towards Gondwana Alive*. Gondwana Alive Society, National Botanical Institute, Pretoria.

Flannery, T. 2001. *The Eternal Frontier*. Atlantic Monthly Press, New York.

Johnson, MR, CR Anhaeusser, CR and Thomas, RJ. 2005. *The Geology of South Africa*, Council for Geoscience, Pretoria.

MacRae, C. 1999. *Life Etched in Stone – The Fossils of South Africa*. Geological Society of South Africa, Johannesburg.

Moore, AE and P A Larkin, PA. 2001. 'Drainage Evolution in South-Central Africa since the Breakup of Gondwana.' *South African Journal of Geology*, vol. 104, pp 47–68.

Partridge, TC and Maud, RR. 2000. *The Cenozoic Geology of Southern Africa*. Oxford Monographs on Geology and Geophysics, New York.

Redfern, R. 2000. *Origins*. Cassel and Company, London.

Retallack,GJ. 2001. 'Cenozoic Expansion of Grasslands and Climatic Cooling.' *Journal of Geology*, vol. 109, pp 407–426.

Tyson, PD, Fuchs R, Fu C, Lebel L, Mitra AP, Odada E, Perry J, Steffen W and Virji H. 2002. *Global-Regional Linkages in the Earth System*. Springer, London.

Viljoen, MJ and Reimold, WU. 1999. *An Introduction to South Africa's Geological and Mining Heritage*. Mintek, Johannesburg.

10
THE ARRIVAL OF HUMANS

The San people of southern Africa are not only unique in their culutral heritage and language, but study of their mitochondrial DNA has revealed that they are the most ancient genetically modern people on Earth.

Ariadne van Zandbergen / IOA

ROUTE MAP TO CHAPTER 10

Age (years before present)	**Event**
65 million	• Dinosaur extinction led to radiation and diversification of mammals, which became the dominant large animal group. Placental mammals became dominant among mammals.
34 million	• An extinction event affected mammals.
25 million	• The heyday of the apes occurred.
14-10 million	• Orang-utans split from the ancestral hominoid lineage.
10-8 million	• African apes split from the hominin line.
ca. 7–6 million	• The oldest hominins appeared: *Orrorin tugenensis* in Kenya and *Sahelanthropus tchadensis* in Chad.
ca. 5 million	• *Ardipithecus ramidus* appeared in Ethiopia.
ca. 4 million	• *Australopithecus anamensis* appeared in Ethiopia.
ca. 3.8–3 million	• *Australopithecus afarensis* appeared in Kenya and Tanzania.
ca. 3.3 million	• *Kenyanthropus platyops* appeared in Kenya.
ca. 3 million	• *Australopithecus africanus* appeared in South Africa.
ca. 2.6 million	• Robust australopithecines appeared, as did the earliest stone tools.
ca. 2.5 million	• The first representatives of the genus *Homo* appeared in East Africa.
ca. 1.8 million	• The first use of bone tools.
ca. 1.7 million	• *Homo erectus* and Acheulian stone tools made their appearance. *H erectus* migrated to Asia.
ca. 1 million	• The last australopithecines became extinct.
ca. 800 000	• Archaic *Homo sapiens* appeared.
ca. 200 000	• *Homo sapiens sapiens* appeared and global dispersion of the species occurred.

THE AGE OF MAMMALS

Since the post-Karoo break-up of Gondwana and the subsequent uplift of southern Africa, the region has largely been an erosional rather than a depositional landscape and environments suitable for fossil preservation are rare. Some sedimentary traps favouring fossilisation occurred in deposits formed along large river systems in the interior, but the dolomitic caves of the Cradle of Humankind northwest of Johannesburg have produced the greatest abundance of fossil mammalian remains, most dating to less than three million years ago. These fossils have given South Africa one of the best records of life in Africa over the past three million years. However, the relatively recent age of these deposits means that only the later stages of mammalian evolution are preserved in the fossil record of South Africa; to discover the big picture it is necessary to look elsewhere.

Mammals, varieties of animals that include humans, are a diverse group. Broadly, an animal is classified as a mammal if it has, among a long list of traits: a single bone comprising the lower jaw; thermoregulation (controlling body temperature); mammary glands for producing milk; body hair and sometimes horns; sweat glands; and internal fertilisation. However, the actual list of characters that defines a mammal exceeds 38 – and that is just for those we know in living mammals. As the fossil record for mammals and near-mammals continues to grow, the number of traits used to define mammals grows accordingly.

As seen in Chapter 8, the earliest mammals arose from therapsids at the end of the Triassic. These earliest mammals were small, shrew-like creatures, probably existing as nocturnal insectivores. Their main competitors were the dinosaurs, and early mammals existed on the periphery of the great dinosaur world throughout much of the Jurassic and Cretaceous. They probably managed to survive because of their small body size and reclusive behaviour. One can imagine that any mammal large enough to provide a decent meal was probably selected out of the gene pool by voracious dinosaurs. After the end-Cretaceous extinction event, however, when all dinosaur species disappeared from the scene, mammalian populations exploded in parallel with a similar expansion among flowering plants, now freed from the overwhelming shade of coniferous forests.

The modern class of mammals resulting from this expansion includes three major sub-classes: **Marsupialia** (pouched mammals in which new-born young spend the early part of their developmental life in a skin pouch on the belly of their mother); **Eutheria** (placental mammals); and the relic group **Prototheria** (monotremes, or egg layers, the best-known of which is the duck-billed platypus, now found only in Australia and adjacent major islands).

POST-CRETACEOUS MAMMALS

Early in the Palaeocene, the epoch immediately following the Cretaceous (which ended 65 million years ago), mammals of all types became abundant in the fossil record. They had an advantage over surviving reptiles because of their ability to control their body temperature. As they expanded into niches vacated by the dinosaurs, new genera and species rapidly evolved, leading to the greatest ever species diversity in mammalian history.

At the end of the Eocene Epoch (about 34 million years ago), there was an extinction event that affected mainly mammals. Its cause is unknown. Terrestrial life was possibly influenced by major meteorite impacts at Chesapeake Bay off the North American coast and Popigai in Siberia, both leaving craters more than 80 km in diameter. Another possible reason is global cooling, as the Drake Passage between South America and Antarctica opened at about this time, causing the oceans to cool (*see* Chapter 9).

So in effect a series of events, geological and otherwise, led to a major extinction event among mammals. This was of great importance as it provided the platform for the evolution of modern forms of mammals. The beginning of the Miocene Epoch, approximately 24 million years ago, saw the ancestors of almost every major form of mammal we are familiar with today evolving into the many niches left vacant by the Eocene extinction event. Among the more successful mammal groups to radiate following this extinction were the primates.

PRIMATE ORIGINS

The earliest primates seem to have been adapted to an arboreal, and probably nocturnal lifestyle, but many soon became diurnal (active during the day). Today the Order Primates is divided into two suborders: the primitive **Prosimians**, which include lemurs,

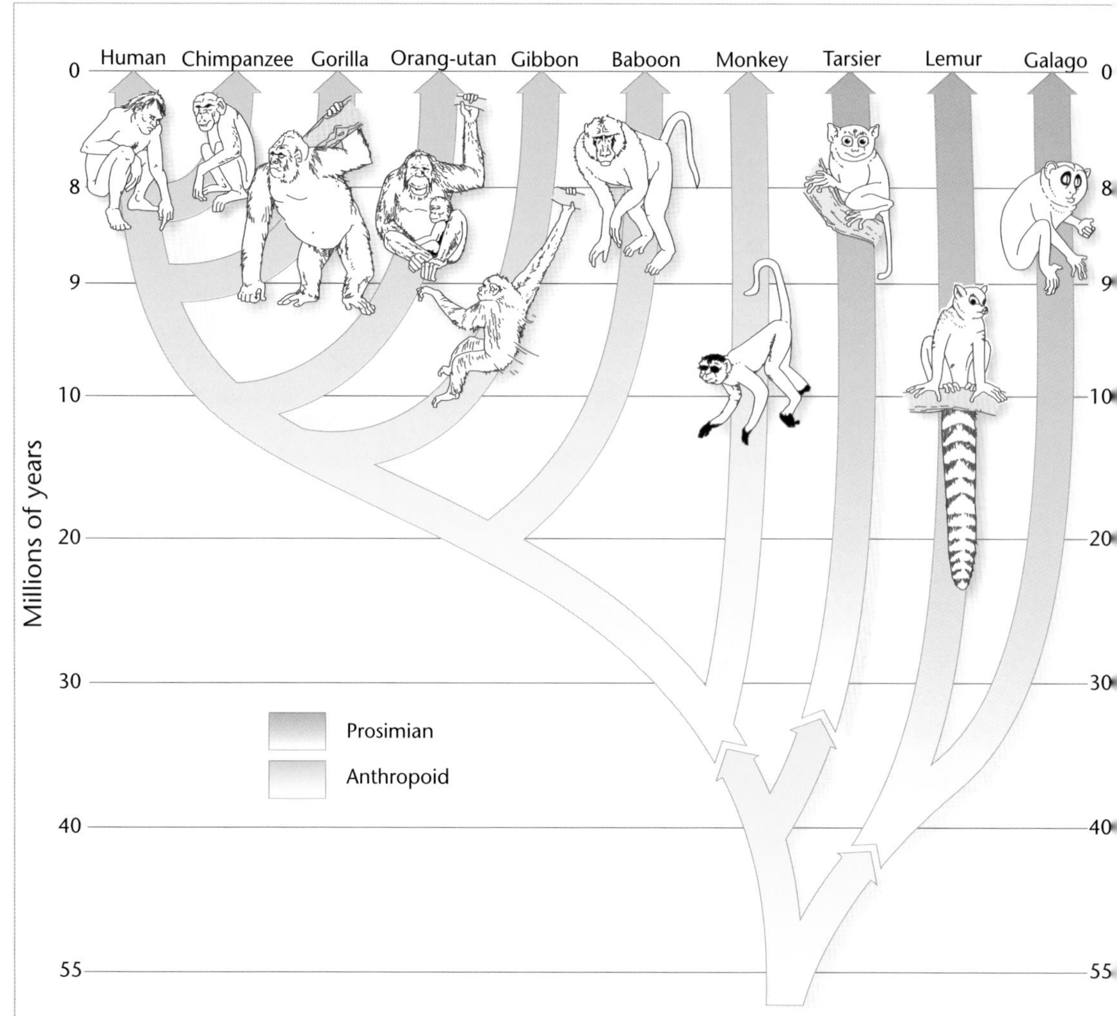

Figure 10.1 *Evolutionary tree showing the relationships between primates. Diversification of primates started about 55 million years ago, producing the various forms known today. The family tree illustrates, for example, that chimpanzees are closely related to humans whereas lemurs are only distantly related.*

tarsiers and bush-babies; and the **Anthropoids** (higher primates), which include the New World and Old World monkeys, as well as apes and humans.

Anthropoids differ from prosimians in fundamentally important ways: they generally have flat nails as opposed to claws (for manipulative ability, especially in feeding and grooming), well-developed stereoscopic and colour vision, and shorter snouts reflecting a reduced reliance on smell. These features in the anthropoids signal a change to diurnal activity and shift in diet from primarily insects to fruits and leaves. Concomitantly, sexual dimorphism emerged and group social behaviour probably became important. The anthropoid line leading to the **hominids** (humans and our **bipedal** ancestors, i.e. that walked on two legs) split from the prosimians in the Eocene Epoch (55 to 34 million years ago) (**figure 10.1**). This led to the Miocene heyday of the apes, beginning about 24 million years ago, when the relatively warm global climate favoured the spread of large-bodied, tailless apes. During the Early Miocene (about 20 million years ago) apes became widespread throughout Africa, Asia and Europe.

Based largely on genetic evidence and supplemented by a sparse fossil record, it appears that

CLASSIFICATION OF PRIMATES

The taxonomic classification system devised by Linnaeus in 1758 – the Linnaean classification of Primates – is still used in a modified form today. Animals are identified in descending order as belonging to a Kingdom, Phylum, Class, Order, Family, Genus, and finally a Species. This classification system is based largely on the animal's physical characteristics; things that look alike are grouped together.

In the Linnaean system, humans would be categorised first as Animalia; then Vertebrata because we have a backbone; Mammalia because we have hair and suckle our young; Primates because we share with apes, monkeys and lemurs certain morphological characteristics such as grasping hands, good vision, depth perception; Hominidae because, among other criteria, we are separated from other apes by being bipedal and having large, complex brains; *Homo* (meaning man) being our generic classification as human; and finally *sapiens*, a species name meaning wise.

The Linnaean system also recognises groupings such as superfamilies and subfamilies. In the human lineage, the most often recognised superfamily is the Hominoidea (hominoids), which includes all living apes and humans. It is from this point that most of the present human origins classification debate begins.

The traditional view has been to recognise three families of hominoid: the Hylobatidae, the Hominidae and the Pongidae. Hylobatidae include the so-called lesser apes of Asia, the gibbons and siamangs. Hominidae include living humans and their fossil ancestors that possess characteristics such as bipedalism, reduced canine size and increasing brain size, as observed in the australopithecines. Pongidae include African Great Apes (gorillas, chimpanzees), and the Asian orang-utan.

Modern-day genetic research suggests humans are more closely related to the chimpanzee and bonobo than either are to the gorilla. Chimps and humans share some 98% of their genes, indicating a common ape ancestor. Divergence times between the two groups based on a molecular clock suggest the chimpanzee-human split occurred about eight million years ago. In turn, the African apes and humans are more closely related to each other than any are to the orang-utan. Recognising these and other genetic relationships, some argue that we must overhaul the present morphologically based classification system for one more representative of our true evolutionary relationships as evinced by our genes.

REWORKING THE FAMILY TREE (HOMINID VS. HOMININ)

This is where the term hominin comes into play. Under the new classification model, Hominoidea would remain a primate superfamily, as has always been the case. Under this hominoid umbrella would fall orang-utans, gorillas, chimps and humans, all in the Family Hominidae.

In recognition of their genetic divergence some 10 to 14 million years ago, orang-utans would be placed in the sub-family Ponginae, and African apes and humans in the sub-family Homininae. Bipedal apes – humans and their fossil ancestors – would fall into the tribe Hominini (thus hominin). All of the fossil genera, such as *Australopithecus, Ardipithecus, Kenyanthropus* and *Homo*, would fall into this tribe.

A few evolutionary biologists want a more extreme classification, which would include humans and chimpanzees within the same genus, the genus *Homo*.

Members of the sub-family Homininae.

Terence McCarthy

Photo Access

Photo Access

Photo Access

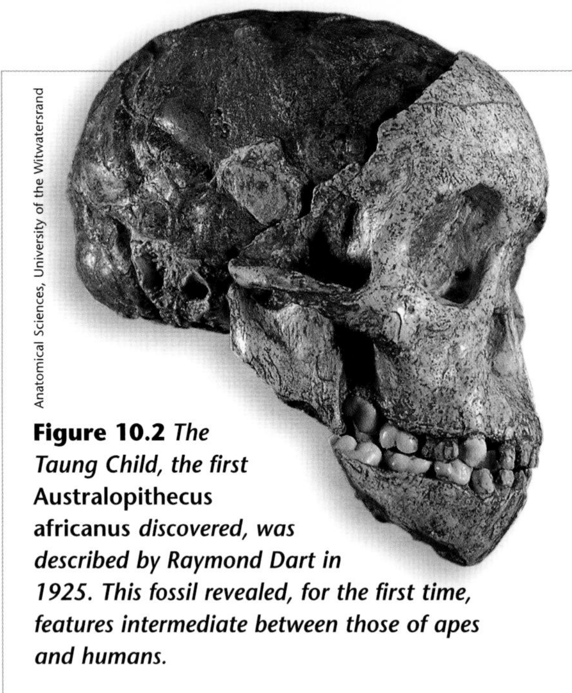
Anatomical Sciences, University of the Witwatersrand

Figure 10.2 *The Taung Child, the first* **Australopithecus africanus** *discovered, was described by Raymond Dart in 1925. This fossil revealed, for the first time, features intermediate between those of apes and humans.*

between 14 and 10 million years ago the orangutan group split from the ancestral hominoid lineage, which includes African apes, humans and our ancestors (*see* 'Classification of primates' on page 279). Between nine and seven million years ago the ancestral lineage of the living gorillas diverged, and the split between the chimpanzee and **hominin** lineages occurred approximately eight million years ago, or perhaps a bit earlier. It appears that the great apes favoured the tropical evergreen forests, while hominins adapted to the more broken woodlands and savannas that were emerging in Africa at that time.

At the same time in Africa, many ancestors of modern African mammalian forms were beginning to gain a foothold. During the late Miocene, different evolutionary lines were evolving in response to changing environmental conditions in Africa, especially the spread of grasslands. One of the finest examples of the fauna of the end of the Miocene can be found at Langebaanweg in the Western Cape, where hundreds of thousands of remains from a rich coastal community have been found. The fossils are between seven and 3.5 million years old, but even though this spans the critical time of the evolution of hominins, no fossil hominin remains have yet been found. For fossil evidence of the evolution of the bipedal ape lineage, it is necessary to move further north in Africa.

THE EARLIEST HOMININS

The complete skull of *Sahelanthropus tchadensis* was discovered by Michele Brunet in Chad and described in the scientific journal *Nature* in 2002 as the world's oldest hominin, dating to between six and seven million years ago. *Sahelanthropus* has many traits characteristic of hominins, including smaller canines and thicker tooth enamel than apes. Also, the area at the back of the skull where neck muscles attach suggests that it walked upright. As with most new finds, there is controversy around its status as a hominin, with some arguing that it is closer in structure to a female gorilla than a hominin.

In 2001 the discovery was announced of a fossil ape-like creature named *Orrorin tugenensis*, dating to around six million years ago. It was reported to be bipedal, one of the most critical characteristics for being designated a hominin. Found at Kapsomin in the Tugen Hills in Kenya's Baringo District, the remains include a thigh bone, pieces of a jaw, some teeth, arm bones and a finger bone, representing five individuals.

Slightly younger is another recently discovered and rather enigmatic species, *Ardipithecus ramidus*, which is now known to have existed from at least five million years down to around 4.4 million years ago. The remains of about 50 individuals – mostly consisting of teeth, jaw and skull fragments – have been found in Ethiopia, but there are also reports of a more complete skeleton encased in **breccia**. The very primitive teeth differentiate it from other hominins, so there is some debate as to whether this is in fact a hominin or an early hominoid more closely related to living chimpanzees.

AUSTRALOPITHECINES

From about four million years ago, the story of hominin evolution becomes somewhat clearer with the emergence of the genus *Australopithecus* (meaning southern ape), a name coined by Raymond Dart some 80 years ago when naming the fossil hominin infant found at Taung in the Northwest Province (the Taung Child) (**figure 10.2**). This child was considered to be the missing link because it shows a mosaic of features intermediate between those of apes and humans (*see* 'Apes versus humans', right).

APES VERSUS HUMANS

There are many similarities between apes and modern humans – a consequence of their similar genetic make-up. Both have a similar anatomy and physiology, reproductive strategy (widely spaced births and extensive parental investment in offspring), and capacity for learned behaviour, especially relating to tool use, infant care and diet.

The differences are relatively minor: humans habitually walk on two legs (bipedalism), have larger and more complex brains, slower growth rates, live longer, and use complex language and symbolism. Many of these differences are reflected in their bones and teeth and can be recognised in fossils (*see* diagrams below). Compared to a chimpanzee, a human has the following skeletal features: a more expanded and rounded skull to accommodate a large brain; a relatively small flat face; smaller canine teeth; rounded as opposed to rectangular arrangement of teeth; a chin; a hole at the base of the skull (*foramen magnum*) placed in the centre of the skull rather than at the back; bowl-shaped rather than elongated pelvis; and big toe close to and aligned with the rest rather than divergent. The last three features are adaptations for upright walking.

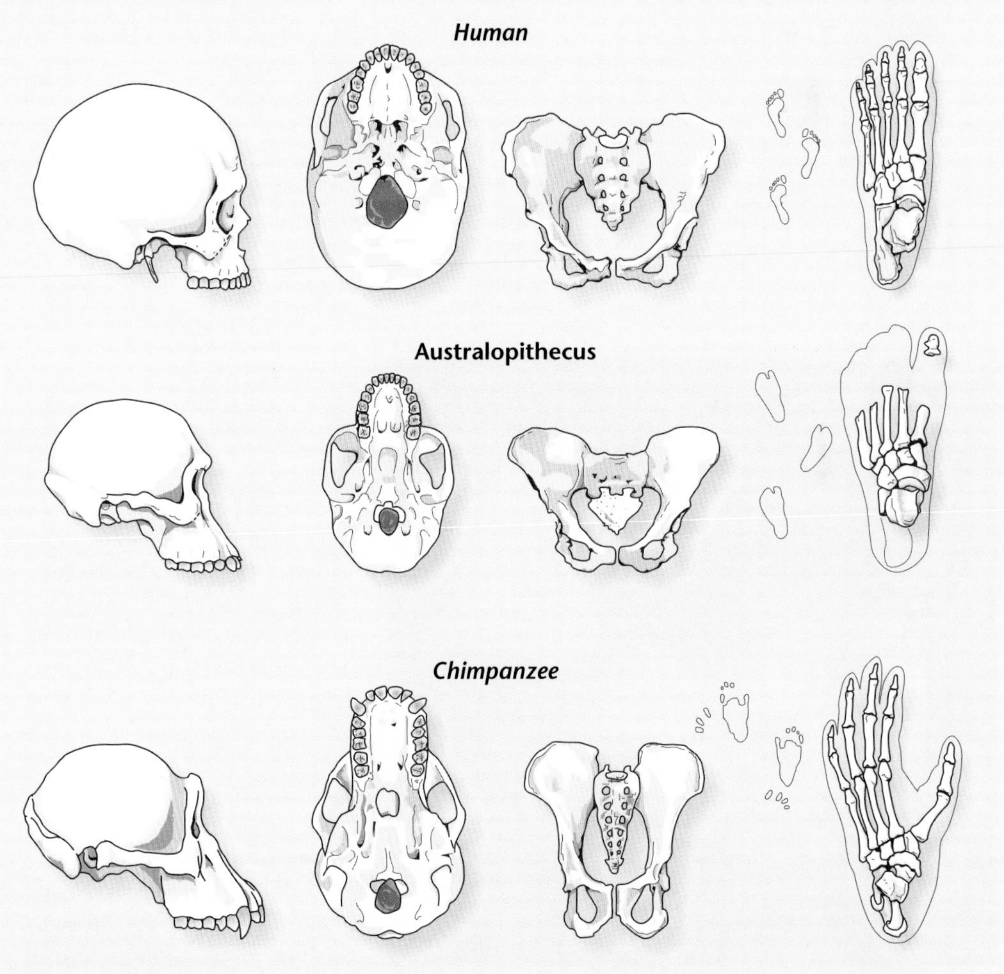

These diagrams show the important anatomical differences between chimpanzees, australopithecines and humans.

Several species of australopithecines lived in Africa between four and one million years ago. The so-called gracile australopithecines, defined by the relatively light build of their jaws, teeth and chewing muscles, seem to have disappeared around 2.5 to two million years ago. The more robust australopithecines, sometimes placed in the genus *Australopithecus* but now more commonly placed into the genus under which they were originally named, *Paranthropus*, survived until about one million years ago before becoming extinct. They are labelled robust because of the tremendous size of their jaws and teeth, and other features of the skull, rather than because of body size. Both gracile and robust forms stood about waist high, relative to an average modern human. They had human-like bodies but their skulls were ape-like.

The most significant adaptation that distinguished the australopithecines from the chimpanzees and other apes was considered to be bipedalism. But as already mentioned, this trait seems to predate the origin of the australopithecines in the genera *Ardipithecus, Orrorin* and *Sahelanthropus*. Other traits that define *Australopithecus* are a slight increase in brain size and retention of a relatively prognathic (pushed out or jutting) face. It now seems that there were also a number of other important adaptations in the rest of the skeleton within this group: some were adapted to climbing trees, while others were more adapted to a terrestrial lifestyle.

The earliest of the australopithecines so far recognised is *Australopithecus anamensis*. Dating to about four million years ago, this species is based on a relatively small number of specimens from the Lake Turkana region of East Africa. A large tibia (shin bone) attributed to *A. anamensis* indicates that this species was well adapted to bipedalism, while the parallel tooth rows are more ape-like, indicating a very primitive head compared to the hominins that came later. Given the present fossil record, this species may be a good candidate ancestor for all later hominins.

Following *A. anamensis* in time is *Australopithecus afarensis* (**figure 10.3**), which lived between 3.8 and three million years ago. This species is the most widespread of the early hominins and probably the best known, including specimens such as the famous Lucy – a relatively complete skeleton discovered in the Hadar region of Ethiopia in the early 1970s by Don Johanson and Tim White. The species is characterised by males who are larger than females, much like the sexual dimorphism observed in gorillas, and a cranial capacity of around 415 cm^3, only slightly larger than that of chimpanzees (400 cm^3). This species was small in stature, adults typically standing waist high, relative to an adult modern human.

Found in Tanzania, Ethiopia and possibly Chad, the species was clearly wide-ranging and lived in diverse habitats. Given its distribution in space and time, it also shows a great deal of variation in body form, leading some researchers to suggest that what is currently regarded as a single species may in fact include several species. There has also been a great deal of debate as to whether *A. afarensis* was a climber, but recent discoveries point to the species being largely terrestrial. The famous Laetoli fossil footprint trail in Tanzania is attributed to this species, and confirms that it walked upright, although its big toe was slightly divergent, reminiscent of chimpanzees.

Christine Steininger

Figure 10.3 ***The skull of* Australopithecus afarensis, *which lived between 3.8 million and 3 million years ago.***

Figure 10.4 ***View over the Cradle of Humankind, with Johannesburg in the distance. The topography and vegetation are probably much the same as they were at the time of* Australopithecus. *Some 35% of African hominin fossils have been found in this region.***

Another new member in the field of human evolution at 3.3 million years ago is the newly named *Kenyanthropus platyops* (flat-faced man-ape from Kenya), which was described in 2001. Although based on a crushed cranial specimen, it seems to have a diversity of features that link it to the australopithecines and possibly to some specimens attributed to early *Homo*. It is really too early to tell how this new form fits into the human family tree.

Another East African australopithecine is *Australopithecus garhi*, which was discovered in Kenya in the late 1990s. This species, dating to 2.5 million years ago, is one of the newly added members to the genus *Australopithecus*. Based on fragmentary evidence of its skeleton it has been suggested that *A. garhi* had both long arms and long legs, giving it unusual body proportions compared to other hominins.

HOMININS ON THE HIGHVELD

The South African record of hominin evolution begins approximately three million years ago. Most of the fossil evidence of early hominins from southern Africa is found in the dolomitic caves of the Gauteng region, the Cradle of Humankind World Heritage Site (**figure 10.4**), where some 35% of the total record of human evolution in Africa has been recovered. Sterkfontein alone has produced in

THE FORMATION OF DOLOMITIC CAVES

Five basic stages of cave formation can be recognised in the dolomite caves within the Cradle of Humankind near Krugersdorp in Gauteng. Dolomite is a calcium and magnesium carbonate, and is slightly soluble in water, especially water that has been acidified by dissolving traces of carbon dioxide.

In stage 1 (*see* figures) a cavern forms along a fracture or crack in the dolomite through the solution of dolomite in what is known as the **phreatic** or **saturated zone**, the zone beneath the ground-water level or water table (the zone above the water table is the **vadose** or **unsaturated zone**). The cavern's original shape is usually determined by the fractures or planes of weakness in the rock. Fragments of insoluble chert, which occurs in the dolomite, litter the floors of the chambers.

In stage 2 the water table drops, usually through natural erosion of the land surface or by down-cutting of a nearby valley, so that the cave becomes air-filled. **Stalactites** and **stalagmites** now begin to form in the cave as lime-rich water drips from the roof of the cave, and lime is deposited on the cave walls and floor, known as **flowstone**. While this filling process is in progress, new cavities are forming below the water table beneath the floor of the cave.

Stage 3 sees the formation of avens, or shafts, that begin to approach and eventually break through to the surface. At this stage a **debris cone** or scree deposit begins to form in the cave as soil, rocks, organic debris and some bones of animals derived from the surface enter the cave through the opening. Animals like hyaenas, leopards and possibly hominins may at this time start living in the cave, as a result of which the discarded bones of their prey begin to accumulate in the cave. If this sedimentary cone becomes calcified (cemented) by lime-bearing water dripping from the cave roof, it forms what is termed **cave breccia**. While this filling is taking place, surface erosion is steadily removing the upper parts of the cave, widening entrances to the cave.

Stage 4 is when the cave is completely filled with cave sediment. Stage 5 is the final stage where erosion has de-roofed the cave entirely and the bone-bearing breccia is exposed on the surface. The cave deposits start to decalcify, mixing the fragments released from the breccia. The process of cave formation and filling is very slow, probably taking tens to hundreds of thousands of years.

This is a generalised scheme and there are many possible variations. For example, material filling an upper cave may collapse into a younger cave that has formed below it; or prolonged rise in the ground-water level may cause decalcification of breccia and the formation of new caverns within older cave fill. One of the important consequences of these various processes is that cave deposits tend to be rather chaotic and as they contain no suitable radioactive substances, it is very difficult to determine their age.

Many of the old, eroded caves in the Gauteng region were mined for the thick seams of lime that were deposited during Stage 2 of cave formation. In following these lime seams underground, the miners exposed the breccia deposits – and as a result, their fossil contents became known to science.

excess of 500 hominin fossils and Swartkrans across the valley has produced a similar number.

The dolomites form part of the Transvaal Supergroup, deposited some 2 500 million years ago (*see* Chapter 4). What makes the dolomitic area such a repository of ancient fossil treasure? The dolomites were partly dissolved by ground water, creating pockets or caves. Once drained of water and opened to the surface by erosion, these cavities were filled by calcium carbonate precipitating from water passing through the cave from the surface, and soil and other material washed into the cave from outside.

Among the external debris that entered the caves were the remains of animals. When calcium carbonate was deposited within the mixed debris that accumulated on the cave floor, it cemented

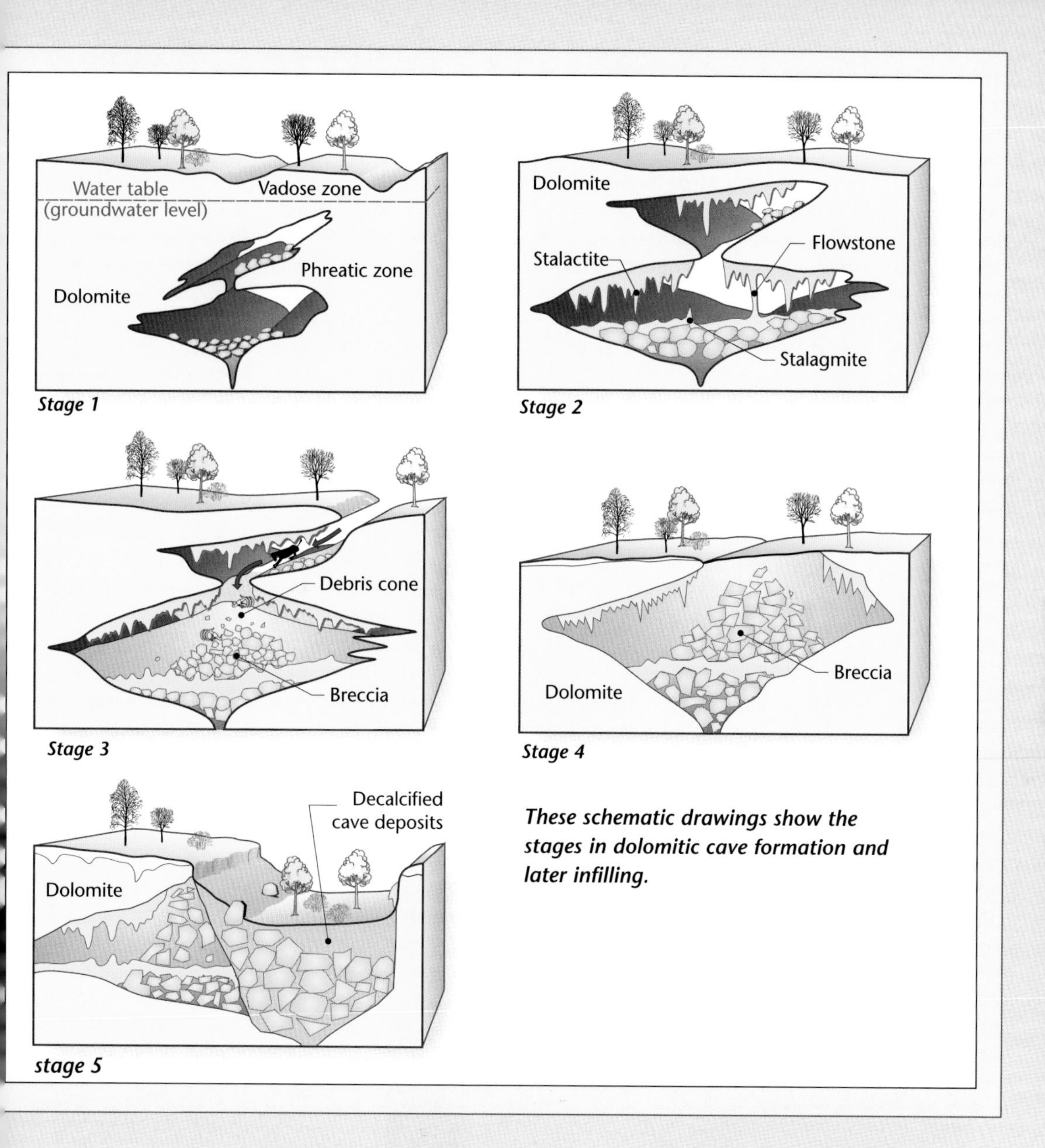

These schematic drawings show the stages in dolomitic cave formation and later infilling.

the material, forming a concrete-like breccia (*see* 'The formation of dolomitic caves', above).

This hard breccia has ensured the preservation of the remains of animals, including so-called ape-men, which found their way into the caves. Extensive research, most notably by Bob (CK) Brain in South Africa, has shown that the fossilised remains of animals, including those of early hominins, accumulated in the older deposits in the caves largely as a result of predation. Many bones amassed in caves from leopards feeding in trees that overhung cave entrances, and from hyaenas, porcupines and owls living in caves. Lee Berger and Ron Clarke have shown that even the famous Taung Child was the victim of predation, in this case by an eagle. These early hominins were clearly the hunted, not the hunters.

Accumulation of bones in cave deposits has ensured that the quality of preservation of the fossils is very good, if rather fragmentary. However, the jumbled nature of cave debris has made the dating of the fossils a difficult and often controversial task. For this reason the dating of South African sites depends largely on comparison of animal remains with East African sites, where accurate dating is made possible by the presence of volcanic deposits associated with similar fossils (*see* 'Measuring the age of a rock', page 68). New technological developments are starting to make direct dating of cave deposits possible, but the methods are not yet universally applicable.

Sometime after three million years ago an animal with a blend of ape and human characteristics occupied the Gauteng Highveld, seeking shelter in riverine forests and scavenging for food in the surrounding broken woodland. This ape-man, known as *Australopithecus africanus,* may well have been the ancestor of our own genus, *Homo,* or may have belonged to a closely related side-branch. Standing some 1.3 m tall, with a brain about the size of a chimpanzee's, *A. africanus* lived in small social groupings in a subtropical landscape dominated by predators such as the false sabre-toothed cat and hunting hyaena, as well as other now-extinct creatures like giant monkeys and small, short-necked giraffes.

The southern African gracile ape-man, *A. africanus* – of which the Taung Child and the famous Mrs Ples, discovered by Robert Broom at Sterkfontein (**figure 10.5**), are well known examples – represents a species that is slightly more advanced than *A. afarensis. A. africanus,* which has only been found in South Africa, had a larger brain and larger teeth. Recently the suggestion was made that *A. africanus* had longer arms and shorter legs than *A. afarensis,* as in apes. This raises doubt as to whether *A. africanus* arose from *A. afarensis,* as had been thought for many years. *A. africanus* possibly represents an independent line of evolution from a more ancient common ancestor like *A. anamensis.*

Still embedded in the rocks of Sterkfontein is a fossil that may give us far greater insight into the early evolution of hominins. Possibly as old as three to four million years, the so-called Little Foot skeleton may well be the most complete early hominin yet discovered, and is therefore certain to shed light on the family tree of our earliest ancestors. This remarkable skeleton, discovered by Stephen Motsumi, Nkwane Molefe and Ron Clarke, has a foot with a slightly divergent big toe, indicating a position between apes and humans (*see* 'Apes versus humans', page 281). Many of the bones of this skeleton are still articulated and show no evidence of carnivore damage (**figure 10.6**), suggesting that this hominid became trapped in the cave and died there.

Global cooling and consequent aridification of Africa intensified between three and two million years ago. The Highveld at that time was probably little different from the open grassland environment we know today (**figure 10.4**). At the same time as these changes were taking place, new forms of proto-human appeared – forms that may well have evolved

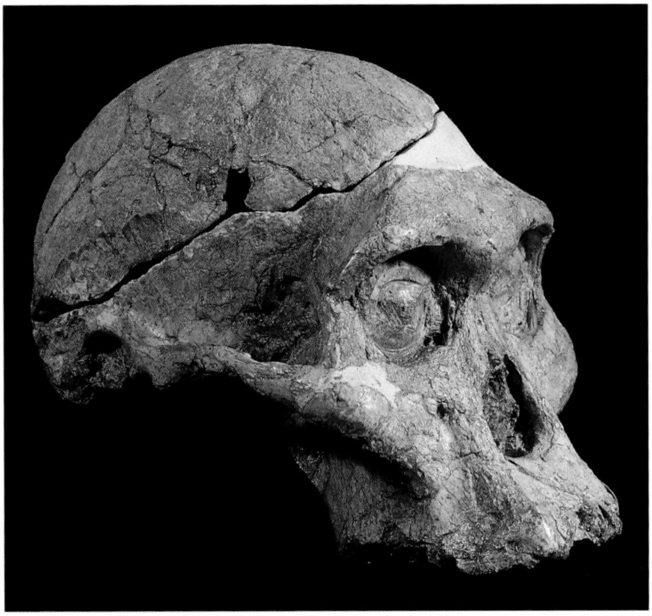

Transvaal Museum

Figure 10.5 *Mrs Ples, the most complete skull of an adult* **Australopithecus africanus,** *was discovered by Robert Broom at Sterkfontein. Classified by Broom as* **Plesianthropus transvaalensis** *(hence Mrs Ples), the specimen was subsequently recognised as* **Australopithecus africanus.**

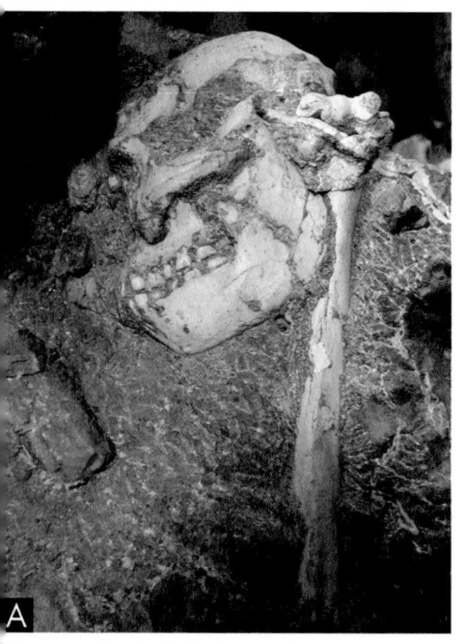

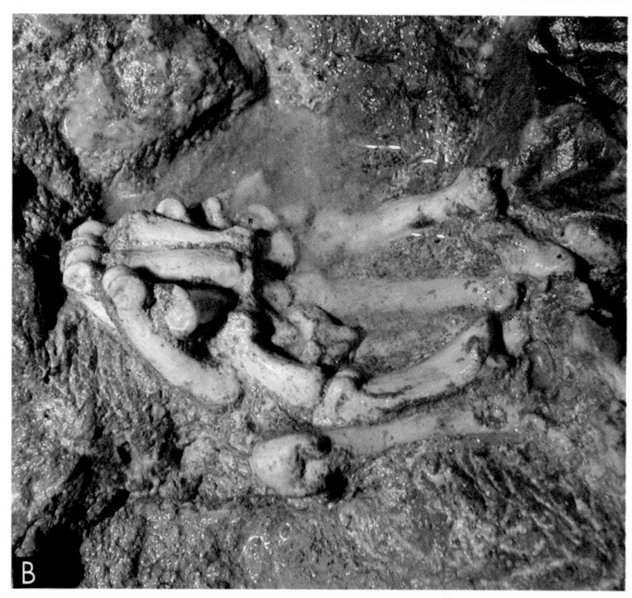

Figure 10.6 *Little Foot, the most complete skeleton of an* Australopithecus, *was found by Stephen Motsumi, Nkwane Molefe and Ron Clarke at Sterkfontein in 1998. The skull* (**A**) *and many of the bones are still preserved in their natural position, as shown by the very well-preserved hand* (**B**).

from *A. africanus* or another gracile australopithecine. Around this time, *A. africanus* may have started to take on characteristics associated with both the robust ape-man (a flatter-faced, larger-toothed australopithecine) and the earliest members of our own genus, *Homo*. Soon afterwards *A. africanus* seems to disappear from the fossil record.

PARANTHROPUS

Emerging about 2.6 million years ago, almost certainly in response to the environmental changes taking place in Africa at the time, was a new form of ape-man. This form was first recognised by Robert Broom from discoveries made at Kromdraai near Sterkfontein. Originally classified as a new genus, *Paranthropus*, then reclassified as a robust form of *Australopithecus*, it is now usually placed back in the genus *Paranthropus*. Fossils of this ape-man are known for their massive teeth and the bony ridge (sagittal crest) on the top of the skull for the attachment of powerful jaw muscles, which suggested that they had a coarse vegetarian diet.

Recent examinations of the hand bones of the southern African robust ape-men reveal that they had the physical capability to make and use tools. Relative to body size, they have the smallest hominin braincases on record, usually less than 420 cm^3, although some individuals have cranial capacities in the low 500 cm^3 range. The body size of *Paranthropus* was similar to that of *Australopithecus africanus*, both being approximately 1.3 m tall.

The earliest paranthropines are known from Kenya and Ethiopia. They have been found in deposits about 2.6 million years old and are known as *Paranthropus aethiopicus*. This species is the suspected common ancestor of later paranthropines. The Black Skull, discovered in Kenya by Alan Walker in 1985, is the best example. This is the first species in the hominin fossil record to exhibit the massive teeth and chewing muscles that suggest a largely vegetarian lifestyle.

Approximately 2.5 million years ago the species *Paranthropus boisei* lived in East Africa. With some of the largest members of the robust australopithecines in this category, *P. boisei* is the epitome of a robust ape-man. It had huge teeth and massive jaws. The best-known example is certainly Nutcracker Man (also sometimes called Dear Boy), which was discovered by Mary and Louis Leakey at Olduvai Gorge in Tanzania in the late 1950s.

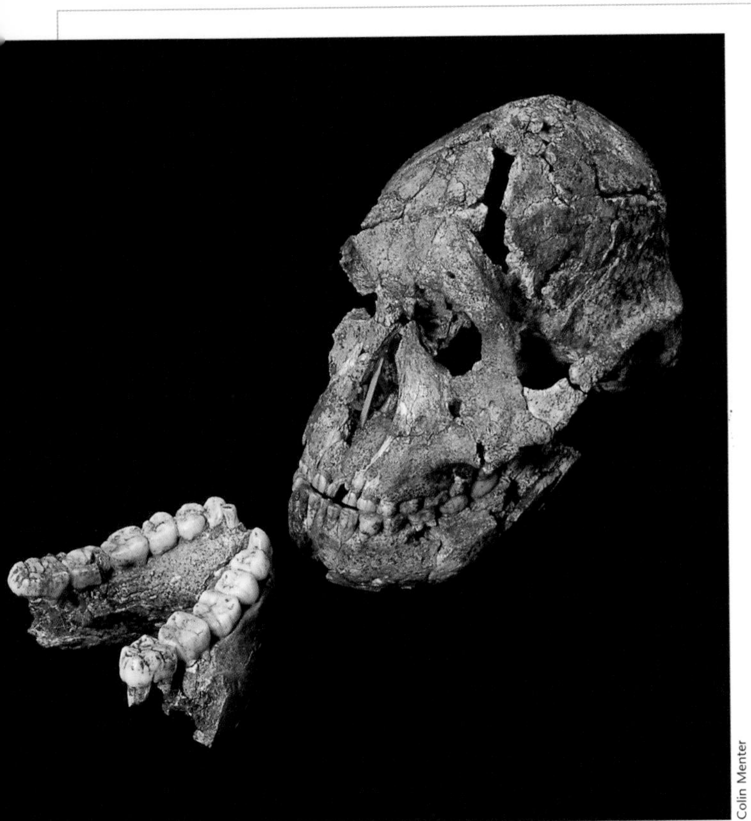

Colin Menter

Figure 10.7 ***A female skull and male mandible of*** **Paranthropus robustus.** ***The specimens were discovered by Andre Keyser at Drimolen in the Sterkfontein Valley.***

Paranthropus robustus (**figure 10.7**) is the southern African version of the robust ape-men. Often called the flat-faced ape-man because of its dished face, it is the best-represented hominin in southern Africa and probably the best-represented fossil hominin in the entire African record. It is found in cave deposits in the Cradle of Humankind at Swartkrans, Drimolen, Kromdraai, Coopers and Gondolin. However, by a million years ago, the robusts had followed the earlier ape-men into extinction.

Studies by Bob Brain, Lucinda Backwell and Francesco d'Errico of fossil bones associated with those of the South African robust ape-man suggest that *Paranthropus* used bone and horn tools, and did so for at least one million years (**figure 10.8**). These were used to dig out bulbs of plants and open termite nests, and have been found at the Swartkrans and Drimolen sites.

HOMO APPEARS

Appearing about 2.5 million years ago, the arrival of *Homo* represents a new direction in hominin evolution. Many evolutionary forms have generally been lumped together as early *Homo*, but the key difference between them and the australopithecines is a much larger and more complex brain that Phillip Tobias, Louis Leakey and John Napier linked to tool-making abilities. Typically, early *Homo* is characterised by its less prominent brow ridges, a reduction in facial prognathism (flattening of the face) and the absence of a ridge on the top of the skull.

The earliest known *Homo* fossil specimens have been found throughout Africa and many have not yet been classified to species. *Homo habilis* and *Homo rudolfensis* are two of the better-known types of confirmed early *Homo* species, but even their status is sometimes questioned, with some palaeoanthropologists wanting to place them in the genus *Australopithecus*. In South Africa, early *Homo* fossils are known from cave deposits at Sterkfontein, Coopers, Swartkrans, Drimolen and possibly Kromdraai.

Homo rudolfensis is the name given to a small number of fossils from Kenya showing affinities with many other hominins, including *Homo habilis, Homo ergaster* and *Homo erectus*. It has been tentatively associated with simple flaked stone tool technology.

With a larger brain than the australopithecines (around 650–800 cm^3), *Homo habilis* (which is Latin for handy man) was the earliest species of hominin to be placed in the genus *Homo*. With its rounder head, reduced prognathism, more human-like teeth and less pronounced brow ridges, it was considerably more human-like in appearance (**figure 10.9**). Based on a fragmentary skeleton from Olduvai Gorge in Tanzania, it has been suggested that *Homo habilis* had relatively long arms and short legs, possibly linking it to *Australopithecus africanus*.

Excavations by Robert Broom at Swartkrans revealed, for the first time, the co-existence of *Homo* and *Paranthropus*. Intriguingly, the early *Homo* species and the robust ape-men appear to have co-existed for hundreds of thousands of years, each grappling with rudimentary tool technology, but apparently occupying different ecological niches. Whereas the robust ape-men possibly used bone tools, early *Homo* was probably responsible for the manufacture of more complex stone tools. Thousands have been found within the Cradle of Humankind, the earliest dating back to around two million years ago (*see* 'The Stone Age', page 290).

The body size of early *Homo* is uncertain, but individuals were larger-brained than the robust australopithecines and had smaller premolar teeth, indicating that they may have been omnivorous, apparently adding substantially more meat to their diet than earlier hominins. A large brain is an expensive organ requiring a disproportionate share of the body's energy to maintain it. Meat is nutritious and for this reason was exploited by *Homo*. Exploiting this food source requires hunting or scavenging, so *Homo* developed tools and also probably collaborative foraging strategies. Tool use, language and co-operative social behaviour are interdependent, together promoting and maintaining a large and intelligent brain.

Homo habilis has long been associated with the crude Oldowan tool industry, the earliest stone tool technology consisting of hammerstones and simple flaked stone tools and choppers. Sites with Oldowan industries often occur close to water sources and have large quantities of butchered bones associated with artefacts. These occurrences have been interpreted by Mary Leakey as the first home bases. *Homo habilis* appears to have been a scavenger and not a hunter.

Beginning at about 1.7 million years ago, a dramatic shift in hominin cranial and skeletal structure took place. This new species had thick-walled bones with large muscle attachments, suggesting

Figure 10.8 *These bone tools are from the Swartkrans site.* **Homo erectus** *or* **Paranthropus** *used them for digging for food (bulbs and termites).*

THE STONE AGE

The Stone Age refers to the earliest human technology, when metals were unknown and tools were made of stone, wood, bone and horns. Stone tool technology is associated with prehistoric and modern humans, and is divided into three periods that together span the last 2.6 million years.

The **Early Stone Age** (ESA) dates to between 2.6 million and 200 000 years ago, and incorporates two distinct types of stone tools known as the Oldowan and Acheulian industries. The Oldowan is the older and more primitive industry dated to between 2.6 and 1.4 million years ago. Oldowan tools are typically pebble cores with a few flakes removed to form simple choppers. These are associated with *Homo habilis* (**A**). The Acheulian dates to between 1.5 million and 200 000 years ago and consists mainly of hand-axes and cleavers. Acheulian artifacts occur at sites with *Homo erectus* and early (archaic) *Homo sapiens*. Spanning more than a million years in Africa, the Acheulian changed remarkably little through time (**B**).

The **Middle Stone Age** (MSA) dates to between 200 000 and 35 000 years ago, and marks a transition to the production of flake tools struck from prepared cores. Thus in the Early Stone Age the central core was used as the tool, whereas in the Middle (and Later) Stone Ages flakes that were struck off the core were used as the tools. The tools are smaller than those of the Acheulian industry (**C**). Principal tool types of the Middle Stone Age industry are side scrapers and points. Middle Stone Age industries are associated with modern humans (*Homo sapiens sapiens*) who sometimes hafted the tools to make spears and knives.

Later Stone Age (LSA) industries emerged with the Cultural Revolution of *Homo sapiens sapiens* some 40 000 years ago. Artifact assemblages from neighbouring areas differed from each other because of varying cultural practices and resource availability. In a short period a rapid succession of industries occurred. These are characterised by a range of task-specific artifacts, many of which were hafted. Tools included stone blades, points and borers, generally small in size (**D**). Bone and ivory were regularly crafted into spear points, awls, needles and fishing hooks.

Oldowan stone tools.

Acheulian stone tools

Middle Stone Age tools

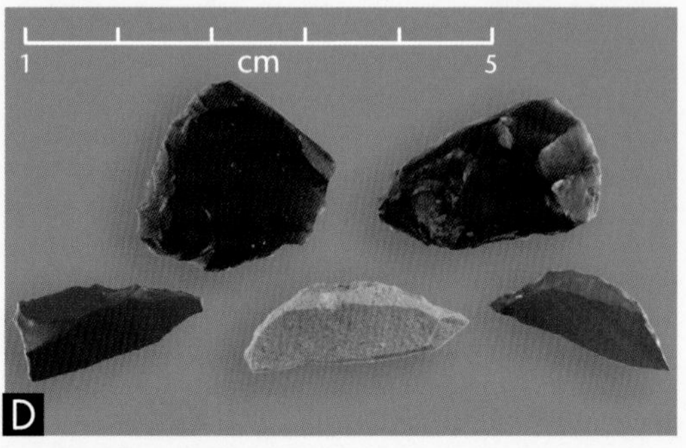

Later Stone Age tools

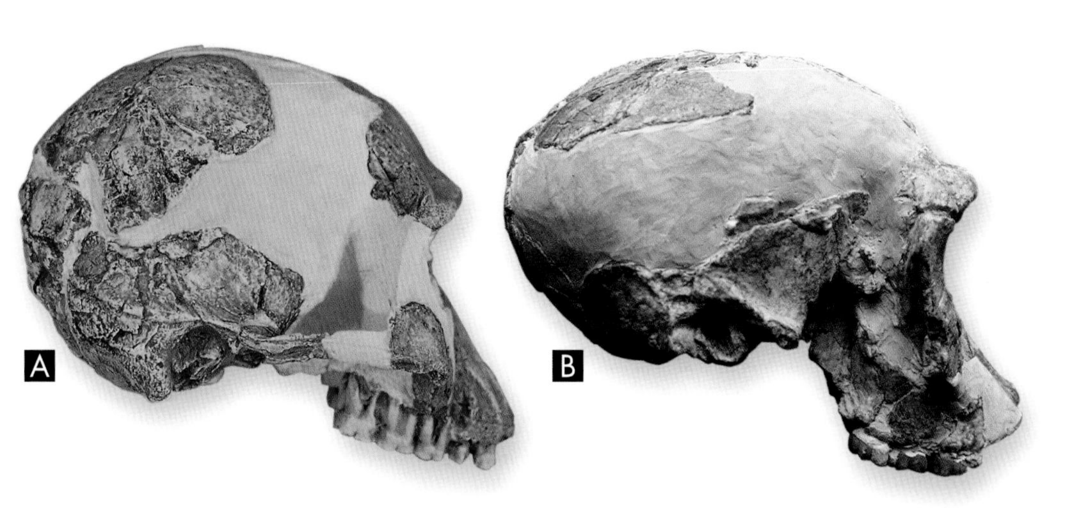

Figure 10.9 *The genus* Homo *appeared about 2.5 million years ago, becoming widespread throughout Africa. Well-preserved specimens have been found at Sterkfontein (***A***) and Olduvai (***B***).*

strong muscles and an active lifestyle. At 1.8 m tall and with a powerful build, it was the antithesis of the australopithecines. With the appearance of this species, *Homo erectus*, a true human ancestor is readily identifiable (*Homo ergaster* is often used as a synonym for an early African *Homo erectus*). Although slightly larger than modern humans, its brain was approximately three-quarters the size. The technology of *Homo erectus* is epitomised by the hand-axe, an often large, teardrop-shaped bifacial tool collectively referred to as the Acheulian culture (*see* **figure 1.1** and 'The Stone Age', left). This design template remained unchanged for a million years.

This species was the first to develop the controlled use of fire about a million years ago, a discovery made by Bob Brain in the younger deposits at the Swartkrans site. It is not clear if this was captured or manufactured fire, but most feel that it probably represents the first evidence of the ability to steal fire from natural bush fires caused by lightning strikes in the area and to transport the flames into caves for warmth, protection from predators, and possibly the cooking of food. This is evidence that hominids had now taken up residence in caves. The controlled use of fire marked a pivotal change in human behaviour as it signalled their ability to adapt the environment to suit their requirements. Bob Brain has proposed that fire was central to the development of complex social behaviour as it extended daylight hours and encouraged conversation and teaching. These advances were undoubtedly the product of a larger and more complex brain.

Increased brain size had a cost. The restricted diameter of the female birth canal requires that human infants are born before their brains are fully developed, unlike other primates. Based on the rate of brain growth, palaeoanthropologist Alan Walker has suggested that the gestation period for a human should be 21 months. Thus human infants are, in effect, born 12 months premature. The trade-off is prolonged dependence and parental investment in the infant. This trait was already in place in *Homo erectus* and must have appeared early in the evolution of the species. Increased dependency of infants on mothers would also have required a more coherent social structure given the diversity of diet of the species. This social structure may be linked to the establishment of home bases as envisaged by Mary Leakey for *Homo habilis*.

EARLY *HOMO* EMIGRATES

Within the period between 1.6 million and 500 000 years ago, *H. erectus* began to master the harsh and competitive conditions of the African veld. Their geographic range increased; some migrated out of Africa, crossing land bridges during periods of lower sea levels, caused by ice ages, and occupied

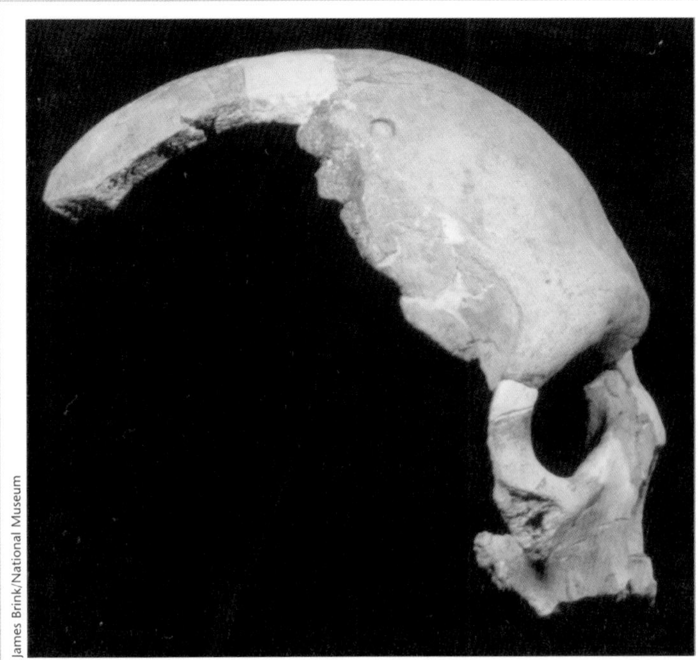
James Brink/National Museum

Figure 10.10 ***The skull of archaic* Homo sapiens *discovered by Thomas F Dreyer in 1932 at the warm water spring site of Florisbad near Bloemfontein is about 260 000 years old. Archaic* Homo sapiens *was replaced by anatomically modern* Homo sapiens *between 100 000 and 200 000 years ago.***

habitable northern hemisphere environments in Europe and Asia – the first hominins to do so. It is now even suggested that in Indonesia, *H. erectus* developed some level of sea-faring or rafting capability to travel between islands. The ability to make stone tools and eventually harness fire gave *Homo* the edge to master and transform the environment.

There is convincing evidence that an archaic form of *Homo sapiens* evolved from *Homo erectus* populations that remained in Africa 400 000 to 90 000 years ago. Rare fossils of this new form, possibly mastering the rudiments of human language and taking on the trappings of modern human behaviour, have been found in Africa. They had features intermediate between those of *Homo erectus* and modern humans. These include the *H. erectus*-like absence of a chin and *H. sapiens*-like expanded forehead to accommodate enlarged frontal lobes of the brain.

Remains of archaic *Homo sapiens* have been found at Florisbad in the Free State (**figure 10.10**), Hoedjiespunt in the Western Cape and Kabwe in Zambia. Evidence also comes from a number of sites along South Africa's southern Cape coastline. It is interesting to note that the Acheulian technology persisted among archaic *Homo sapiens*. These pre-modern humans were the forerunners of modern humans who appeared between 200 000 and 100 000 years ago.

HOMO SAPIENS SAPIENS ARRIVES

A major cultural transition occurred about 200 000 years ago when the Middle Stone Age flake tool industry replaced the classic Acheulian hand-axe culture of a million years standing (*see* 'The Stone Age', page 290). For the first time flake tools were hafted to form spears. Spear hunting of large game requires group co-ordination, which is indicative of complex social structures and language. These technological changes coincided with the appearance of anatomically modern humans.

Very few human skeletal remains from this time period are known. The oldest, dating back to 160 000 years ago, were discovered by Tim White at Herto in Ethiopia in 1997. Other old occurrences are from the South African sites of Klasies River Mouth (115 000 years) and Border Cave (90 000 years).

Modern humans are characterised by having an extremely large brain and small, flattened face arising from a reduced dependence on smell, and smaller teeth. Interestingly, *Homo sapiens sapiens* is the first hominid to have a chin, a feature that created an enlarged space below the tongue which became necessary to accommodate increasingly sophisticated language. The occurrence of cut marks on the hominid bones from both Herto and Klasies River Mouth suggest ritual or cannibalism.

Genetic evidence, examining both mitochondrial DNA (from the female line) and Y chromosome DNA (from the male line), as well as the fossil evidence, adds support to the theory that modern humans arose in Africa between 200 000 and 100 000 years ago and spread rapidly across the entire Old World. All modern humans appear to be descendants of a single woman, the so-called Mitochondrial Eve. She was one of a group of about 10 000 archaic *Homo sapiens* with advantageous traits (perhaps a genetic predisposition for language, or perhaps greater intelligence) that gave her descendants a selective advantage.

This population appears to have spread rapidly and did not interbreed with other archaic *Homo* species that existed at the time. Modern humans entered the Near East, Europe, Asia and Australia between 90 000 and 50 000 years ago, probably via coastal routes. By 30 000 years ago hominid diversity had apparently vanished and *Homo sapiens sapiens* was evidently the only survivor. Migration continued and humans arrived in the Americas as recently as 13 000 years ago and even later on islands in the Indian and Pacific oceans.

The earliest evidence of symbolism and personal ornamentation in the form of engraved ochres and shell beads (**figure 10.11**), dates back to 77 000 years at Blombos Cave in the Western Cape in South Africa. These developments heralded a global cultural revolution that swept through *Homo sapiens sapiens* with the transition to the Later Stone Age, commencing around 40 000 years ago. Arising from this revolution was a versatile toolkit, bows and arrows, appreciation of art and music, burial of the dead and evidence of trade.

EMERGENCE OF MODERN HUMANS: DID CLIMATE CHANGE PLAY A ROLE?

Accelerated speciation among the hominids, which started with *Homo erectus* and culminated with *Homo sapiens sapiens*, occurred over the last 1.8 million years. During this time brain capacity increased dramatically and as a result hominid lifestyle and tools became progressively more sophisticated. This period coincided with a major global climatic upheaval in the form of the Pleistocene ice ages, which began 1.8 million years ago (*see* Chapter 9).

Although the ice was concentrated on the continents of the northern hemisphere, the climate of Africa experienced drastic change. Southern Africa in particular became extremely arid. Because of the harsh desert-like conditions, it is likely that the ranges of *Homo* became restricted and populations became isolated, confined to more habitable areas, particularly coastal sites. Geographic isolation is recognised to be an important factor in the evolution of new species. The limited size of an isolated gene pool means that particular genetic traits that may emerge are not diluted and can even become amplified. It is therefore possible that the ice ages of the Pleistocene accelerated speciation and selected for intelligence among hominids.

The evidence of early modern culture (symbolism and personal ornamentation) from southern Cape coastal sites suggests that cultural modernity originated there. During the Pleistocene ice ages a vast sand desert lay in the interior of southern Africa, extending from the Atlantic in the west to the Drakensberg in the east, and from the southern Karoo to the Equator (*see* Chapter 9). This vast sand sea could have isolated early *Homo sapiens* communities living along the southern Cape coast from their kin to the north, except possibly for a strip along the east coast. But this route too may have been closed by dunes such as those in the Port

Figure 10.11 ***The earliest evidence of personal ornamentation, in the form of shell beads, was found at Blombos Cave in the Western Cape.***

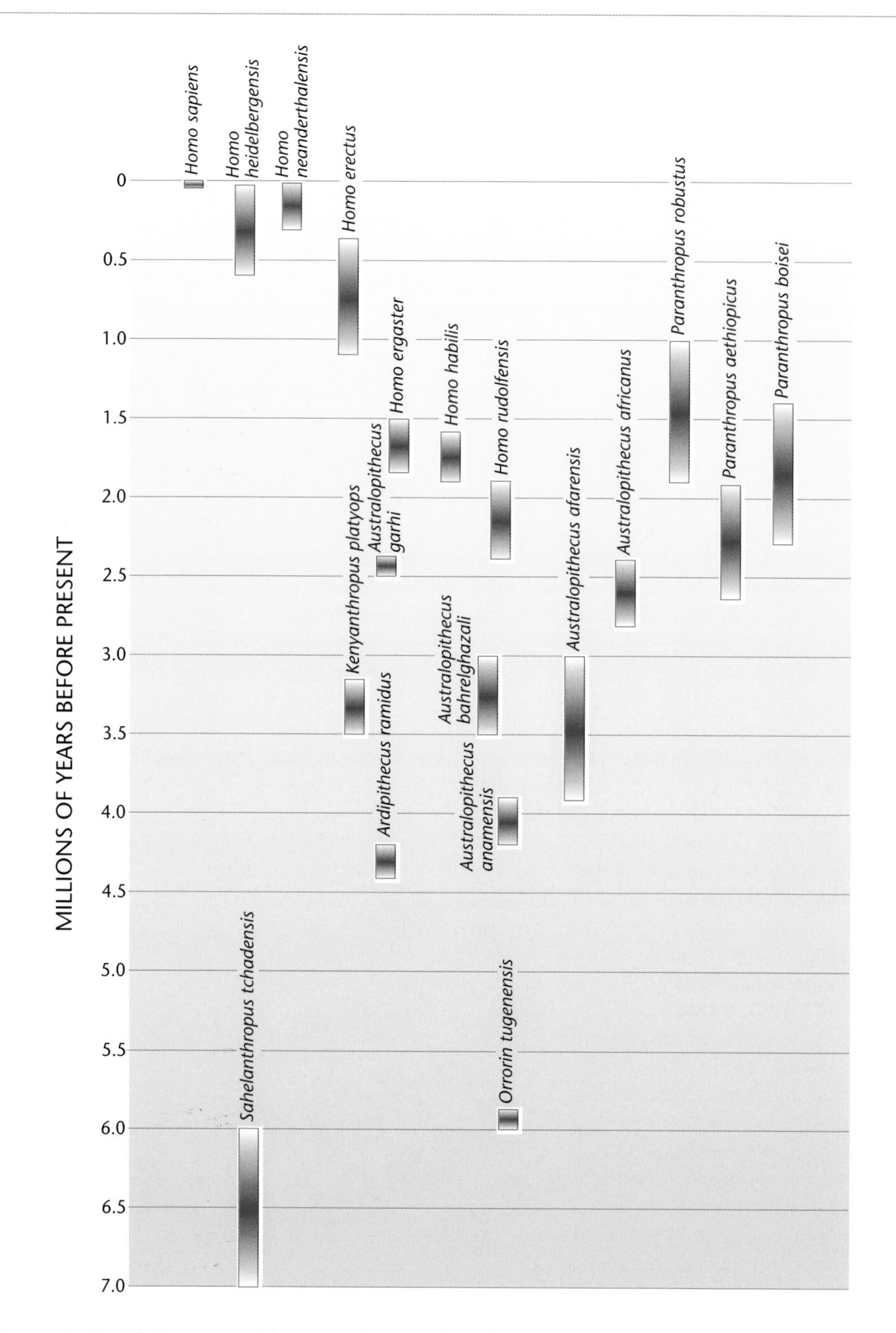

Figure 10.12 *This diagram illustrates the periods in the past during which different species of hominids lived.*

Elizabeth area today, which result from strong onshore winds. In this isolated community human culture may have evolved.

The dry conditions were periodically broken by brief globally wetter and warmer interludes (**figure 9.21**) at approximately 100 000-year intervals. Perhaps during the warm, wet period between 140 000 and 120 000 years ago the originators of this new culture were enabled to take their first steps to colonising the world – carrying with them the genetic signature of Mitochondrial Eve.

The end of the ice age about 10 000 years ago resulted in a globally more equable climate. This in turn heralded another cultural revolution characterised by a change from hunting and gathering to a sedentary lifestyle. Domestication of wild plants and animals enabled people to develop permanent settlements, and the first villages arose. The sedentary lifestyle made it possible for communities to support specialist craftsmen who developed pottery, metal smelting, irrigation systems, use of fibres for clothing, transport and trade. Populations increased rapidly, leading to a need for increased social organisation and record-keeping in the form of writing, and the first city-states ruled by kings and priests followed. These developments first appeared in the Middle East and later emerged independently in other parts of the world, heralding the dawn of our civilisation.

Early hominin evolution was an African phenomenon that took place on the emerging savannas of the continent. It seems likely that global cooling during the late Miocene – possibly accentuated by uplift of the African Super Swell and the associated drying – created the environmental pressures that triggered the origin of the early hominin family, along with a host of other savanna-adapted mammalian species. Unfortunately, there is an enormous gap in the fossil record between 14 and six million years ago. Exactly who gave rise to the hominin line and what they looked like remains a mystery.

Following the Miocene, hominins emerged as a highly successful African evolutionary line and radiated into a wide variety of niches, culminating in *Homo sapiens sapiens* as a globally distributed species. Episodic ice ages during this time and the associated aridification of parts of Africa may have played an important role in this radiation and in selecting more intelligent hominid forms.

Today we have a pretty good understanding of the big-picture of human evolution (**figure 10.12**), and new discoveries are being made all the time, shedding light on the details of the human family tree. Changes are occurring in the understanding of human origins because of fossil discoveries and technological innovation. These are so rapid that any attempt to be up to date is almost always thwarted by a new discovery, such as *Homo floresiensis*, a possible descendant of *Homo erectus*, discovered in 2004 in Indonesia and which may even still be living today on a remote island in the Indonesian archipelago.

SUGGESTED FURTHER READING

Brain, CK, 2004 (2nd ed). *Swartkrans: A Cave's Chronicle of Early Man*. Transvaal Museum Monograph no. 8, Pretoria.

Brain, CK. 1981. *The Hunters or the Hunted? An Introduction to African Cave Taphonomy*. University of Chicago Press.

Dawkins, R, 2004. *The Ancestor's Tale: a Pilgrimage to the Dawn of Life*, Weidenfeld & Nicolson, London.

Diamond, J. 1991. *The Rise and Fall of the Third Chimpanzee*. Radius Publishers, Great Britain.

Fleagle, J. 1999 (2nd ed). *Primate Adaptation and Evolution*. Academic Press, New York.

Johanson, D and Edgar, B. 2001. *From Lucy to Language*. Witwatersrand University Press, Johannesburg.

Klein, RG. 1989. *The Human Career. Human Biological and Cultural Origins*. University of Chicago Press, Chicago.

Leakey, R and Lewin, R. 1992. *Origins Reconsidered. In Search of What Makes Us Human*. Abacus Publishers, Great Britain.

MacRae, C. 1999. *Life Etched in Stone – the Fossils of South Africa*. Geological Society of South Africa, Johannesburg.

Walker, A and Shipman, P. 1996. *The Wisdom of Bones – in Search of Human Origins*. Weidenfeld & Nicholson, London.

11
THE FUTURE

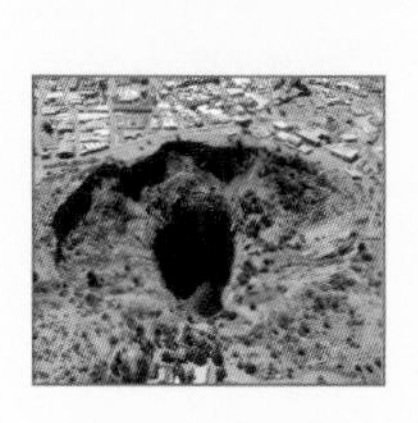

De Beers

The discovery of diamond-bearing kimberlite pipes in 1870–71 in Kimberley and elsewhere created the entrepreneurial and financial impetus for geological exploration of southern Africa.

WHY IS GEOLOGICAL KNOWLEDGE SIGNIFICANT?

This is a book about history, the history of the southern African region of the planet. To many of us, a history is of interest in itself. In the same way as it is of interest to understand how the social structure of South Africa has emerged through its political past, it is somehow satisfying to know how the physical character of the region came to be as it is. Such knowledge provides an added dimension to our perspective of our surroundings.

While knowledge of geological history is satisfying in itself, it also has practical uses. Perhaps one of the most important arises from the mineral wealth contained in rocks – and South Africa is particularly well-endowed in this respect. Study of the distribution and origin of South Africa's rock formations over the years has paid huge dividends to those who have taken the trouble to investigate them, and the discovery and development of the region's mineral wealth has provided great prosperity for the country as a whole. The investigations, and more importantly, discoveries, are far from over, notwithstanding more than a century and a half of mining activity. Changes in world demand for commodities, changes in technology, and improvements in our understanding of geology will ensure the continued development of new mining enterprises in the decades to come.

Knowledge of geological history can also provide some insight into the future. Although we can never predict the future, knowledge of conditions that prevailed in the past and what created these conditions, can give us some insight into what the future might hold.

Our geological history has a bearing on certain important philosophical issues. Perhaps central to these is our place in the Universe. During the Dark Ages, scientific thought fell under a religious dictatorship, echoes of which still reverberate today. The accepted view of the world was based on that proposed by the Greek natural philospher Ptolemy. The Earth lay at the centre of the Universe, and the Sun, Moon and planets revolved around it close-by, while the stars lay further away. Man had dominion over Earth, and was accountable only to God.

Nicolaus Copernicus displaced Earth from its imagined central place in the Universe, through his discovery that the Sun, not the Earth, lies at the centre of our Solar System. His heliocentric theory was supported by Johannes Kepler, who worked out that the planets move in elliptical rather than circular orbits. Their theories were supported by Galileo, who first observed the orbiting moons of Jupiter through his newly invented telescope. But Galileo paid a heavy price for his affirmation of these then heretical ideas, and had to publicly withdraw his support for a heliocentric solar system at his trial before the Holy Inquisition in 1633 (he was exonerated in 1992).

The truth is irrepressible. Since Galileo's time, astronomers have further reduced the status not only of Earth, but of the entire Solar System, through their investigations of the cosmos. Our Sun is now known to be a relatively insignificant star in a galaxy containing hundreds of billions of stars, many of which appear just like the Sun. Moreover, our Milky Way galaxy (**figure 11.1**) is one among millions of galaxies that populate the Universe. Our little Solar System is small fry indeed when viewed in the context of the vastness of Space.

While astronomical discoveries continue to erode our imagined status and importance in the Universe, it was the discovery of the techniques to read and date the messages contained in rocks that has perhaps had an even more profound effect on our worldview. We now know that the geological record is immensely long, and that our time on this planet has been extremely short. The Earth has been around for 4 600 million years, and has supported life for at least 4 000 million. In contrast, our species emerged on this planet only 200 000 years ago. To put these huge numbers in perspective, think of the 4.6 billion years of Earth history as a line 1 m long. The length of our history on this time line is just four hundredths of a millimetre.

Evolution and extinction

There is a school of thought that maintains that the evolution of life and intelligent life in particular – that is, us – was inevitable. This is known as the anthropic principle. The study of the physics of matter and the cosmos has revealed that the Universe is extremely finely tuned, down to the level of the properties of atomic particles such as the magnitudes of their charges and masses. Had these properties been

Figure 11.1 *The Sun is one of billions of stars that make up the Milky Way galaxy.*

just slightly different, the history of the Universe would have been quite different, and life would not exist. John Gribbon and Martin Rees, in their book *Cosmic Coincidences*, go so far as to say:

> *The conditions of our Universe really do seem to be uniquely suitable for life forms like ourselves.*

There are two facets to the anthropic principle. The strong version maintains that we are here because the Laws of the Universe made it inevitable that we should appear, a version that has been used to support the notion of a creative force. The general theme of these arguments is that the Universe was designed especially for our arrival. The weak version maintains that had the Laws been different, we would not be here to marvel at the fine-tuning of the Universe, and our being here has nothing to do with the tuning of the Universe, although this is a necessary condition for our existence.

The geological record has something to say about our being here. We now know that 99.9% of all species that have ever lived are extinct, that they failed to meet the challenges of an ever-changing world. Some species became extinct for no apparent reason, perhaps simply losing out in the ceaseless competition of daily life. The same thing happens to businesses all the time.

But many species disappeared in mass extinction events (**figure 11.2**), when environmental conditions changed abruptly – so fast that there simply was not enough time to adapt. The extinction toll in many of these events is frightening, especially in the so-called Big Five mass extinctions: 85% of species lost at the end of the Ordovician, about 80% at the end of the Devonian, 96% at the end of the Permian, 76% at the end of the Triassic, and 70–80% at the end of the Cretaceous.

The causes of most mass extinctions are not known. However, let us consider just one, the end-Cretaceous event, where the trigger was most probably an impact by a projectile from Space. The reign of the dinosaurs ended abruptly as the global ecosystem collapsed following the collision. As a group, they had dominated the world for

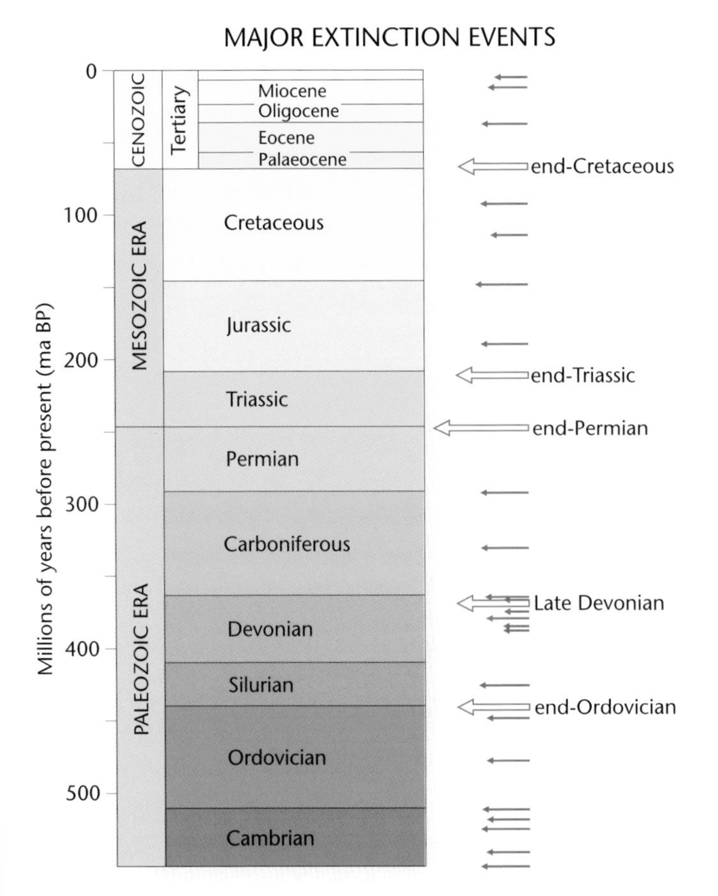

Figure 11.2 ***This diagram shows the principal extinction events over the last 540 million years.***

What makes humans different?

The evolution of intelligence itself is quite another thing. It is conceivable that had they not been destroyed, intelligence may have emerged among a species of dinosaur – and they were bipedal animals, too. There are some biologists who believe that animal evolution has direction, leading to ever-more complex forms, and inevitably an intelligent species will ultimately appear. Perhaps purely by chance that culmination happened to fall on a mammalian species – us.

The long record of both normal and mass extinctions warns us that in the grand scheme of things, *Homo sapiens* may not be very special, and that extinction could await our species as well. Strangely, this seems a terrible thing to contemplate, even though we as individuals only live for 70-odd years. Such is the power of our genetic programming; self-preservation is important to us, but equally important is the survival of our species. Genes may be selfish, as British biologist Richard Dawkins maintains, but that selfishness extends beyond the individual and the family group to the entire species.

Despite the odds, we are here, even though it may simply be a consequence of a cosmic accident 65 million years ago, coupled with several other chance events that subsequently impacted on the terrestrial environment. Geological history tells us that we are very different from anything that has ever lived before. The fossil record is full of repeated themes or strategies that species evolved time and again in the relentless arms race of life. We see in species after species over geological time the evolution of wings to enable flight (**figure 11.3**), of sharper eyesight to be better equipped to catch prey or avoid becoming prey; we see the evolution

more than 100 million years. Mammals co-existed with them, but dinosaur dominance of most ecological niches was such that mammals were forced to eke out a precarious living on the fringes of the dinosaurian world.

This changed radically after the Cretaceous, and mammals rose to dominate the new world in which we live today.

Now, suppose that that projectile from Space had missed Earth instead of colliding with it. What would the world be like today? Chances are that dinosaurs would still be the dominant species on Earth, and we would not be here to notice.

And this raises a fundamental question: how inevitable was the evolution of our form of mammalian intelligent life?

of thicker armour plate, bigger teeth, stronger legs, faster legs, and many more themes repeated over and over again in insects, therapsids, dinosaurs, reptiles, fish, birds and mammals. Invariably the outcome of these evolutionary strategies is more and more specialisation and ultimately extinction when environmental conditions change.

We seem to be different – a radical new experiment in evolution. As a mammalian species, we are not particularly strong or particularly fast. We have no in-built self-protection devices such as a set of horns or a thick hide. We have hairless bodies and no natural protection against the cold. We walk upright, exposing our soft bellies to predators. Our young are born weak and utterly defenceless, incapable of the simplest act of self-preservation, and need to be nursed and protected for years before they can fend for themselves. On the face of it, in fact, we actually seem to be severely physically disadvantaged.

But we have an arsenal of compensating abilities. We have hands with opposed thumbs that can grasp and manipulate objects, which, coupled with an upright posture, creates many advantages. We also have intelligence and have evolved the ability to use complex language.

Of course, all animals possess some degree of intelligence and are able to communicate with other members of their species. They can make decisions and anticipate actions of other animals such as prey or predators. Some can work as teams, behaviour often seen among prides of lions or marine mammals when hunting. But in *Homo sapiens*, evolution of the brain has honed intelligence and communication to unprecedented levels of sophistication, and coupled these with unique physical abilities.

It is this combination that makes us different from any species that has gone before. These abilities have made us extremely adaptable as a species, easily the most adaptable on Earth, and we survive in a wider range of habitats than any other mammalian species. Physically weak we may be, but our special abilities have made us the most formidable species ever to have walked the planet.

ANTICIPATING THE FUTURE

Our combination of specialised skills represents a completely new strategy in evolution, never tried before. It was a while in the making, five million years or so, but this is still short in the context of geological time. We are unique, so the geological record provides no parallels as to how a species like us has coped with life in the long term. But given our strong desire for preservation of our species, there are things we can and must learn from the geological record, especially that there are events that happen from time to time that could seriously affect us, so we need to be prepared.

Philosopher George Santayana remarked: *Those who fail to learn the lessons of history will be forced to repeat them.* This applies as much to geological history as any other kind. The scale of such geological events varies from the local to the global, but whatever the scale, the one outstanding fact is that we are powerless in the face of geological events. This does not mean, however, that we cannot take evasive action; advance warning is all that is needed.

Leonard Hoffmann / IOA

Dietmar Nil / Naturepl.com / Photo Access

Figure 11.3 ***The ability to fly has evolved independently in birds, insects and mammals.***

Predicting earthquakes

Perhaps the most destructive common geological events are earthquakes. Although earthquakes happen all the time, very few, perhaps only about 1 000 per year, are capable of inflicting serious damage, and many of these occur away from inhabited areas. Most earthquakes occur at the margins of tectonic plates (*see* Chapter 2), where plates move relative to each other, usually in the upper 20 km or so of the crust. Deeper-seated earthquakes, as deep as 700 km, occur occasionally at subduction zones.

An earthquake is generated when stresses in rock build up to the point where the rock ruptures, releasing pent-up energy as a shock wave that spreads outward at supersonic speeds, gradually dissipating. When the shock wave reaches the surface, it creates ground vibration that, if sufficiently strong, can cause severe damage to human constructions (**figure 11.4**), with consequent loss of life. As a Japanese philosopher once observed: *It is not earthquakes that kill people, but their possessions.* Aftershocks invariably follow a major earthquake, and although of lesser magnitude, can nevertheless still be very destructive.

Earthquakes are estimated to have killed 27 million people between 1949 and 1986 in China alone. The worst occurred on 27 July 1976 in Tangshan, killing between 655 000 and 800 000 people. Unfortunately, some of the most densely populated areas on Earth happen to be along seismically active tectonic plate margins, such as California, the Japanese islands, central China and the Mediterranean region.

Destruction due to ground motion is not the only threat posed by earthquakes. Massive landslides are a common associated feature, as are **tsunamis**, large waves triggered by undersea earthquakes (**figure 11.5**) or associated landslides. These waves travel at speeds of between 800 and 1 000 km per hour and are a major threat to coastal communities.

The worst disaster on record occurred on 26 December 2004, when a magnitude 9 earthquake off the northwest coast of the island of Sumatra in Indonesia triggered a massive tsunami. Wave surges up to 10 m high devastated coastal settlements around the northern Indian Ocean, completely swamping many of the smaller islands in the Indonesian archipelago. Villages and coastal resorts in Indonesia, Thailand, Sri Lanka and India were particularly badly affected. More than 280 000 people died, but the full death toll will probably never be known as entire villages were obliterated. Hundreds of tourists from around the globe, including several South Africans, perished in the catastrophe. Tsunamis are predictable, and much of this tragic loss of life could have been avoided had an early warning system been in place in the region.

Predicting the earthquakes that cause tsunamis is quite another matter. Considerable advances have been made in recent years in monitoring the build-up of stresses that presage an earthquake, pinpointing the exact spot where an earthquake starts, and how the rupture spreads. But earthquake zones are characterised by complex patterns of faults, making prediction impossible at this stage.

Terence McCarthy

Terence McCarthy

Figure 11.4 ***Extensive damage was caused to buildings in Tulbagh and Ceres in the Western Cape by 6.5 and 6.2 magnitude earthquakes in September 1969 and April 1970.***

Figure 11.5 *The 5-m-high fault scarp left by the 7.1 magnitude Madison Canyon earthquake in Montana, USA, in August 1959. When the sea floor is displaced in this way, it acts like a giant piston, creating a tsunami.*

So, although careful monitoring of seismic activity and measuring accumulating stress can reveal the stress build-up, it is not yet possible to predict when the rock will fail, where exactly the failure will occur, or the likely magnitude of energy release. Nevertheless, there is much at stake, and research in this field continues.

Better progress has been achieved in the field of earthquake engineering. Improved understanding of the ground motion associated with earthquakes and how this affects structures has resulted in safer building design, which has undoubtedly reduced the loss of life in earthquake-prone areas. While it may never be possible to reliably predict earthquakes, it probably will be possible to design buildings that can withstand all but the most extreme earthquakes.

Dealing with volcanic eruptions

Many areas at risk from earthquakes also face the threat of volcanic eruptions. More than 500 million people live close to the 550 active volcanoes on the Earth. The term active is misleading, though, as volcanoes thought to be extinct can spring to life. Mt Pinatubo in the Philippines erupted in 1991, having shown no signs of activity for 600 years, and 12 of the 16 largest explosive eruptions in the 19th and 20th centuries occurred at volcanoes with no known historical activity. About 60 volcanic eruptions occur on land each year, but this represents only about 20% of total activity, as most occur at ocean ridges on the sea floor. These undersea volcanoes pose little threat.

In contrast, those on land, and especially those associated with subduction zones, are extremely

Figure 11.6 ***Both the eruptions and the pyroclastic flows of volcanoes associated with subduction zones are extremely destructive.***

explosive (*see* Chapter 2). The eruptions of these volcanoes are dangerous in themselves, particularly the pyroclastic flows (**figure 11.6**) that they release. In addition, the steep sides of these volcanoes and the very fragmental nature of the ejected material make the volcanoes prone to severe debris avalanches or mud-flows brought on by heavy rain. Modern monitoring techniques have enabled geologists to achieve significant success in averting tragedies associated with these volcanic eruptions, and undoubtedly many lives have been saved. The loss of property will, however, never be avoided.

Dangers from earthquakes, landslides, volcanic eruptions and mud-flows really only affect people living at the margins of tectonic plates. Southern Africa is far from this type of activity and its inherent dangers. Things may well change in future if the hot blob beneath the African Super Swell rises nearer the surface. But that is probably millions of years away and for the moment the region is reasonably safe.

However, the danger arising from volcanic eruptions is not zero. The threat from explosive volcanoes is not always confined to those in the immediate vicinity of the eruption. Some eruptions have widespread effects, even global. The 1815 eruption of Tambora left 12 000 dead, but a further 80 000 people starved to death as a result of the ensuing famine. The real figure is probably much higher as 1815 was recorded world-wide as the year without a summer, when summer frosts blighted crops around the world. The reason for this was the dispersal of dust and gas, especially sulphur dioxide, into the upper atmosphere, which blocked out sunlight and lowered global temperatures.

Explosive volcanic eruptions of this type create another associated long-distance risk – that of tsunamis. Many explosive volcanoes occur on islands along the Ring of Fire that girdles the Pacific Ocean (*see* Chapter 2) and, being at sea level, explosive eruptions can create huge tsunamis capable of crossing oceans. Some 36 000 people died in Java and Sumatra as a result of the tsunamis generated by the explosion of Krakatoa in 1883.

Volcanoes related to mantle plumes, such as Hawaii or St Helena, produce basaltic lava, which is very fluid, allowing rapid gas escape. These volcanoes are not explosive and the very fluid nature of the lava results in relatively flat shield volcanoes. While the eruptions are often spectacular, producing fountains of lava hundreds of metres high, they pose little risk to the safety of nearby inhabitants, although they are a risk to property.

But there is a far more sinister danger associated with these volcanoes, one that today poses possibly one of the greatest threats on Earth in terms of potential loss of life. Although these volcanoes do not have steep sides, most are in the oceans where they build extremely high, conical mountains. The highest point on Earth's surface is Mt Everest, but the tallest mountain, measured from base to summit, is Mauna Kea in Hawaii, which reaches a height of 10 000 m from its base 5 500 m below the surface of the Pacific Ocean. These volcanic islands have a tendency to collapse under their own weight, creating huge landslides. Collapse is triggered when magma heats water that is trapped in the rock in the interior of the islands or by earthquakes associated with eruptions. The sea floor around volcanic islands like Hawaii is littered with the debris of these immense avalanches.

An island recently identified as potentially unstable is La Palma in the Canary Islands off the northwest African coast. If it collapses, it will send a catastrophic tsunami across the Atlantic, which will devastate the coasts of parts of Europe, the eastern seaboard of the Americas and the Caribbean. Investigations on islands in the Bahamas have revealed the presence of 2 000-tonne blocks of coral reef lying up to 20 m above sea level, washed there by tsunamis in the past. Such a tsunami today would cause incalculable damage and loss of life, and careful monitoring is essential. La Palma is not the only island threatening to collapse; on the Hawaiian islands there are also signs of impending failure. Past landslides in this archipelago washed marine deposits up to 375 m above sea level.

Attack from Space

Potentially one of the most serious threats we face on a global scale is the impact of a body from Space. This threat was not taken seriously by most until 1994, when the world watched live TV coverage of the Shoemaker-Levy 9 comets crashing into Jupiter (**figure 11.7**). Although these bodies were relatively small (100–200 m in diameter), each caused gigantic impact features, scars in the atmosphere of Jupiter larger than the Earth.

Had they hit Earth, the damage would have been less spectacular, perhaps making craters 1–3 km in diameter, similar in size to the 200 000-year-old Tswaing impact crater near Pretoria (**figure 4.33B**),

NASA

Figure 11.7 ***The vulnerability of Earth to impact by an extraterrestrial object was vividly brought home by the collision of Shoemaker-Levy 9 comets with Jupiter in 1994.***

or Roter Kamm in Namibia (**figure 11.8**). The shockwave through the ground would have been equivalent to a major earthquake, but the air blast and associated impact debris would have done most of the damage. Loose rocks, debris from buildings, cars and even trees would have become lethal projectiles and within a radius of 50 km around the site all life would have been destroyed. Had these bodies landed in the ocean – a more likely scenario given the fact that water covers 70% of the Earth's surface – the resulting tsunamis would have caused widespread devastation.

A somewhat larger (1 km in diameter) body would form a crater 10–20 km in diameter, but would destroy everything within a radius of about 500 km if it fell on land, and more extensive devastation if it fell in the ocean because of the resulting tsunamis. The impact of such large bodies has the added effect of throwing up a massive, globe-encircling dust cloud, which would severely affect Earth's climate, possibly for several years. It is this aspect that makes impacts potentially so destructive to life on Earth.

Calculated flux rates of extraterrestrial bodies onto Earth and the known impact crater record of planets and satellites in the Solar System provide a basis for probability estimates of how often impact events of different magnitudes could be expected to happen. According to these calculations an event leading to a global catastrophe, a mass extinction, will happen once every 100 million years or so. Smaller magnitude events, producing craters in the 20–50 km size range and capable of causing devastation on the continental scale, could occur once every 500 000 to a few million years. Smaller events of the kind that produced the Tswaing crater may occur as often as once to three times per hundred years.

The 20th century probably saw two of these smaller impacts: in 1908 in Siberia, where the Tunguska event destroyed 2 000 km^2 of tundra forest and killed reindeer herds and their keepers; and in 1930 when a Tunguska-like airburst of a 10- to 50-m object caused significant ground damage in Brazil's rainforest.

The type of object responsible for these two events has not been identified; in both cases it

Dion Brandt

Figure 11.8 *Roter Kamm is a 2.5 km diameter impact crater in southern Namibia. It is 3.4 million years old, and only the rim protrudes through the desert sand.*

appears that they exploded before hitting the ground, leaving no trace of what they were made of. This suggests they may have been comets (composed mainly of ice) rather than asteroids. Between 1990 and 1999, 23 objects between 2 m and 39 m in size are known to have hit the Earth. The two largest fell in the Pacific and caused no damage.

Awareness of the potential threat posed by an extraterrestrial collision has led to increased efforts to identify and plot the orbits of asteroids, in addition to the constant vigil maintained for comets (mainly by amateurs). There are several thousand asteroids known in the size range between 100 m and 1 000 m. The orbits of several classes of these, called the Apollo and Aten asteroids, are now known to cross that of the Earth. Some come quite close: in May 1996, a 400-m asteroid passed within 450 000 km of Earth (about as close as the Moon); and a 1.5-km asteroid, 1997 FX11, is destined to pass within 42 000 km of Earth on 26 October 2028.

We have a fair idea about the abundance and orbits of the larger asteroids, and the orbits of about 700 of the estimated 1 100 asteroids larger than 1 km in diameter are known. The smaller ones number in the millions, however, and will possibly never all be tracked. But they pose a lesser threat, and it is likely that such bodies would mostly break up on entering the Earth's atmosphere.

Figure 11.9 ***This Hubble telescope image of Mars shows a desolate planet. Detailed investigations indicate that frozen water may be present below the Martian surface.***

PANSPERMIA

It is clear from the rock record that life is almost as old as the oldest rocks on Earth. All life forms, from the most primitive archaebacteria to humans, perhaps the most complex life form, use the same molecular code to store genetic information. The inescapable conclusion is that all life on Earth has a common ancestry and that it arose only once (*see* Chapter 6). All living things are descendants from that first ancestor. Or so it presently appears.

The most essential life-supporting commodity is water. We once believed that life required water, an energy source and living conditions neither too hot nor too cold in order to survive – more or less in the range 5–50°C. Over time, the limits of too hot and too cold have been extended as we discover increasingly bizarre places that support life. We now know that life forms exist that can live comfortably at temperatures below the freezing point of pure water, while others live normally above its boiling point.

Organisms derive energy by using the electron transfer associated with changing the oxidation state of a substance. There seems to be no end to the range of substances that organisms can extract energy from – to name just a few: oxygen, hydrogen, hydrogen sulphide, methane, sulphur dioxide, sulphur, iron, iron oxides, and assorted organic compounds. Wherever a local chemical disequilibrium exists, life moves in to exploit it. Life can exist in what we would consider the harshest of places, provided there is water and a suitable energy source.

While we know that there is no intelligent life on other planetary bodies in our Solar System, or indeed within a radius of 40 light years of Earth, there may well be primitive life forms on other bodies in space. The current quest is to search for life on Mars (**figure 11.9**), which seems the most likely place in our immediate vicinity to look for it, especially now that

Figure 11.10 *The surface of Mars shows evidence that liquid water – a prerequisite for life – existed there.*

we are beginning to understand where water is likely to be found on Mars (**figure 11.10**). Meteorites derived from Mars contain microscopic features that have been interpreted – quite controversially, though – to be fossils of micro-organisms, suggesting that life was once present there.

If life is found on Mars, it may provide a test for the notion of panspermia, the idea that life first appeared somewhere in Space and that Earth was seeded from this extraterrestrial source. The litmus test will certainly be the genetic make-up of any extraterrestrial life forms we might find out there. Should it be the same as ours, panspermia becomes a viable explanation for the origins of life on Earth.

LIFE AND PLATE TECTONICS

Life and surface processes on Earth are inextricably linked. Opinions vary on how close the linkage is. We understand the interconnectedness of life and the atmosphere, but the linkage may run much deeper. The Earth alone among the inner or terrestrial planets exhibits plate tectonic activity. If plate tectonics were purely a geological process, related only to cooling of the planet, we might expect to see it on our neighbouring planets. But it is not there. What these planets lack is oceans; no oceans, no plate tectonics.

Plate tectonics may ultimately be driven by the interaction between the oceans and the hot interior of the Earth. The oceanic crust becomes hydrated at ocean ridges and dehydrated at subduction zones, forming dense rock that pulls the plates forward into the subduction zone. The released water escapes, and helps to induce melting to produce granite magma, which forms continents. Without oceans there would be no continents.

The surface temperature of the Earth has remained within a narrow range capable of sustaining oceans for more than four billion years, despite a 30% increase in solar radiation. Some believe the Earth's surface temperature is regulated by life (*see* Chapter 6) and in a sense, therefore, plate tectonics may be sustained by life. And, of course, it was at the hydrothermal vents on the ocean floors where life probably first began. Thus there may be a feedback loop between life and plate movement that has been operating for millions of years.

The movement of the continents and the formation of mountain ranges has constantly changed global climate and fragmented the ranges of species. This constant change has played a fundamental role in driving evolution. Nothing ever stays the same, neither the Earth, nor its climate, nor its life. Life's diversity is a product of plate tectonics, which may in turn be sustained by life, albeit indirectly. Life, the atmosphere, the oceans and the solid Earth may be operating as a single, integrated system.

SEX AND PLATE TECTONICS

Why do we have sex? Most of us think we know the answer, but the question has deep significance to evolutionary biologists, who have been pondering it for decades.

Organisms reproduce in a variety of different ways. Some, like ourselves, reproduce sexually, with separate male and female individuals. Others reproduce by cloning themselves in asexual reproduction.

Between these two extremes is a variety of intermediate reproductive strategies. For example, some organisms are able to alternate between sexual and asexual reproduction whereas others are both male and female at the same time, capable of mating with themselves or others.

Viewed in an evolutionary context, sexual reproduction exists because it offers some advantage over asexual reproduction. But herein lies a puzzle. Imagine two co-habiting species of organisms, identical in all respects except that one reproduces sexually and the other asexually. Among the asexually reproducing variety all individuals are capable of reproduction, so numbers will expand fast. Through normal, slow mutation, coupled with natural selection, the asexuals will become perfectly matched to their environment. Since the offspring are exact copies of the parent, successful adaptations will be passed on from one generation to the next.

In contrast, only the females of the sexually reproducing organism will be able to reproduce, so there will be fewer offspring. Each newborn has a scrambled mix of genes from both parents, and some will not be well matched to their environment and will not thrive. The phenomenon of upwardly mobile parents with downwardly mobile offspring is well known to all. In time, therefore, the asexually reproducing organism should completely out-compete and swamp the sexually reproducing variety.

So why is asexual reproduction not the norm? American biologist Christopher Wills believes environmental change may have something to do with it. In a sexually reproducing population there is wide genetic diversity in each generation. If environmental conditions change, provided the change is not too drastic, there will be some individuals who can cope and carry the species forward. In contrast, asexually reproducing organisms have limited genetic diversity and have only the slow normal mutation rate to adjust to environmental change. For them, environmental change could spell catastrophe.

The Earth's surface environment is constantly changing, partly as a result of plate tectonics. It is therefore possible that plate tectonics has played a role in ensuring the dominance of sexual reproduction among species with longer life spans.

HOW STABLE IS OUR WORLD?

We know that life has profoundly affected the composition of Earth's atmosphere over time, removing carbon dioxide, which became fixed in carbonate rocks and as carbon in sediments, and more importantly, releasing oxygen – ultimately producing an atmosphere rich in free oxygen and capable of supporting life forms such as ourselves. Oxygen is continually being added to the atmosphere by plants to replace that lost as a result of combustion, decay and respiration, and the plants maintain an oxygen concentration at a constant level of around 21%.

If the oxygen concentration rose much higher, to 25% or more, some believe there is a real risk of spontaneous combustion of organic material exposed to the atmosphere. But this is still a hotly debated topic. There are indications that oxygen abundance in the atmosphere may have exceeded 30% during the Carboniferous Period, when most of the northern hemisphere coal deposits formed in vast swamps.

If oxygen abundance were much lower than at present, we would be in trouble. At oxygen levels below about 15% we would suffocate. Nitrogen, too, is continually cycled by organisms, a process essential to life on Earth. We are an integral part of this atmospheric cycling system and live in a symbiotic relationship with plants in general. They supply our food and oxygen, and we supply them with carbon dioxide in return. Notwithstanding this total dependence on plants for the air we breathe and the food we eat, our behaviour as symbiotic partners on the planet has not always been exemplary. In fact, until relatively recently, we have considered ourselves separate from Nature, masters of the world, and have shown disrespect for our environment.

Global climate change

Destruction of forests, combustion of fossil fuels and other forms of pollution not only affect our immediate quality of life, but they may pose a long-term threat. Excessive emission of gases such as carbon dioxide, methane, nitrous oxide and chlorofluorocarbons (CFCs) has altered the composition of the atmosphere, and is changing some of the parameters that control climate. These gases appear to be contributing towards global warming.

Terence McCarthy

Figure 11.11 ***Sedimentary deposits such as these exposed by donga erosion near Burgersfort in Mpumalanga reflect major climatic changes in the region over the past 100 000 years. During dry periods, vegetation was sparse, and debris eroded from hillsides collected in valleys to form sand deposits (lighter layers). During wetter periods, soil formed (darker layers), only to be buried when the climate changed. Wet and dry periods lasted for many thousands of years.***

In addition, they appear to be destroying the ozone layer that protects terrestrial life from ultraviolet radiation from the Sun.

Since the beginning of the Industrial Revolution, the concentration of carbon dioxide and other greenhouse gases in the atmosphere has increased substantially. In addition, average global temperatures increased by approximately 0.5°C in the last century, and the sea level is rising. This realisation sparked international concern in the 1980s. Some projections are alarming: global temperatures may rise by as much as 5°C in the next century. Sea level may rise by up to a metre, swamping many low-lying regions such as Bangladesh, and posing threats to cities such as London and New York. Rainfall and vegetation patterns may alter substantially and ecosystems may be threatened. Already evident in South Africa is an increase in the severity of storms, which is likely to worsen further. Diseases like malaria could spread to Europe and North America.

Global climate change is a complex subject and depends on many factors in addition to the composition of the atmosphere, as we have seen in previous chapters. These include: variations in the energy emitted by the Sun itself; cyclical variations in the eccentricity (deviation from a circular orbit) of the Earth's orbit around the Sun, and in the angle of tilt of its axis of rotation; changes in the reflectivity of the Earth's atmosphere and surface; and changes in the positions of landmasses and ocean currents due to plate tectonic processes. Because of these complexities, understanding the climate and making predictions are fraught with problems. However, it is possible to reconstruct past climates, which gives us some perspective on current warming.

Meteorological records extend back for little more than 150 years, except in China where they go back several centuries. It is necessary to use other methods, known as **proxy methods**, to evaluate earlier climates. Tree rings store information on local rainfall and temperature conditions during the life of the tree and have been extended back as much as 6 000 years. Geological data provide important information on climatic conditions at the time different rock sequences were deposited and on the extent of ice sheets in the past. Analysis of pollen grains extracted from sedimentary rocks provides information on the types of vegetation growing in the past, and thus provides climatic information, as does the nature of the sediment itself (**figure 11.11**).

Study of the crater fill at the Tswaing meteorite impact crater near Pretoria, for example, has provided insight into climates over the last 200 000 years. The relative abundances of different isotopes of oxygen and hydrogen vary with temperature, and can be used to study past climate. Samples collected from stalagmites and stalactites provide particularly long records as they form very slowly. Ice cores from polar regions have also been used as a sample source. These have the added advantage of providing samples of the atmosphere at the time, trapped as small gas bubbles in the ice, and extend back about 800 000 years. Sediment cores recovered from the oceans provide even longer records.

The Tswaing crater sediments suggest that an important influence on climatic variation in the interior of southern Africa over the last 200 000 years has been the variation in radiation received from the Sun, caused mainly by variation in the angle of Earth's axis of rotation – the so-called precession of the equinoxes. These have resulted in variations in annual rainfall from less than 500 mm per annum to almost 900 mm per annum. The current rainfall of about 650 mm per annum is part of a rising trend. Antarctic ice core data and analysis of pollen data from the southern African interior show a marked rise in temperature in our region between 15 000 and 10 000 years ago, as the last ice age ended. Temperatures then fluctuated around a generally decreasing trend, reaching a low during the Little Ice Age between about 1500 and 1700 AD, a time when the River Thames would freeze over in winter.

Since then the temperature has been rising slowly. Current global warming may therefore be a consequence of a combination of factors, and only a component may be due to the rising concentrations of greenhouse gases in the atmosphere. The ice cores show that there is a clear relationship between temperature and carbon dioxide concentration in the atmosphere, with interglacial periods having higher carbon dioxide concentrations. Carbon dioxide concentration has varied widely over the past 420 000 years, due to unknown natural causes. Moreover, the concentration was already rising before the Industrial Revolution, but the rate of increase has accelerated dramatically since then.

Facing another ice age

Global warming is a source of concern because of its possible effects on climate, and especially on food production and sea level. Where it will lead in the long term, should it continue, is also of concern. Some have suggested that it may actually result in an ice age by upsetting global heat transfer systems. Under present conditions, cold water in the north Atlantic Ocean is transferred southward into the Indian and Pacific Oceans by deep ocean currents that originate off the coast of Greenland as the sea freezes during winter. As ice forms, the sea water becomes denser due to its higher salt content. This dense, cold water sinks and flows southward. The movement of this water draws in warmer surface water from the south via the Gulf Stream, which transfers heat to the north Atlantic.

Scientists argue that global warming will increase the amount of fresh water entering the north Atlantic from rivers in Canada and Greenland, which will dilute the surface ocean water. In addition, less ocean ice will form in the winter. The net effect will be a weakening of the deep ocean conveyor, and consequently a weakening of the Gulf Stream. Heat transfer from the Equator into the north Atlantic will fall, and Europe and North America will begin to freeze, initiating an ice age.

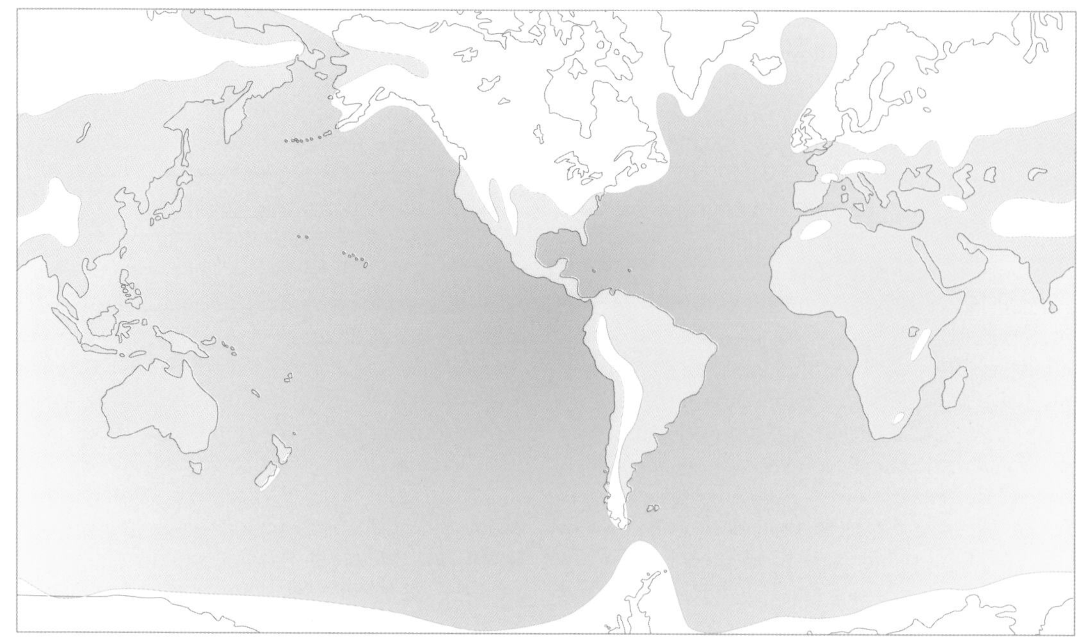

Figure 11.12 ***During the last ice age, which ended about 10 000 years ago, much of Europe and North America were buried under ice sheets hundreds of metres thick, as illustrated here. At that time, the sea level was about 130 m lower than now, and continental outlines would have differed somewhat from those shown here. A recurrence of such an ice age would be catastrophic for many nations.***

Should an ice age similar to the most recent one occur, it will be catastrophic for our civilisation, and is likely to cause immense social and political upheaval as displaced nations try to find somewhere to live (**figure 11.12**). In the event of an ice age, the climate in South Africa will become hyper-arid and agriculture will probably not be possible over most of the interior. So the region will not be immune to its physical effects, quite apart from the likely global social upheaval that will follow in its wake.

Throughout the whole debate about global warming and the greenhouse effect is an underlying notion that the world's temperature and climate should stay exactly as it is. That is the way we want it to be. The geological record tells us that this is a vain and naive hope. Instead, it tells us that dramatic change is inevitable. We have been living in an abnormally warm and climatically stable period and cannot expect it to last indefinitely. We may be contaminating the atmosphere in a way that could influence global climate, but there are other forces at work that we do not yet fully understand, which have caused major changes in climate in the past.

IS THE EARTH'S MAGNETIC SHIELD IN JEOPARDY?

The Earth is one of several planets that possess a magnetic field (*see* 'Earth's magnetism' on page 36), behaving as if it had a bar magnet inside aligned more or less along the axis of rotation. The field is actually very weak; in fact it is several hundred times weaker than that exerted by a toy horseshoe magnet. Nevertheless it is very important to us. The field extends about 60 000 km into Space on the sunlit side of the Earth, and much further on the dark side. It is believed to arise in the liquid outer core of the Earth some 3 000 km below the surface from flowing streams of molten metal.

The Earth is bathed in a constant stream of high-energy particles emanating from the Sun, known as the **Solar Wind**. Many of these particles are electrically charged; in fact, most consist of protons (positively charged) and electrons (negatively charged). This stream of particles deforms the Earth's magnetic field into the shape of a tear-drop, the tail pointing away from the Sun. A magnetic field exerts a force on a moving charged particle.

You can demonstrate this by bringing a magnet up to a TV or computer screen (CRT or cathode-ray tube); the presence of the magnet causes distortion of the image on the screen. This happens because the screen is illuminated by a stream of electrons fired from an electron gun at the back of the tube: the magnetic field exerts a force on the electron stream, altering its path and causing distortion.

The Earth's magnetic field similarly influences charged particles in the Solar Wind, deflecting them away from the Earth. In effect, the Earth's magnetic field acts as a giant shield. Every now and then, activity on the surface of the Sun increases dramatically (**figure 11.13**), causing a sudden increase in the intensity of the Solar Wind. This in turn can cause disturbances in the Earth's magnetic field, so-called **magnetic storms**. These are rapid variations in the strength of the field as it changes shape in response to gusts in the Solar Wind. Minor disturbances produce aurorae in the sky at high latitudes. More severe storms cause disturbances in radio communications, while very severe storms can cause major disruption.

The worst case on record occurred in March 1989, when a severe magnetic storm cut off electricity to six million people in Quebec, Canada, as a result of its effect on power transmission lines. Such storms can also cause problems with radio and telephone communications and with radar and radio navigation systems. But only extreme magnetic storms have this effect, because the Earth's magnetic field shields us.

We know from studies of rocks that the Earth's magnetic field has periodically switched around (*see* 'Earth's magnetism' on page 36) in events known as magnetic reversals. During these events, the Earth's magnetic field changes direction: the magnetic North Pole becomes the South Pole, and *vice versa*. It is not that the Earth flips around; rather it is as if the imaginary bar magnet inside the Earth has flipped around. If such a reversal were to occur now, the North Pole of compasses would point south. The last such reversal occurred about 780 000 years ago and there is growing evidence to suggest we may be on the brink of another such reversal.

It has been known for many years that the strength of the Earth's magnetic field is not the same everywhere: in particular, there is a region over the South Atlantic where the field is very weak – essentially a hole in the magnetic shield. According to Pieter Kotze of the Hermanus Magnetic Observatory in the Western Cape, charged particles approach Earth much closer in this region than anywhere else on Earth, resulting in intense radiation. Although this does not affect the Earth's surface, it could potentially cause severe problems for astronauts and satellites in low orbits, and as a result it is the subject of intense study.

The cause of the hole in the magnetic field has now been pinned down: the Earth's magnetic field in a localised patch below the southern Cape and the adjacent Atlantic has actually reversed, and has become a local north pole. Compasses in the southern Cape still point in the right direction, though, because the overall magnetic field of the Earth is still dominant over this small patch. But the patch is growing and the strength of the Earth's

NASA

Figure 11.13 ***Occasionally, violent flares of hot gas rise from the Sun's surface and are channelled by its magnetic fields into great arcs. At these times, the Solar Wind fluctuates wildly, causing chaotic oscillations in the Earth's magnetic field.***

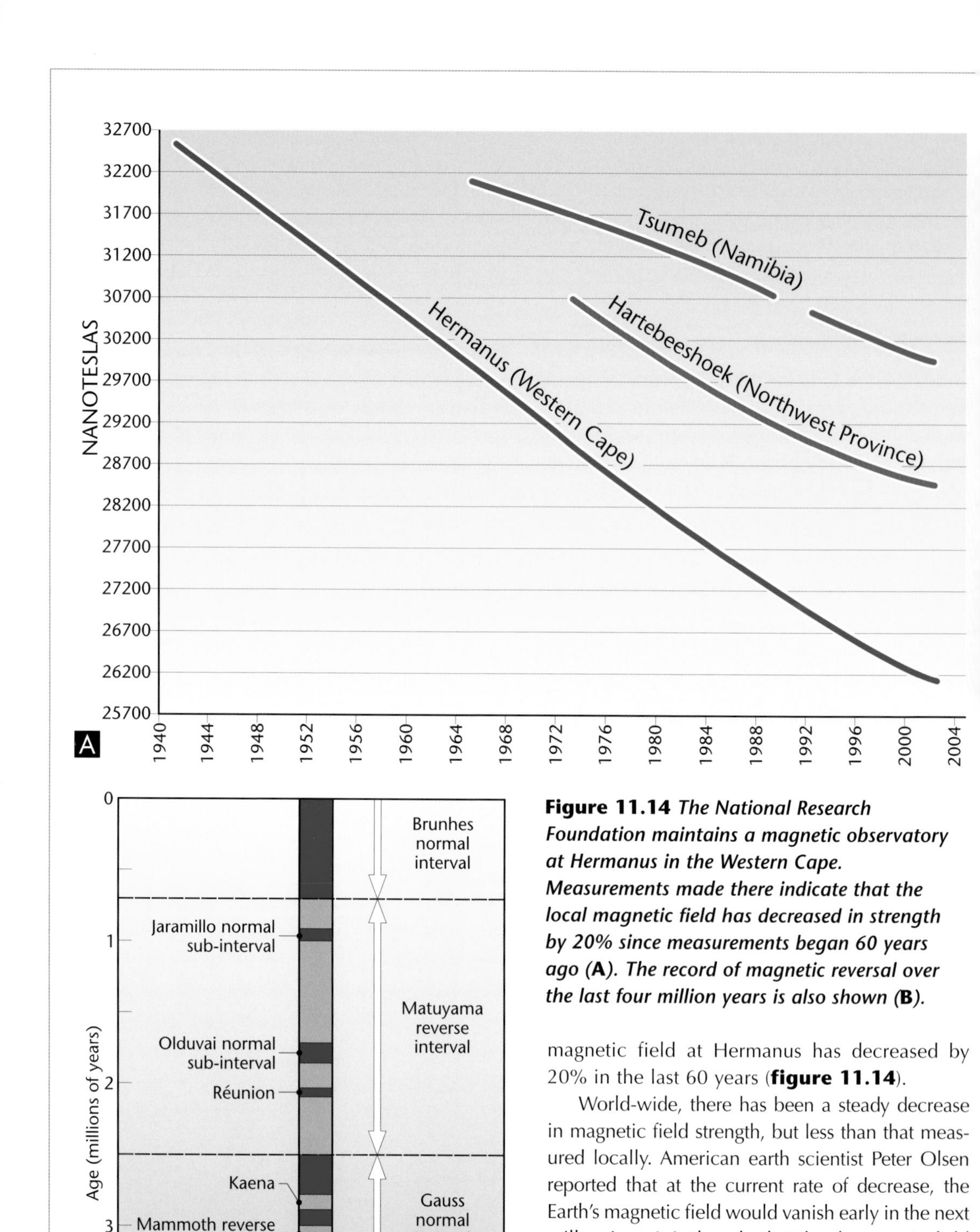

Figure 11.14 *The National Research Foundation maintains a magnetic observatory at Hermanus in the Western Cape. Measurements made there indicate that the local magnetic field has decreased in strength by 20% since measurements began 60 years ago (**A**). The record of magnetic reversal over the last four million years is also shown (**B**).*

magnetic field at Hermanus has decreased by 20% in the last 60 years (**figure 11.14**).

World-wide, there has been a steady decrease in magnetic field strength, but less than that measured locally. American earth scientist Peter Olsen reported that at the current rate of decrease, the Earth's magnetic field would vanish early in the next millennium. It is thought that the decrease in field strength may be a reflection of an incipient magnetic reversal. It is hypothesised that reversals come about as patches of reversed magnetism such as that below the southern Cape form and grow, weakening the overall field. Ultimately they coalesce and form a new field, but in the reverse direction. The time required to form the new field is not known. Of course, this decrease in field strength may simply be a natural oscillation in the field, and the field may recover. We have insufficient data to be able to predict which way events are likely to turn out.

Effects of a weakened magnetic field

As matters stand, we are faced with the prospect of a steady weakening of the Earth's magnetic field, especially in South Africa. What are the consequences of this likely to be? We know that there have been many reversals in the past, and the palaeontological record tells us that life was not unduly affected – there were no mass extinctions associated with these reversals. The atmosphere will in all probability provide some protection from the solar radiation.

Weakening and disappearance of the field may result in the loss of the ozone shield. This possibility arises from observations made during intense solar activity in October and November 2003. At that time, an intense flux of electrons from the Solar Wind was funnelled down into the upper atmosphere over the North Pole by the Earth's magnetic field lines. The electrons ionised nitrogen, which reacted with oxygen to produce nitrous oxides. These in turn destroyed ozone, resulting in a massive ozone hole over the northern hemisphere. A globally weakened magnetic field will allow widespread penetration of electrons from the Solar Wind into the upper atmosphere where nitrous oxide formation may occur, resulting in extensive loss of the ozone shield.

But loss of the ozone layer and an increase in solar radiation may be the least of our worries. Ours is a very technological world, and we have become dependent on sophisticated electronics. How will this technology shape up as the protective magnetic field steadily weakens? Will power grids become increasingly sensitive to even mild magnetic storms? How will communications and computers be affected? Will satellites still be able to operate effectively over our region? Compasses will be useless for navigation, but will the Global Positioning System (GPS) still be able to function? How will high-flying aircraft be affected as the hole in the magnetic field deepens?

These are issues that will need to be addressed in the near future because of the possible risks to our way of life. At the same time we need to intensify our research on the magnetic field: study of the Earth's magnetic field, especially in South Africa, should be on the agenda alongside or even ahead of global warming and the ozone hole. We need to know what is going to happen because our very survival may depend on it. If we understand the dangers we can take the necessary evasive action.

THE SIXTH EXTINCTION

Our great advantage in evolutionary terms has been our intellect and consequent ability to manipulate the world around us. But will this turn out to be our undoing? The release of greenhouse gases into the atmosphere is one of many environmentally damaging acts committed by humankind, but most serious is the destruction of natural habitats as the world's population grows and consumption of resources steadily increases.

While extinction is a natural, ongoing process, our activities have accelerated the natural rate of extinction, some estimate by more than a thousand-fold. The International Botanical Congress predicts that at current rates about two thirds of all bird, mammal, butterfly and plant species will be extinct by the end of this century. The problem appears to have begun way back in the Pleistocene, during the time that hominins (*Homo erectus*) acquired the ability to make and use tools, opening the way for a change from being scavengers to being predators.

Coincident with the development of tool use and the spread of *Homo* from their birthplace in Africa, we see in the fossil record the decline and eventual extinction of virtually all of the mega-fauna on the planet – animals such as mammoths, woolly rhinoceros, *Deinotherium* (an elephant with downward-pointing tusks in the lower jaw) and countless others.

The extinction of this mega-fauna passed like a wave around the globe, striking different continents at different times, as *Homo erectus* and especially *Homo sapiens* radiated from Africa. Australia began to experience extinction about 60 000 years ago, while it commenced in the Americas about 12 000 years ago, Madagascar 2 000 years ago and New Zealand only 1 000 years ago, in each case co-incident with the arrival of humans.

The mega-faunal extinction in North America was particularly severe and 75% of large herbivore genera were lost. Those apparently quintessential large beasts of North America – the bison, grizzly bear, moose and elk – in fact only arrived along with the first humans 13 000 years ago from Asia, when low sea levels (caused by the ice age) created a land bridge between Alaska and Asia. Having co-habited with humans for perhaps 100 000 years in Asia, these large animals had defence strategies against human hunting hard-wired into their brains, and they rapidly expanded in numbers to fill the vacuum as the indigenous fauna was wiped out.

The reason large animals are particularly vulnerable to extinction, even at modest rates of predation, is their very long gestation period and the fact that they generally produce one young at a time. But even so, how was it possible for early humans to kill large animals like woolly mammoths and mastodons with the primitive weapons at their disposal? We are conditioned to think of wild animals as being timid because they flee when approached by humans. This was probably not the case when humans first arrived on a new continent. There, the local animals had no experience of humans and would probably have been totally unafraid.

Annie Dillard, in her collection of essays *Teaching a Stone to Talk,* describes this phenomenon on a visit to the Galapagos Islands. Here, animals such as the famous marine lizards (**figure 11.15**) have no fear whatsoever of humans. Their only natural enemies live in the ocean and when they are afraid they will not enter the water. They do not seem to be able to acquire a fear of humans either. Charles Darwin describes an encounter with one of these lizards in his book *Voyage of the Beagle.* He picked up a large lizard lazing on a rock and threw it into the sea. The lizard promptly swam back and climbed back onto the rock. Darwin continued to throw the reptile back repeatedly, but it simply did not develop any fear of him. He wrote:

> *Perhaps this singular piece of apparent stupidity may be accounted for by the circumstance that this reptile has no enemy whatever on shore, whereas at sea it must often fall prey to the numerous sharks. Hence, probably, urged by a fixed and hereditary instinct that the shore is its place of safety, whatever the emergency may be, it there takes refuge.*

Anyone who wished to kill one of these lizards could simply walk up and dispatch it. This complete lack of fear of humans led to the extinction of many animal species across the Pacific islands, starting with the arrival of the Polynesians, and later, sailors from European countries.

Perhaps it was the same with the mega-fauna of the continents. Early man may have simply walked up to these large animals and stabbed them to death with spears. Africa was to a significant degree spared this extinction (only 13% of genera were lost), perhaps because the mega-fauna of this continent co-evolved with humans and developed a healthy respect for our hunting abilities. Sadly, the

Figure 11.15 *These marine lizards of the Galapagos Islands have no instinctive fear of humans. Similar behaviour probably accounts for the extinction of larger animals that coincided with the migration of humans across the globe.*

invention of the rifle has tipped the balance and African mega-fauna, too, are now threatened.

While our ancestors were purely hunter-gatherers, it was primarily the mega-fauna that bore the brunt of our activities, but this changed with the development of agriculture. Thereafter, the onslaught took on a more insidious form as bush was cleared to make farmland and fences were erected. Hunting continued to take its toll, but the real destruction was to the natural habitat, and the fragmentation of what remained. In these new circumstances, the majority of species affected were not those used as a food source, and included small mammals, insects, birds and reptiles. As the human population has expanded and industrialisation spread, the problem has intensified, and today we are experiencing what amounts to a global extinction event, which anthropologist Richard Leakey has termed the **Sixth Extinction**.

Survival or extinction?

Will we be among the survivors of this extinction, given our symbiotic existence on Earth? Or will we destroy mammalian life and make way for a new age, the Age of Insects? We urgently need to heed the warnings contained in the geological record and understand our symbiotic dependence on life at large – and our inherent vulnerability in the face of geological change. Unless we establish a more sustainable way of living and co-existing with our fellow species, we – and many of them as well – are destined for early extinction.

Our evolutionary success is a result of our extraordinary intelligence coupled with our manual dexterity and the fact that we are generalist feeders. In the past, these traits have stood us in good stead through some severe environmental crises, notably a succession of ice ages. But in recent times we have embarked on a road of increasing specialisation, a specialisation rooted in technology. This technology has been developed during a protracted period of terrestrial and astrophysical stability.

As we proceed down this road of technological specialisation, we need to be constantly aware of two very important messages from the geological record. The first of these is that species that become super-specialised become vulnerable, especially during environmental change. The second is that there have been environmental changes in the past that could seriously affect our technology, especially reversals in the earth's magnetic field.

Research in the earth sciences has made a major contribution to human development by providing the raw materials on which our society depends. It has also contributed to our intellectual progress. It has played a significant role in our understanding of life, especially evolution. Through geological research, we are beginning to gain deeper insights into the world around us, how it came to be the way it is and how it may change in the future. We no longer have to appeal to capricious gods to protect us from natural disasters such as earthquakes and volcanic eruptions: we understand them and are slowly developing the means to predict them.

Earth science is a relatively new discipline, and the revolutionary plate tectonic paradigm, which has given us a unifying framework in which to comprehend the functioning of the Earth we live in, is only about 40 years old. There is still a lot to learn. New techniques are constantly being developed with which to probe and examine the Earth, and our understanding is growing rapidly. In the longer term, research in the earth sciences offers the prospect of providing a comprehensive knowledge and understanding of our environment, knowledge that can be used to develop a more sustainable way of living.

SUGGESTED FURTHER READING

Barrow, JD. 2003. *The Constants of Nature*. Vintage Press, London.

Diamond, J. 2002. *The Rise and Fall of the Third Chimpanzee*. Vintage Press, London.

Flannery, T. 2001. *The Eternal Frontier*. Atlantic Monthly Press, New York.

Leakey, R and Lewin, R. 1996. *The Sixth Extinction*. Phoenix, London.

Newson, L. 1998. *Atlas of the World's Worst Natural Disasters*. Penguin, Toronto.

Steel, D. 2000. *Target Earth: the Search for Rogue Asteroids and Doomsday Comets that Threaten our Planet*. Readers Digest, New York.

Glossary

Abyssal plain Flat areas of the ocean floor

Acheulian period Division of the Stone Age, extending from about 1.65 million to about 200 000 years ago

Agate Very fine-grained quartz, usually showing concentric colour banding; formed in cavities in rock (often gas bubbles trapped in lava)

Alluvial plain Broad, flat region formed by coalesced river floodplains

Ammonite Extinct diverse group of floating or swimming marine molluscs with coiled, chambered shells used for regulating buoyancy; closely related to modern squids, cuttlefish, octopus and *Nautilus*

Amphibian Four-limbed vertebrates that remain dependent on water for egg-laying and breeding, including modern frogs, newts and salamanders; the ancestral group to reptiles

Amygdales Former gas bubbles in lava (vesicles) that have become filled by minerals

Anapsid Member of a group of reptiles having a skull structure in which there are no accessory openings behind the eyes to accommodate jaw muscles

Andesite Fine-grained igneous rock (volcanic) consisting mainly of amphibole (calcium, aluminium, iron and magnesium silicate) and plagioclase (calcium, sodium, aluminium silicate)

Apatite Mineral consisting of calcium phosphate

Arachnid Member of the group of arthropods to which spiders and mites belong, characterised by having four pairs of legs and an external skeleton of chitin

Archaea, Archaebacteria Primitive form of bacteria

Archaean Eon Period of Earth history extending from 4 600 million to 2 500 million years ago

Archosauria The so-called Ruling Reptiles that include the ancestors of all crocodiles, dinosaurs and birds, characterised by two accessory openings in the skull behind the eye to accommodate jaw muscles; many showed a strong trend towards a bipedal (two-legged) running gait

Arctica Early continent that consisted of parts of North America, Siberia and Greenland, believed to have formed between 2 500 million and 1 500 million years ago

Arthropod Member of the great group of invertebrate animals that includes crustaceans (shrimps, lobsters, crabs, etc.), arachnids (spiders, mites) and insects, all characterised by having an external skeleton made of a hard substance called chitin, and a variable number of jointed legs; many insects also develop wings in the adult stages and can fly

Asteroid Small bodies (less than 1 000 km diameter) orbiting the Sun between Mars and Jupiter

Asthenosphere Zone within the Earth between depths of about 70 and 200 km possessing relatively lower seismic velocities, probably due to its very plastic nature; also referred to as the Low Velocity Zone (or Layer) or LVZ

Atlantica Early continent that consisted of parts of eastern South America and western Africa, believed to have formed about 2 000 million years ago

Banded iron formation Sedimentary rock consisting of alternating layers of fine, white quartz (silicon dioxide) and black and red oxides of iron

Basalt Fine-grained igneous rock (volcanic) consisting of plagioclase (calcium aluminium silicate) and pyroxene (calcium, magnesium, iron silicate)

Basement Rocks, usually granite or related rocks, that lie beneath the oldest layered rocks (usually sedimentary) in a region

Basin Local depression in the earth's surface

Batholith Large body (greater than 100 km^2) of intrusive igneous rock

Bioherm Mound-like or reef-like structure at sea built up by marine organisms (e.g. algae, corals)

Biosphere Thin envelope around the earth, at or near its surface, in which life can exist

Blastoid Extinct members of the group of marine invertebrates known as the Echinodermata, which had an external shell

Blastular embryo Hollow ball of cells formed in the early stages of the development of Metazoan animals

Brachiopod Member of the group of two-shelled marine invertebrates commonly known as lamp shells, in which the two shells or valves are of unequal sizes; superficially similar in appearance to mussels (which are molluscs)

Bract A modified, usually reduced leaflike structure, normally associated with a flower or reproductive part of a plant

Braided river River with a complex pattern of channels that diverge and rejoin, separated by sand or gravel bars; they form in cases where rivers are supplied with more sediment than they can effectively transport

Breccia Rock consisting of angular fragments set in a matrix of finer material

Calcite Mineral consisting of calcium carbonate

Chert Sedimentary rock consisting of very fine-grained quartz (silicon dioxide) formed by chemical precipitation from water

Chromite Mineral consisting of an oxide of iron and chromium

Clastic sedimentary rock Sedimentary rock consisting of discrete particles that were moved as separate grains to the site of deposition

Clubmosses Group of early land vascular plants with microphylls (primitive leaves) growing in whorls from the stems; they have spores in cone-like structures. Today only herbaceous moss-like clubmosses survive but during the Carboniferous some grew to tree size. Also known as lycophytes, lycopods and formally as Lycopsida

Coal Sedimentary rock consisting mainly of carbon, formed by the accumulation and chemical alteration of plant matter

Columbia Supercontinent that may have existed on Earth around 1 800 million years ago

Comet Large body consisting mainly of ice laced with dust that orbits the Sun in a very eccentric orbit

Conglomerate Coarse-grained sedimentary rock consisting of pebbles, cobbles or boulders with sandy material filling the spaces between larger particles

Conifers One of the groups of gymnosperms or naked seed plants; e.g. pines, spruces, firs and yellowwoods; all are woody plants, frequently trees, and produce cones

Continental crust The Earth's crust beneath continents; consisting mainly of granite and related rocks, it is usually about 35 km thick, but may be thicker beneath mountain ranges

Continental drift Theory that continents move with respect to each other

Continental shelf The fringes of continents that are submerged below sea level; the edge of the shelf is usually marked by an abrupt increase in slope onto the Continental slope that passes down to the abyssal plain

Cordaitales Group of extinct spore-producing plants with long, strap-shaped leaves; probably ancestral to the conifers

Core Innermost portion of the Earth, believed to consist of nickel and iron; the inner core is solid and the outer core liquid

Cratons The stable, ancient cores of continents; usually made up of granodiorites and greenstone belts

Crinoid Member of the group of attached, stalked echinoderms commonly known as sea lilies

Cross bedding Stratification (layering) in a sedimentary rock that is inclined to the surface on which the sediment was deposited; formed by the migration of ripples or dunes across the sediment surface

Crust Outermost layer of the Earth situated above the Mohorovičić Discontinuity

Cuticle Protective outer layer secreted by the outer skin of plants and arthropods

Cyanobacteria Group of micro-organisms capable of oxygen-producing photosynthesis

Cycadeoid Group of extinct plants that physically resembles the cycads but differs in the reproductive cycle

Cynodont Group of advanced mammal-like reptiles (therapsids) from which mammals are considered to have evolved

Cynognathus Flesh-eating cynodont mammal-like reptile, which was the size of a large wolf

Cystoid Extinct members of the group of marine invertebrates, known as the Echinodermata, which had an external shell

Dacite Fine-grained igneous rock (volcanic) consisting of feldspar (sodium and potassium aluminium silicate), quartz and amphibole (calcium, magnesium, iron, aluminium silicate)

Diamictite Sedimentary rock whose particles are completely unsorted by size, consisting of both very fine-grained and large particles; often deposited under glacial conditions

Diapirism Slow, upward movement of a less-dense mass of material through a denser medium (e.g. a hot air balloon)

Diapsid Member of the group of reptiles characterised by two accessory openings in the skull behind each eye, to accommodate jaw muscles; includes all the Archosauria, as well as the ancestors of all modern lizards and snakes

Dicroidium Generic name for the leaves of the extinct seed fern that has bifurcating fronds and was dominant during deposition of the Molteno Formation (Upper Triassic Period)

Dicynodont Large and diverse group of mammal-like reptiles (therapsids) which lived from the middle Permian to the Late Triassic; these animals were the dominant land-living herbivores of the Permian Period and were characterised by a horny beak rather than many teeth

Differentiation Processes by which crystals are separated from magma as it crystallises, resulting in the formation of layers of rocks of different mineral composition

Dinocephalian Diverse group of carnivorous and herbivorous mammal-like reptiles (therapsids) that lived during the Permian; they were the first large vertebrates to live on land, and formed an important component of the earliest therapsid faunas of South Africa and Russia

Diorite Coarse-grained igneous rock (plutonic) consisting mainly of amphibole (calcium, aluminium, iron and magnesium silicate) and plagioclase (calcium, sodium, aluminium silicate)

Dolerite Medium-grained igneous rock consisting of plagioclase (calcium aluminium silicate) and pyroxene (calcium, magnesium, iron silicate); occurs as dykes or sills

Dolomite Mineral, calcium magnesium carbonate, or rock entirely comprising grains of this mineral

Dome Uplift that is more or less circular in plan, with rock layers dipping away from its centre

Dyke Tabular intrusive rock that cuts across other rocks

Echinodermata Group of marine invertebrates that have a five-fold body symmetry, members of which develop armour in the form of hard shells and spines (the group name means spiny skins); includes such familiar forms as starfish, sea urchins, sea cucumbers and sea lilies

Echinoid Member of the group of marine invertebrates known as the Echinodermata, which bears an external shell and an armour of moveable spikes; includes the familiar forms known as sea urchins, sand dollars and pansy shells

Ediacaran fauna Late Precambrian multicellular animals constituting earliest macroscopic fauna

Eubacteria Primitive form of bacteria

Eukaryote Organisms whose cells have a complex structure, including a nucleus and organelles

Fault Surface along which a rock mass has been broken and displaced

Feldspar Group of minerals consisting of sodium, calcium and potassium aluminium silicates; plagioclase is the name given to the sodium, calcium variety

Fold Bend in a layer of rock

Foraminifera Small marine organisms that usually have calcareous shells with minute holes; belong to a subdivision of the Protozoa

Foreland basin Depression formed on the landward side of the mountain range lying along the collision zone between oceanic and continental plates

Fractional crystallisation *See* differentiation

Fracture zone Linear zone along which many faults are developed; the zone of fracture extending from a transform fault across the ocean floor

Gabbro Coarse-grained igneous rock consisting of plagioclase (calcium aluminium silicate) and pyroxene (calcium, magnesium, iron silicate)

Gastropod Member of the diverse group of molluscs that includes the snails, slugs and their relatives; some live on land (land snails and slugs), but the majority are aquatic (fresh water: mainly snails, and marine – e.g. whelks, periwinkles, marine snails, etc.)

Ginkgo Generic name of the group of gymnospermous plants that were abundant in the past; today only *Ginkgo biloba* survives; the formal name for the group is Ginkgoopsida

Glacial striations Grooves and scratches on a rock surface formed by glaciers dragging rock fragments over the surface

Glossopterids Members of the family to which the extinct seed fern, *Glossopteris,* belongs

Glossopteris Generic name for the leaves of a group of seed ferns that dominated the floras of Gondwana during the Permian Period; remains of this plant make up the bulk of southern hemisphere coal deposits today

Gneiss Coarse-grained metamorphic rock with a distinct banding of lighter and darker layers; its mineral composition is usually similar to granite

Gondwana Ancient continental landmass made up from the present landmasses of Africa, South America, the Falkland Islands, India, Australia, New Zealand, Antarctica and Madagascar that is believed to have formed about 500 million years ago

Gorgonopsian Group of flesh-eating mammal-like reptiles (therapsids) that lived during the late Permian and became extinct at the end of the Permian Period

Graben Elongated block of rock that has been lowered by faulting relative to the surrounding blocks

Granite Coarse-grained igneous rock consisting mainly of quartz (silicon dioxide) and alkali feldspar (sodium and potassium aluminium silicates)

Granodiorite Coarse-grained igneous rock consisting of feldspar (mainly sodium aluminium silicate), minor quartz (silicon dioxide) and amphibole (calcium, iron, magnesium aluminium silicate)

Granophyre Fine-grained igneous rock consisting mainly of quartz (silicon dioxide) and alkali feldspar (sodium and potassium aluminium silicates); quartz and feldspar crystals form a distinctive interlocking texture

Greenstone Metamorphic rocks composed of minerals with a distinctly greenish colour, usually a product of metamorphism of basalt

Greenstone belt Large, linear mass of greenstone; a characteristic feature of the geology of ancient cratons

Grenville Belt Linear region of metamorphic rocks in eastern Canada formed about 1 100 million years ago

Group Term used in stratigraphy to denote a related group of layered rocks

Gutenberg Discontinuity Sudden change in seismic velocity that occurs at the boundary between Earth's mantle and core

Gymnosperm Seed plant with seeds not enclosed in an ovary (e.g. conifers, cycads, cycadeoids and ginkgos); means naked seed (from Greek)

Hadean Era Geological period forming the earliest part of the Archaean Eon, including final stages of formation of Earth by meteorite bombardment (4 600 million to 3 800 million years ago)

Hominin Group of primates that includes all the bipedal forms as well as humans, i.e. all the fossil genera such as *Australopithecus, Ardipithecus, Kenyanthropus* and *Homo*

Horsetails Group of seedless vascular land plants with primitive leaves growing from nodes (resembling bamboo); today only the genus *Equisetum* survives but in the past some plants grew to tree size

Horst Elongated block of rock that is elevated by faulting relative to the surrounding rocks

Hotspot The expression at the Earth's surface of a mantle plume (column of rising, hot material from the deep mantle), usually taking the form of localised, intense volcanic activity

Hydrosphere Total layer of liquid water at and near the Earth's surface

Hydrothermal fluid Hot groundwater containing various dissolved substances

Hydrothermal vent Point of discharge of hydrothermal fluid, usually on the sea floor

Iron formation Sedimentary rock consisting mainly of iron oxide minerals

Island arc Arc of volcanic islands (e.g. Japanese archipelago, Aleutian Islands) formed where one plate containing oceanic crust is subducted beneath another such plate

Isostasy Balance of all large portions of the Earth's crust as if floating on a denser, plastic sub-layer

Joint Fracture in a rock mass along which no significant relative movement has taken place

Kaapvaal Craton Ancient core of the southern African continent, formed between 3 600 and 3 100 million years ago; consists mainly of granodiorite batholiths and greenstone belts

Kibaran Belt Linear zones of metamorphic rocks on the African continent formed about 1 200 to 1 000 million years ago

Kimberlite Rare igneous rock consisting of silicate minerals rich in magnesium, iron and alkali metals; it is the source rock of diamonds

Komatiite Fine-grained igneous (volcanic) rock consisting of silicate minerals rich in magnesium and iron (mainly olivine and pyroxene)

Lahar Avalanche of debris consisting of rocks of various sizes in mud

Lava Molten rock erupted from a volcano; the term is also used to describe the solidified form

Layered intrusion Igneous intrusion consisting of layers of rock of different mineralogical composition

Lithification Process by which loose sediment is converted into sedimentary rock; cementation and compaction are the two most important processes involved

Lithosphere Relatively rigid outer layer of Earth, including the crust and upper part of the mantle

Liverworts Group of non-vascular, seedless land plants resembling broad-leaved mosses that grow in damp places

Low velocity layer (zone) *See* asthenosphere

LVZ *See* asthenosphere

Lystrosaurus Genus of dicynodont mammal-like reptiles (therapsids) living at the time of the Permo-Triassic extinction; one of the few land-living vertebrates to have survived this calamity

Magma Molten rock before it erupts on surface

Mammal-like reptile Common name used for the group of primitive synapsid tetrapods that are the distant ancestors of mammals; comprises two groups, pelycosaurs and their descendants the therapsids

Magma chamber Large space below the Earth's surface occupied by molten rock

Mantle Region of the Earth's interior between the base of the crust and the core

Mantle plume Hot, buoyant mass of material that rises from deep within the Earth to the base of the lithosphere

Matrix Relatively finer-grained material occupying the spaces between larger particles in a rock

Meandering river River channel pattern characterised by broad loops

Relatively small (30-cm-long) anapsid marine reptile that lived during the Early Permian Period; its remains are known only from Brazil, Namibia and South Africa

Mesosphere Zone of the Earth's mantle below the asthenosphere

Metamorphic belt Linear zone on the Earth's crust underlain by metamorphic rocks

Metamorphic rock Rock formed from pre-existing rocks that have been subjected to elevated temperature and pressure (and usually involving hot fluids), such that their mineral composition has undergone change

Metamorphism Alteration of mineral composition of rock caused by exposure to elevated temperature and pressure (and possibly hot fluids)

Metazoa Sub-kingdom of animals whose bodies are made up of many specialised cells grouped together into tissues, and that have a centralised, co-ordinating nervous system

Meteorite Particle of natural material that has fallen to Earth from Space

Microfossil Remains of tiny plants or animals that require the use of a microscope to be studied; includes fossil bacteria, algae, pollens, spores, protozoans, microscopic crustaceans, etc.

Mid-ocean ridge Mountain range on the sea floor where tectonic plates separate

Migmatite Rock consisting of an intimate mixture of metamorphic and igneous rocks (usually granite)

Mineral Naturally occurring chemical compound

Mobile belt Term often used synonymously with Metamorphic belt

Moho *See* Mohorovičić discontinuity

Mohorovičić Discontinuity Shallowest major seismic discontinuity below the Earth's surface, which marks the base of the Earth's crust; it arises from the difference in chemical composition between the crust and the mantle

Mountain belt Long mountain chain (e.g. Andes)

Nebula Mass of dust and gas in Space

Nucellus Tissue in a plant seed that comprises the chief part in which the embryo sac develops, e.g. the edible part of a peanut

Nena Continent consisting of what are today the Baltic, Ukraine, Europe, Antarctica, Greenland and a portion of North America, which is believed to have formed between 1 600 and 1 300 million years ago

Oceanic crust Crust that underlies the oceans (excluding continental shelves); it is 5–7 km thick and consists of basalt and gabbro

Ornithischia Herbivorous branch of the two great dinosaur lineages, possessing a four-pronged (tetra-radiate) pelvic structure, and including such well-known forms as *Stegosaurus, Iguanodon, Triceratops, Parasaurolophus*, etc.

Palaeomagnetism Magnetism that was imprinted in rocks when they formed; its measurement enables determination of the orientation and location of the rock mass relative to the Earth's magnetic poles at the time the rock formed

Pangaea Supercontinent consisting of all of the Earth's present continents; believed to have formed about 300 million years ago

Panspermia Notion that life on Earth was seeded from Space

Pareiasaurid Family of plant-eating early anapsid reptiles that formed an important component of the Permian reptilian fauna; their bodies were covered by large bony scutes (scales) and they had characteristically serrated teeth

Pediment Gently sloping erosion surface at the base of a mountain

Phanerozoic Eon Period of Earth history extending from 545 million years ago to the present

Photosynthesis Metabolic process in which light or heat energy is used to convert carbon dioxide and water to carbohydrate

Pillow lava Pillow-shaped masses of lava formed during undersea volcanic eruptions, or when erupting lava issues into a lake or other body of water

Placoderm Group of early fish that had jaws

Plagioclase Mineral consisting of sodium and calcium aluminium silicate

Planetesimals Small bodies that grow into planets by a process of accretion

Plate Tectonics Theory that proposes that the outer layer of the Earth (lithosphere) consists of separate, rigid plates (in which the continents are embedded) that move relative to each other and in the process move the continents

Plate boundary Region where the rigid plates that form the outer layer of the Earth come into contact; at these boundaries plates may converge, diverge or slide past each other

Plesiosaur Group of reptiles that were adapted to an aquatic life, having a long neck and paddles; they lived during the Mesozoic Era

Pluton A body (less than 100 cm^2) of intrusive igneous rock

Procaryote Life form consisting of simple, single cells, such as cyanobacteria, that are not specialised and do not have organelles or a nucleus inside them

Procolophonid Family of plant-eating early anapsid reptiles common during the Triassic Period in Antarctica, Argentina, China and South Africa

Prosauropod Primitive plant-eating member of the early Saurischian dinosaurs, including the ancestors of the later gigantic brontosaurs

Proterozoic Era Period of Earth history extending from 2 500 million to 545 million years ago

Pseudotachylite Dark, fine-grained rock, often containing fragments of other rocks, which occurs in the form of dykes; believed to form by frictional melting of the surrounding rock

Pterosaur Member of a group of bat-like flying reptiles that co-existed with the dinosaurs, and also became extinct at the end of the Cretaceous; often referred to as pterodactyls, but this name properly applies to only one branch of the pterosaurs

Pyroclastic flow Avalanche of hot, often incandescent, volcanic rock and dust, together with superheated gases, emanating from a volcano

Pyroxene An iron, magnesium silicate; calcium-bearing and calcium-free varieties exist

Quartzite Sedimentary rock comprising sand grains (composed of the mineral quartz) in which the grains interlock due to recrystallisation (usually a result of heating), making the rock very hard

Rauisuchid Member of a group of primitive land-dwelling crocodile-like archosaurs, which were the top ambush-predators of larger-bodied land animals during the Triassic Period

Red bed Sandstone or siltstone with a red colour, caused by the presence of red oxides of iron

Reef (1) A structure built by marine organisms; (2) a planar body of rock enriched in a mineral of economic importance (e.g. gold)

Refractory brick Brick made from materials having extremely high melting temperatures; used to line the inside of metallurgical furnaces

Rheological properties Properties relating to flow and deformation

Rhynchosaur Member of a group of curious, pig-sized, plant-eating archosaurs with highly specialised teeth; relatively common in some parts of the world during the Triassic Period

Rhyolite Fine-grained igneous rock (volcanic) consisting mainly of quartz (silicon dioxide) and alkali feldspar (sodium and potassium aluminium silicates)

Rift valley Valley of regional extent formed by the collapse of a fault-bounded central zone

Ring dyke Dyke with a circular plan-form

Ripple marks Small waves on a sand surface produced by flowing water or wind; often preserved in sedimentary rocks

Rodinia Supercontinent believed to have formed about 1 100 million years ago

Saurischia Largely carnivorous branch of the two great dinosaur lineages, which possessed a three-pronged (tri-radiate) pelvic structure, and included such well-known forms as *Compsognathus, Allosaurus, Velociraptor,* and *Tyrannosaurus,* etc.; this line includes the ancestors of the birds; one saurischian line consisted of plant-eaters: the prosauropods and sauropods

Sauropoda Group of well-known gigantic herbivorous dinosaurs of the Mesozoic Period, including such familiar forms as *Brachiosaurus, Brontosaurus, Seismosaurus,* etc.

Sea-floor spreading Theory that new ocean floor (oceanic crust) is being created at, and is spreading away from, mid-ocean ridges

Seamount Conical mountain rising from the sea floor; probably submerged volcanoes

Sedimentary rock Rock formed by the accumulation and consolidation of sediment

Sedimentary structure Structure (e.g. ripple marks, cross bedding) that formed in sediment at the time of deposition; often preserved in the sediment after it has been converted to rock

Seismic discontinuity An interface within the Earth at which the velocity of seismic waves suddenly changes

Seismic wave Vibration within the Earth created by release of energy either on or below the surface

Shaman Priest, witchdoctor or diviner

Shatter cone Nested conical joint surfaces in rock

Shield volcano Large volcano shaped like a circular, medieval shield, constructed by repeated flows of fluid lava (usually basalt); the slopes seldom exceed 10°

Silicate mineral Mineral in which silicon is a major constituent

Sill Sheet-like body of igneous rock intruded between layers of older rock

Smoker Hot spring on the sea floor where cooling of the spring water induces precipitation of dissolved solids, producing smoke-like clouds in the water; particles are often black (termed a black smoker)

Spherule Spherical particle, often concentrically layered

Sporangia *(pl.)*, **sporangium** *(sing.)* Hollow unicellular or multicellular structure in a plant, in which spores are produced

Spore Reproductive cell of a plant, usually unicellular, capable of developing into an adult without fusion with another cell; reproduction in seedless plants is by spores, and in seed plants by pollen fusing with an egg cell

Stomata *(pl.)* **stoma** *(sing.)* Minute opening bordered by guard cells in the epidermis of leaves and stems of plants through which gases pass (critical for photosynthesis and respiration)

Stratigraphy Study and documentation of rock sequences

Strato-volcano Conical volcano made up of alternating layers of lava and volcanic ash

Stratum *(sing.)*, **strata** *(pl.)* Layer of sedimentary rock

Stromatolite Mound-like growth of carbonate rock, due to accumulation of calcium carbonate crystals on slimy algal or bacterial colonies

Subduction Subsidence of the edge of a plate into the mantle

Subgroup Term used in stratigraphy to denote a related group of layered rocks: subordinate to a Group

Supercontinent Single continent formed by the amalgamation of several previously separate continental masses

Supergroup Group of rock strata formed during a single, major and widespread episode of rock accumulation

Supernova Tremendous stellar explosion involving the almost total destruction of a star

Suture Line of junction between two adjoining geological terranes

Synapsid Member of the group of reptiles characterised by the presence of only one accessory opening in the skull behind each eye, to accommodate jaw muscles; includes all mammals and their extinct ancestors, pelycosaurs and therapsids

Temporal opening Accessory opening behind each eye socket in the skull of a vertebrate animal, to accommodate more complex and powerful jaw muscles (*see also* Anapsid, Diapsid, Synapsid)

Terrane Region having similar geological formations and history

Tethys Sea Sea that is believed to have once separated Gondwana and Laurasia, the closing of which gave rise to the Alps and the Himalayan mountains

Tetrapod Vertebrate animals possessing four limbs instead of fins, i.e. amphibians, reptiles, mammals and birds

Therapsid Diverse group of mammal-like reptiles from which mammals evolved

Thermal subsidence Subsidence of the lithosphere (including the crust) caused by cooling; occurs where the lithosphere has been thinned by stretching, bringing the hot asthenosphere closer to the Earth's surface. As the asthenosphere cools, its density increases and it subsides, causing subsidence of the lithosphere

Thermophile Heat-loving

Therocephalian Diverse group of long-ranging mammal-like reptiles (therapsids) that lived from the Permian to the mid-Triassic; they varied in size from that of a meerkat to that of a large dog

Theropoda Two-legged, meat-eating dinosaurs belonging to the Saurischia, including the smallest dinosaurs known, but also the largest land-dwelling meat eaters of all time – the tyrannosaurids; this group of dinosaurs included the ancestors of birds

Tidal flat Large, flat area of land that is inundated at high tide; tidal flats are usually covered by mud, silt and sand

Transform faults Strike-slip (wrench) faults usually lying perpendicular to a mid-ocean ridge that compensate for different spreading rates along the length of the ridge

Trench Elongated depression on the sea floor formed above a subducting plate

Trilobite Early form of scuttling marine arthropod, distantly related to lobsters, crabs, etc., and including the first animals to possess eyes capable of true sight; their bodies are characteristically divided into three longitudinal rows – hence the group name

Tsunami Very large ocean wave caused by a major geological disturbance such as an earthquake, volcanic explosion or landslide; tsunamis travel at 800–1 000 km/h; often incorrectly referred to as tidal waves

Turbidite Sedimentary deposit formed by a turbidity current

Turbidity current Sediment-rich slurry that cascades down the continental slope onto the abyssal plain below

Ubendian Belt Belt of metamorphic rocks formed 1 800 million years ago in southern and central Africa

Ur Believed to be the Earth's oldest continent, consisting of cratonic areas of South Africa, Madagascar, India and Western Australia; believed to have formed between 3 000 million and 1 500 million years ago

Vascular system (of plants) System of cells and tissues for the uptake and transport of fluids and dissolved nutrients within a plant (the plumbing); non-vascular plants have to live in moist places because they do not have a well-developed system for the internal transport of fluids

Vein (1) In geology: small, dyke-like intrusive body; (2) in zoology: blood vessel carrying blood towards the heart; (3) in botany: vascular system in leaves

Vermiculite Silicate mineral that readily breaks into flat sheets (micaceous) that greatly expand on heating; used in horticulture and insulation

Volcanic ash Dust-sized particles of lava released from volcanoes during eruptions

Weathering Chemical decomposition of rocks as a result of exposure to the atmosphere

Index

THE MAKING OF SOUTH AFRICA

This series of maps illustrates the growth of southern Africa over time. The present extent of the major rock formations is shown, but each was probably more extensive. The coastline of South Africa is shown for reference only – the present coastline only formed 90 million years ago.

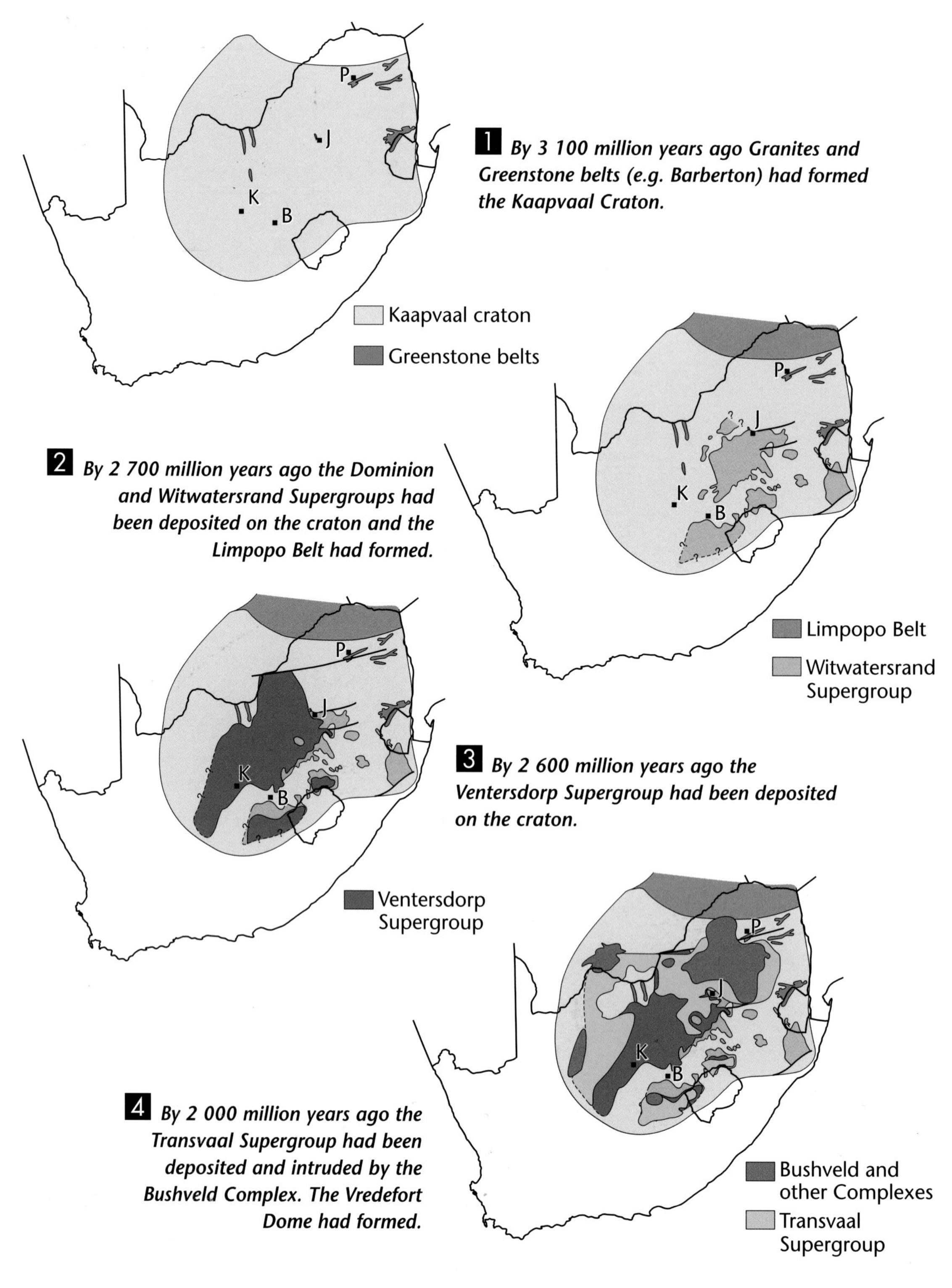